Introduction to Industrial Energy Efficiency

Introduction to Industrial Energy Efficiency

Energy Auditing, Energy Management, and Policy Issues

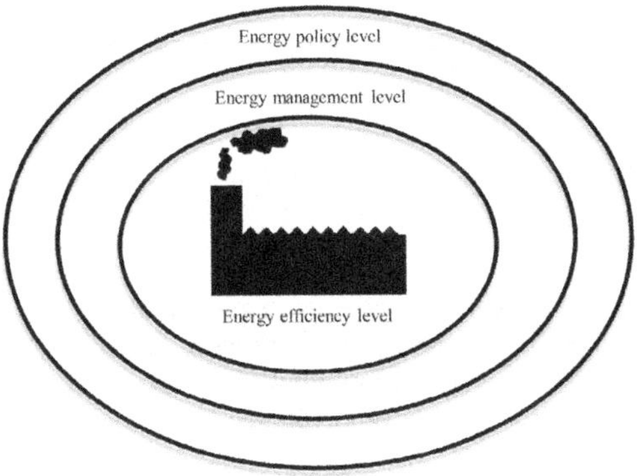

Patrik Thollander
Linköping University, Linköping, Sweden; University of Gävle, Gävle, Sweden

Magnus Karlsson
Linköping University, Linköping, Sweden

Patrik Rohdin
Linköping University, Linköping, Sweden

Johan Wollin
Volvo Construction Equipment, Brussels, Belgium

Jakob Rosenqvist
Tranås Energi, Tranås, Sweden

ELSEVIER

ACADEMIC PRESS
An imprint of Elsevier

Academic Press is an imprint of Elsevier
125 London Wall, London EC2Y 5AS, United Kingdom
525 B Street, Suite 1650, San Diego, CA 92101, United States
50 Hampshire Street, 5th Floor, Cambridge, MA 02139, United States
The Boulevard, Langford Lane, Kidlington, Oxford OX5 1GB, United Kingdom

British Library Cataloguing-in-Publication Data
A catalogue record for this book is available from the British Library

Library of Congress Cataloging-in-Publication Data
A catalog record for this book is available from the Library of Congress

ISBN: 978-0-12-817247-6

For Information on all Academic Press publications
visit our website at https://www.elsevier.com/books-and-journals

Publisher: Joe Hayton
Acquisition Editor: Raquel Zanol
Editorial Project Manager: Andrae Akeh
Production Project Manager: Nirmala Arumugam
Cover Designer: Miles Hitchen

Typeset by MPS Limited, Chennai, India

Contents

Preface

This book would not have been completed without several key persons. First of all, we would like to express our sincere gratitude to Associate Professor Mats Söderström for his detailed and sharp views on earlier drafts of this book. We would also like to thank Professor Sven Werner at the University of Halmstad and Dr. Per-Åke Franck at CIT Industrial Energy AB for their review of the Swedish book manuscript. Their valuable comments helped to improve the book.

We also want to thank the key people who reviewed an earlier version of the book manuscript. We would like to thank Roland Dahlström, Head of Scania Industrial Maintenance AB; Leif Rydell, former Energy Engineer at Xylem Water Solutions; and Henric Dernegård, lead energy controller at Södra Forest Owners Economic Association for their valuable comments from an industrial perspective. We would also like to thank Dr. Nawzad Mardan at the University of Gävle and Associate Professor Curt Björk at Linköping University for their valuable comments from a research perspective.

Also, we would like to address our sincere thanks and gratitude to a number of people who have served us with proof reading of selected parts or major versions of earlier drafts of the book manuscript.

We would like to address our sincere gratitude to a number of people who have served us with proof reading of earlier drafts of the part involving Lean energy management: Dr. Esther Widmer and Anna Wollin Ellevsen. Also, we would like to thank our colleagues at Volvo Construction Equipment such as Frank Bengtsson, Adam Björk, Katarina Heikkilä, Niklas Nillroth, and Boris Radojcic and to Birgitta Sjögren at Swedish Environmental Research Institute.

Moreover, our sincere thanks to Dr. Osamu Kimura, CRIEPI, Japan and Dr. Sandra Backlund, Swedish Ministry of Economics for providing comments on the parts involving policies and programs. Furthermore, our sincere thanks to Hugo Salamanca and Vida Rozite at the International Energy Agency for very valuable comments on an earlier draft of the manuscript, in particular, those related to energy audits. Furthermore, sincere thanks to Dr. Maria Johansson, Simon Johnsson and Elias Andersson, Linköping University for valuable comments on selected parts of the manuscript.

Moreover, we would like to express our thanks to co-authors of previous Journal articles, conference papers, and book chapters that some parts of the book emanate from.

Finally, we would like to thank Fredrick Andersson at the Swedish Energy Agency who many years ago strongly advocated that we seriously should start the book manuscript process and for your views on a previous version of the book related to policy.

Finally, our sincere thanks to our dear families, you are invaluable for us!

Introduction

1

Chapter Outline

1.1 Energy efficiency—"first fuel"

In addition to an increased share of renewable energy, more efficient energy use is by far the most important means for achieving sustainable energy systems—locally, regionally, and globally. Efficient energy use in the manufacturing industry is a key factor in ensuring that individual companies, industries, and industry as a whole continue to be competitive on a global market with scarce resources. According to the Intergovernmental Panel on Climate Change (IPCC),[1] more efficient energy use in industry is the single most important means of preventing global warming. The International Energy Agency is in the same line, calling energy efficiency *first fuel*, that is, the first action that should be taken.

Efficient industrial energy use refers to everything from operational energy efficiency measures in the production process and capital-intensive major investments in new production lines, to technology solutions for controlling supply temperatures in heating systems, procurement of energy services from external energy service providers, implementation of energy audits, internal energy management work, and

[1] The IPCC is the international body for assessing science related to climate change. The IPCC was founded in 1988 by the World Meteorological Organization and the UN Environment Program to provide politicians with regular assessments of the scientific basis for climate change, its effects and future risks, and options for adaptation and limitation.

Introduction to Industrial Energy Efficiency. DOI: https://doi.org/10.1016/B978-0-12-817247-6.00001-8

the establishment of energy strategies at management level in an organization and more. This complexity regarding both technology and organization places high demands on the organization as well as the competence for the energy efficiency work to be carried out successfully. This book aims at increasing knowledge for our readers about, and thus promoting, a more efficient industrial energy use. The focus is on manufacturing industry, but much of the content is applicable in other sectors and industries as well. Efficient energy use, in turn, contributes to the cost-effective improvement of industrial energy systems, the elimination of barriers to energy efficiency, and the improvement of the procurement competency of the industry in terms of energy-related products and services.

1.2 Background

Global climate change due to increased greenhouse gas emissions, which, in turn, is a consequence of primarily the use of fossil fuels for energy supply, is pushing decision-makers in taking important decisions related to climate change mitigation. Within the EU, the EU Energy Service Directive (ESD) was launched in 2006 aimed at improving energy efficiency by 9% to 2016. In 2012 the ESD was replaced by the Energy Efficiency Directive. The aim of the 2012 directive was to reduce energy end-use by 20% by 2020, compared with the 2005 level. The new target within the EU for 2030 was set in 2018 and states a 32.5% improvement in energy end-use by 2030.

1.2.1 *Climate change mitigation*

The human-related, so-called anthropogenic, environmental problems have changed over the centuries; from local and regional phenomenon to today's global environmental issues such as ozone thinning and increased greenhouse effect. These induce entirely new demands on how to pursue sustainable environmental work. Many researchers today consider that global climate change is by far the greatest threat to mankind (IPCC, 2014). The use of fossil fuels is the single largest cause of the emergence of global climate change issues. Therefore, the highest priority in sustainable environmental work is to limit greenhouse gas emissions without sacrificing measures to improve sustainability on regional and local levels. All gases, except for the two-atom molecules containing only one kind of atoms (N_2, H_2, O_2, etc.), are so-called greenhouse gases. The strength of a greenhouse gas is due to the gas's ability to absorb heat radiation and its useful life. Methane is, for example, a 23 times stronger greenhouse gas than carbon dioxide. Gases with absorption bands where the atmosphere is transparent, gives rise to extremely strong greenhouse gases. Examples of gases with these properties are those which contain bonds between carbon and chlorine and carbon and fluorine, the so-called freons. Despite the fact that carbon dioxide is not a particularly strong greenhouse gas, the extensive combustion of fossil fuels means that these emissions, compared with other greenhouse gases, are very large.

Global climate change, EU's efforts to limit greenhouse gas emissions and rising energy prices, means that there are strong incentives for both state and industry to act strongly to achieve more efficient supply and end-use of energy.

1.2.2 Key areas for reducing energy costs

The negative effect of increased energy costs can be reduced in two ways: either the company aims to negotiate a lower price with the energy supplier, or the company can undertake internal measures to improve energy end-use efficiency and also investigate potential own supply of energy. The four key areas in which a company can reduce its energy costs are:

- operation changes, for example, more efficient operating and maintenance routines and changed behavior;
- investments in energy-efficient technologies;
- time control, for example, load management; and
- conversion from one energy carrier to another, for example, from electricity to district heating.

The overall aim of the book is to convey a system perspective on more efficient energy end-use in different types of industrial energy systems.

1.3 Global industrial energy use

Industry accounts for a large share of global energy end-use. Depending on how the term *industry* is defined in different reports, the industry accounts for between 25% and 50% of global energy end-use. In 2014 the industrial sector used 43,000 TWh/year or 36% of the world's total energy end-use (IEA, 2017). The world's most energy-intensive industries are according to the IEA (2017) chemicals and petrochemicals, iron and steel, cement, pulp and paper, and aluminum representing about 69% of the global industrial energy end-use (IEA, 2017). Chemicals and petrochemicals and iron and steel account for about 10% and more than 8% of the world's energy end-use, respectively (IEA, 2017).

About half of the energy end-use in the chemical and petrochemical industry comes from oil, and most of the remaining energy end-use emanates from natural gas. The iron and steel industry mainly uses coal in its processes. Both industries also use a lot of electricity. More than half of the energy end-use in the cement industry comes from coal, while the remainder is relatively evenly distributed between electricity, natural gas, and oil, and a very small part of biomass. For the pulp and paper industry, almost half of energy end-use comes from fossil fuels, while the remaining share is biofuels. Electricity is the dominant energy form for the aluminum industry, while almost half of the energy end-use is fossil fuels. The energy use in the different industries is found in Fig. 1.1, showing the five most energy-intensive industries globally.

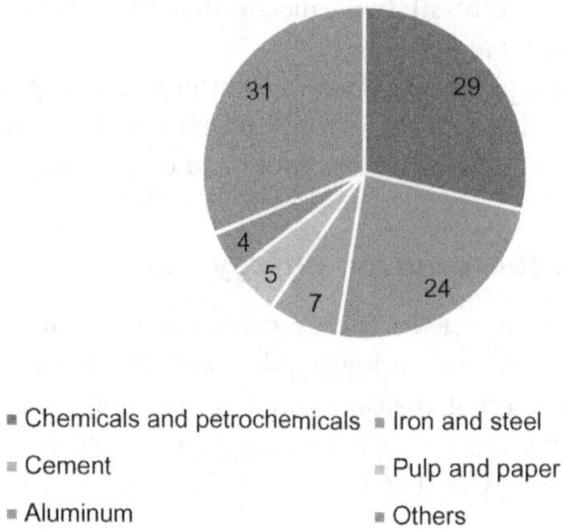

Figure 1.1 Energy use in the different industries (%), in total 43,000 TWh (IEA, 2017).

1.4 Energy end-use among different industrial companies

Among non-energy intensive small and medium-sized enterprises, the support processes, such as ventilation, space heating, cooling, and lighting, stand for the companies' main energy use. On the contrary, among energy-intensive companies, the production processes represent the largest energy use. Table 1.1 shows the energy end-use in a large non-energy intensive engineering industry, while Table 1.2 displays the energy end-use in a large iron foundry, which is energy intensive and where the production process is discontinuous. Table 1.3 displays a large chemical pulp mill, which is highly energy intensive and where the process is continuous.

As shown in Tables 1.1–1.3, energy end-use differs significantly between the different industrial companies. In the engineering company, production processes account for less than 30% of the total energy end-use, while in the iron foundry, about 50% of the energy end-use is found in the production processes. For the chemical pulp mill, only about 16% of the total energy end-use can be found in "Others" in which support processes are included, but also a number of minor production processes. The differences between the companies clearly affect how they can conduct effective industrial energy management, but also their respective potential for more efficient energy use.

1.5 Energy efficiency

The book primarily uses the concept of *energy efficiency*, which should not be considered as energy saving. Energy savings is specified as the amount of kWh saved, that is, it is an absolute measure. Energy efficiency, however, is indicated in

Table 1.1 Energy end-use in a large non-energy-intensive engineering industry (Thollander et al., 2007).

Processes	MWh/year	%
Metal cutting	2370	22.3
Space heating	2310	21.8
Ventilation	1840	17.3
Compressed air	1360	12.8
Lighting	1310	12.3
Metal chip collector	610	5.7
Tap hot water	600	5.7
Phosphating	100	0.9
Engine heaters (parking lot)	90	0.9
Administration (office)	25	0.2

Table 1.2 Energy end-use in a large iron foundry (Thollander et al., 2005).

Processes	MWh/year	%
Melting and holding	13800	42.7
Space heating	5530	17.1
Ventilation	4700	14.6
Compressed air	2100	6.5
Forming	2020	6.3
Ladle heating	1300	4.0
Tap hot water	1200	3.7
Sand preparation	580	1.8
Administration (office)	580	1.8
Lighting	490	1.5

Table 1.3 Energy end-use in a large chemical pulp mill (Klugman, 2008).

Processes	GWh/year	%
Evaporator	710	29.5
Drying	500	20.7
Digesters	450	18.7
Others	390	16.2
Bleaching	200	8.3
Boilers and feed water	160	6.6

relation to something. It indicates how effective a process is or how much more effective it can become, given a certain measure, such as X MWh per ton more efficient. Governmental energy goals are often formulated in terms of energy savings. However, for individual companies, it is more relevant to consider energy use from an efficiency perspective, that is, how a unit to a customer can be produced more efficiently and thus more sustainable.

In Jaegemar (1996) energy efficiency is defined as: "Energy efficiency is a measure of the balance between the energy gained and the sacrifice necessary to bring about this gain." An energy-efficient system is thus a system that provides its function with the lowest possible energy use at reasonable cost. This definition is useful, for example, when discussing indoor environment aspects as when Abel and Ekberg (2002) stress the importance of two aspects when implementing efficiency measures:

1. The implementation of an energy efficiency measure should not have a negative effect on the building (or production) and its function.
2. The primary use of resources when implementing an energy efficiency measure should be related to the total energy reduction due to the measure.

These two requirements are to be fulfilled in order for a measure to be categorized as an efficiency measure. This sorts out measures where the function of the building, such as indoor air quality (IAQ) issued, or the production is reduced or when the measure results in a use of resources not in proportion to the reduction of energy use (Abel and Ekberg, 2002). Another aspect in defining an energy efficiency measure could be to establish an energy payback criteria, where the energy payback index represents the potential saving per year divided by the embodied energy, and thus results in a relation of how long it will take for the measure to save as much energy as it has required in production.

1.5.1 EU's energy efficiency directive, definitions—an example

Energy efficiency means the ratio of output of performance, service, goods or energy, to input of energy.

Energy savings means an amount of saved energy determined by measuring and/or estimating consumption before and after implementation of an energy efficiency improvement measure, while ensuring normalization for external conditions that affect energy consumption.

Energy efficiency improvement means an increase in energy efficiency as a result of technological, behavioral, and/or economic changes.

Source: EU's Energy Efficiency Directive (2012).

1.6 Vertical or horizontal perspective

When aiming to improve energy efficiency in companies, a vertical perspective is usually applied. The company's energy use at a geographically defined site is studied, for example, a factory within a company group or all plants in the company group. The company's energy end-using processes are audited, analyzed, and improved, for example, by energy auditing, energy management, procurement of energy services, and process integration activities.

Another point of departure for improving energy efficiency is the horizontal perspective, where instead the product's processing chain or value chain is studied.

According to this perspective, the energy end-use of the product from cradle to grave is included, for example, via a life cycle analysis.

In this book, an emphasis is placed on the vertical perspective, and these methods are described later on in the book.

1.7 Systems perspective on energy efficiency

In order to realize the full energy efficiency potential, both technology measures, behavioral measures as well as knowledge about both single processes and a number of processes togehter, are needed. This way of viewing a challenge from a number of different perspectives is called a systems' perspective or systems' thinking. With regard to improved energy efficiency, it is crucial to apply a system perspective. When undertaking system changes, it is important, for example, to take into account the type of energy used. A process that requires heating at low temperatures does not necessarily need to use electricity. Electricity should preferably rather be reserved for processes that require high-quality energy, such as electric motors, while more low-quality energy sources (such as excess heat from other parts of the plant) can be used for heating.

Energy savings in a certain part of the system can lead to consequences in other parts of the system, which can counteract or even increase overall energy demand. For example, if there is a heating requirement in a room, the installation of more energy-efficient lighting means that less heat is supplied to the room (through losses in the lighting as all electricity uses in the lighting fixture is later transformed into heat). Thus additional heat must be supplied to the room during the cold winter season, due to the improvement of the lighting. This, of course, applies only if there is a heat requirement in the room. On the other hand, in cooled rooms, a switch to energy-efficient lighting affects the cooling demand positively. However, it is important to consider what energy quality is required, that is, is it wise to use electricity for heating?

When implementing several energy efficiency measures at the same time, it is important to take into account that the actions may interfere with each other and that the respective energy efficiency potentials are not necessarily additive. For example, the energy efficiency potential for reduced cooling demand in a location can decrease if an energy efficiency measure in a process is carried out simultaneously, which reduces the overall heat demand in the room. Adding the saving potential for the cooling requirement with the energy efficiency potential of the process in the room gives an incorrect result.

When analyzing industrial energy systems, the choice of system boundary is very important. As mentioned earlier, different energy efficiency measures may interfere with each other, which may be overlooked if an overly tight system boundary is selected. System boundaries are important in discussions about electricity production and the use of excess heat from industrial processes, that is, heat that turns out as a surplus and could be used for secondary heating delivery in, for example, a district heating system. In a district heating system with cogeneration plants, that is, plants that produce both heat and electricity, the amount of electricity that can be produced is controlled by the amount of heat that can be delivered. Therefore if the industrial excess heat from an industry is delivered to a district heating system with combined heat and power (CHP), the electricity generation in the CHP will decrease. If it is very profitable to produce electricity, it can be unprofitable to use industrial excess heat in these district heating systems. Energy policies that promote renewable electricity generation in district heating systems, for example, the Swedish and Norwegian electricity certificate scheme, which favors biofuel-based electricity generation, may then counteract with the aim of increasing industrial excess heat in these district heating systems.

1.8 Potential for improved energy efficiency

The potential for improved energy efficiency varies between different companies, industries, and countries (Sorrell et al., 2000). As a result of various barriers such as lack of time, of capital and information asymmetries and imperfections that inhibit improved energy efficiency, not all energy efficiency measures are realized, even though they are perceived as profitable. The difference between potential for energy efficiency and what de facto is implemented is referred to as the energy efficiency gap, initially defined by Hirst and Brown (1990).

In a paper by Jaffe and Stavins (1994), they outline a number of different levels of "energy efficiency potentials," or "energy efficiency gaps," stating that the actual energy efficiency potential level is dependent on which view that is applied. On the lowest level, the economist's potential, the potential can be reached by overcoming the so-called market failures in the energy technology markets. The technologist's potential, which is on a higher level, can be reached by overcoming both market failures and market barriers, while the highest level, the hypothetical level, may be reached by also overcoming market failures on the energy markets Jaffe and Stavins (1994).

The energy efficiency gap is based on the view that there are a number of cost-effective energy efficiency measures, primarily technical, which are profitable and lead to more efficient energy use. Energy use is calculated by multiplying power with time. Technologies and processes that use energy have a power demand during operation hours which means that how technologies and processes are operated affect both power demand and operation hours. The part related to how technologies and processes are used refers to as the energy management gap meaning that there is in fact a higher energy efficiency potential if energy management is also included in addition to energy efficiency technology measures. This refers to as the extended energy efficiency gap, initially defined together with the energy management gap by Backlund et al. (2012a). Fig. 1.2 displays this gap.

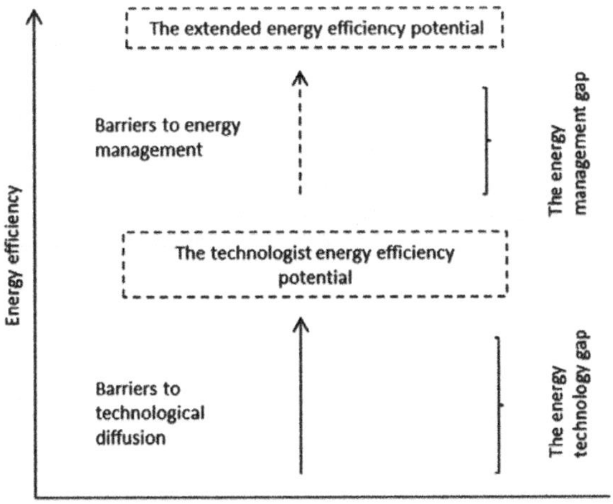

Figure 1.2 The extended energy efficiency gap: the energy efficiency potential level is increased if energy management practices are also included (the energy management gap) (Backlund et al., 2012a).

In a paper by Backlund et al. (2012b), the extent to which the energy management gap exists and if so to what extent, was explored through a case study. Two different groups of companies were studied:

- Category X: Energy-intensive firms that have participated in a voluntary agreements program (VAP) or conducted an energy audit in the last 3 years.
- Category Y: firms that have not participated in a VAP nor conducted an energy audit in the last 3 years. Mostly non-energy intensive and small and medium-sized enterprises.

The result of the perceived energy efficiency and energy management gap is displayed in Table 1.4 and reveals that the potential seems to be quite evenly distributed between technology and management. Notably, the energy-intensive companies perceive a larger relative potential for management (6%) than for technology (4.6%) and for the non-energy intensive and small and medium-sized companies, the potential is relatively larger and also, technology is perceived to have a higher degree of energy efficiency potential than management-related measures.

A consecutive study of the energy efficiency measures implemented within the Swedish VAP, the program for electricity-intensive industry, which was a 5-year policy program for industry involving energy auditing and the implementation of the world's first industrial energy management standard, SS 62000 was carried out. The study found that about 61% of the implemented 1254 energy efficiency measures emanated from the implementation of new technology, and the rest came from management and operational measures (Paramonova et al., 2015).

In a study of the Swedish steel industry, Brunke et al. (2014) found that the extended energy efficiency potential was perceived to be 9.7%, which was 2.4% higher than if only the potential for cost-effective technologies was considered (Brunke et al., 2014).

Table 1.4 Estimated energy efficiency potential (Backlund et al., 2012b).

Category	X	Y	Total
Energy use (TWh/year)	8.8	0.6	9.4
Percentage of accounted energy use (%)	94	6	100
Potential for energy efficiency from technology (%)	4.6	20	6
Potential for energy efficiency from management (%)	6	13	6

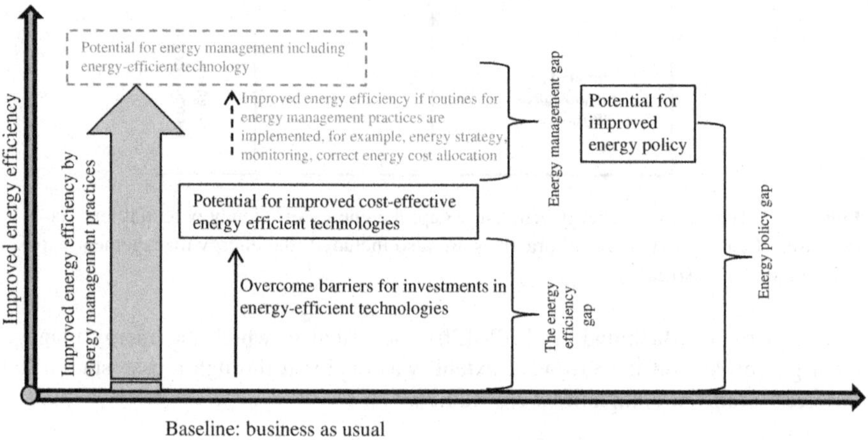

Figure 1.3 The energy efficiency, energy management, and energy policy gaps (Thollander and Palm, 2013).

From the national perspective, energy policies are used to promote improved energy efficiency. Properly designed, these policies can help companies reduce both the energy efficiency and the energy management gaps. The difference between potential for energy efficiency and energy management with the help of policies, and which policies that are put into place, that is, that are currently present in an Economy, is referred to as the energy policy gap, initially defined by Thollander and Palm (2013).

Fig. 1.3 displays the three gaps related to improved energy efficiency.

1.8.1 Energy efficiency and nonenergy benefits or multiple energy benefits

When an assessment is made of the implementation of an energy efficiency measure on an operational level, the cost-effectiveness of an energy efficiency measure is assessed using commonly used tools, for example, net present value, internal rate of return, life cycle cost, or simple payback. These assessments normally include:

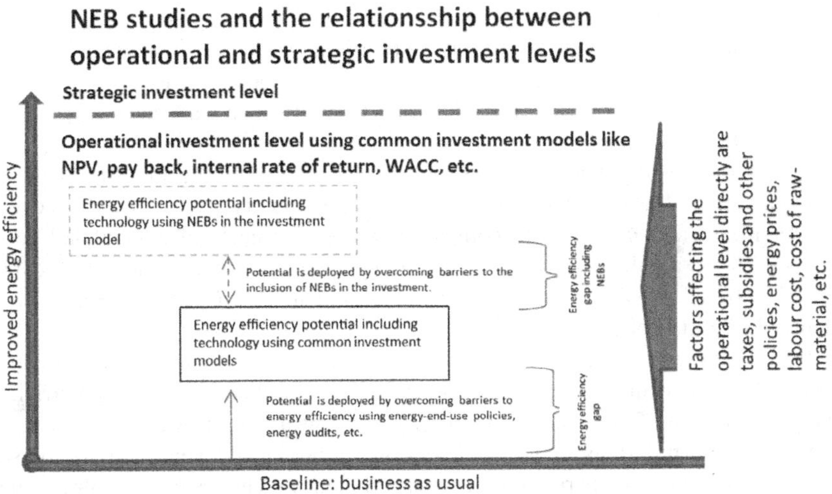

Figure 1.4 Inclusion of NEBs and the effect it has on the energy efficiency potential, and its relationship to the operational and strategic dimensions of investments in an industrial company (Nehler et al., 2014).

- the investment cost of the measure;
- the reduced annual cost for energy; and
- sometimes the maintenance cost.

Research in the area reveals that if other benefits are included as well, for example, increased productivity, reduced waste, and improved working conditions, the cost-effectiveness of the measures increase (Nehler et al., 2014). These additional benefits are referred to as nonenergy benefits (NEBs) or multiple energy benefits.

Research reveals that when larger investments on a strategic level are undertaken, many of the NEBs are included in the assessment (Nehler and Rasmussen, 2016), while on an operational level, these NEBs often remain overlooked. Fig. 1.4 displays the positive effects of including these benefits in an investment assessment.

Fig. 1.4 shows that when including NEBs in the assessment, a number of more measures are likely to pass threshold for what is perceived as cost-effective or not among industrial companies, leading to the fact that the energy efficiency potential is higher than if NEBs remains overlooked (Nehler et al., 2014).

1.9 The disposition of the book

The book is divided into three parts and has its starting point in the energy efficiency, energy management, and energy policy gaps, see Fig. 1.3.

Part I focuses on energy auditing and energy efficiency of companies from a predominantly technical perspective. This part is aimed toward students at, for example, university levels, energy and environmental managers in industrial companies, technical consultants, and various public policy actors and stakeholders, such as regional energy agencies and county administrative boards. Part I has its starting point in the so-called energy efficiency gap. Chapters 3–12 are included in Part I of the book.

Part II focuses on energy management in the manufacturing industry. This part is mainly aimed toward students at, for example, university levels, as well as energy and environmental managers in companies. Even technical consultants and public actors at, for example, national regional energy agencies may benefit from this part. Part II is based on the so-called energy management gap. Chapters 13 and 14 are included in Part II of the book.

Part III focuses on energy efficiency policy instruments and how these can be designed, implemented, and evaluated as well as one chapter covering barriers to energy efficiency. This part is primarily aimed toward actors who administer and operate policy programs at national, regional, and local levels, and for persons interested in studying barriers to energy efficiency. This part may also benefit groups that want to take a holistic approach to energy efficiency-related issues and companies that want to reduce the negative environmental impact of energy use throughout the value chain, that is, including suppliers. Part III is based on the so-called energy policy gap. Chapters 15–17 are included in Part III of the book.

After the book's introductory chapter, general energy theory and a review of commonly used concepts, terms, and definitions is provided in Chapter 2. Chapter 2 is aimed toward readers who need to freshen up their energy skills or lack a technical energy educational background.

References

Abel, E., Ekberg, L., 2002. Energieffektivitet, Rapport Effektiv 2002:01, Borås, Sweden (In Swedish).

Backlund, S., Thollander, P., Palm, J., Ottosson, M., 2012a. Extending the energy efficiency gap. Energy Policy 51, 392–396.

Backlund, S., Broberg, S., Ottosson, M., Thollander, P., 2012b. Energy efficiency potentials and energy management practices in Swedish firms. In: Proceedings of the ECEEE Industry Summer Study, 11–14 September, 2012.

Brunke, J.C., Johansson, M., Thollander, P., 2014. Empirical investigation of barriers and drivers to the adoption of energy conservation measures, energy management practices and energy services in the Swedish iron and steel industry. J. Cleaner Prod. 84 (1), 509–525.

Hirst, E., Brown, M.A., 1990. Closing the efficiency gap: barriers to the efficient use of energy. Resour. Conserv. Recycl. (3), 267–281.

IEA (International Energy Agency), 2017. Tracking Clean Energy Progress 2017, Energy Technology Perspectives 2017, Excerpt Informing Energy Sector Transformations.

IPCC (International Panel for Climate Change), 2014. Climate Change 2014. Impacts, Adaptation and Vulnerability. Intergovernmental Panel on Climate Change. Fifth Assessment Report.

Jaegemar, L., 1996. Design of Energy Efficient Buildings – Applied on HVAC Systems in Commercial Buildings (Doctorial thesis). Chalmers University, Sweden.

Jaffe, A.B., Stavins, R.N., 1994. The energy efficiency gap: what does it mean? Energy Policy 22 (10), 60–71.

Klugman, S., 2008. Energy systems analysis of Swedish pulp and paper industries from a regional cooperation perspective—case study modelling and optimization. Linköping Studies in Science and Technology, Dissertation No. 1194, Division of Energy Systems. Linköping University, Linköping.

Nehler, T., Rasmussen, J., 2016. How do firms consider non-energy benefits? Empirical findings on energy-efficiency investments in Swedish industry. J. Clean. Prod. 113, 472–482.

Nehler, T., Thollander, P., Ottosson, M., Dahlgren, M., 2014. Including non-energy benefits in investment calculations in industry – empirical findings from Sweden. In: Proceedings of ECEEE Industry Summer Study, Arnhem, June 2–5, 2014.

Paramonova, S., Thollander, P., Ottosson, M., 2015. Quantifying the extended energy efficiency gap – evidence from Swedish electricity-intensive industries. Renew. Sustain. Energy Rev. 51, 472–483.

Sorrell, S., Schleich, J., Scott, S., O'Malley, E., Trace, F., Boede, E., Ostertag, K., Radgen, P., 2000. Reducing Barriers to Energy Efficiency in Public and Private Organizations. SPRU's (Science and Technology Policy Research).

Thollander, P., Palm, J., 2013. Improving Energy Efficiency in Industrial Energy Systems – An Interdisciplinary Perspective on Barriers, Energy Audits, Energy Management, Policies & Programs. Springer, London, ISBN 978-1-4471-4161-7.

Thollander, P., Karlsson, M., Söderström, M., Creutz, D., 2005. Reducing industrial energy costs through energy efficiency measures in a liberalized European electricity market – case study of a Swedish iron foundry. Appl. Energy 81 (2), 115–126.

Thollander, P., Palm, J., Söderström, M., 2007. Industrial energy auditing – a key to competitive energy-efficient Swedish SMEs. In: Bergmann, D.M. (Ed.), Energy Efficiency Research Advances. Nova Publisher, New York, pp. 213–238.

Further reading

EC (European Commission), 2006. Directive 2006/32/EC of the European Parliament and of the Council of 5 April 2006 on energy end-use efficiency and energy services and repealing Council Directive 93/76/EEC. Brussels.

EC (European Commission), 2012. Directive 2012/27/EU on energy efficiency, amending Directives 2009/125/EC and 2010/30/EU and repealing Directives 2004/8/EC and 2006/32/EC. vol. 4, pp. 1–56.

Nehler, T., Rasmussen, J., 2016. How do firms consider non-energy benefits? Empirical findings on energy-efficiency investments in Swedish industry. J. Cleaner Prod. 113, 472–482.

General energy theory[1]

2

Chapter Outline

2.1 Introduction

Services of various kinds are necessary to enable society to function satisfactorily. Services, for example, lighting and heating, require energy and are therefore called energy services. The concept of energy services focuses on the benefit, and to achieve the benefit energy is used. Energy is thus a means. An example where the focus is on benefit (function) rather than means is the energy service "reasonable temperature" which may be more suitable to use than the term "heating," because sometimes it also requires "cooling" to get the right temperature. Another example is the use of the term "communication" instead of "travel and

[1] Parts of this section are based on Areskoug and Eliasson (2011).

Introduction to Industrial Energy Efficiency. DOI: https://doi.org/10.1016/B978-0-12-817247-6.00002-X

transport." Communication can take place in a variety of other ways than through travel and transport. Examples of energy services are as follows:

- Lighting
- Reasonable temperature
- Preparing and storing food
- Communication
- Exchange of information
- Manufacturing of goods (molding, joining, melting, and so on)

The last bullet "manufacturing of goods" is central to many industries, but other energy services are also of great importance. "Preparing and storing food" is, for example, an important energy service in the food industry. As indicated in the bullet list above, the division of the energy service "manufacturing of goods" can be further refined.

Energy services change over time. For the past 50 years, energy has increasingly been used for cooling in all sectors. Going far back in time, fire was used to accommodate several energy services simultaneously, such as temperature (in this case heating), lighting and preparation of food.

Energy is thus a means of meeting a benefit. Energy is distributed in different ways depending on which energy carrier is used. The energy carriers have in turn been produced in a process, such as electricity in a nuclear power plant, or electricity and district heating in a cogeneration plant. These processes also require energy of different forms and energy sources.

Thus, to provide an energy service, a certain amount of energy of some kind is required. The energy service could possibly be met with a smaller need for energy. The energy service could also possibly be met with another type of energy carrier or energy source, which in turn means that the supply of energy is influenced.

It is worth noting that it is the use of energy that forces a need for energy supply. The following sequence of events should serve as a starting point for discussions on energy use:

1. What energy services are needed?
2. How to satisfy the need for energy services with the lowest possible energy quantity in order not to misuse energy quality?
3. How does society ensure that different social functions are provided with energy in the most efficient way?
4. Which energy sources should be used?

The last question often gets the most space in discussions about energy use, but focus should instead be placed at the user perspective.

2.2 What is energy?

From a historical perspective, the understanding of the concept of energy has changed. During the 18th century substance theory dominated, and heat was considered a substance, a gas or liquid-like matter, that penetrated materials. The substance was called caloric. Benjamin Thompson Rumford, however, showed at the end of the 18th and early 1800s that ice blocks, although absorbing large amounts of heat when melted, did not increase in mass. James Prescott Joule conducted experiments in the 19th century where he confirmed Rumford's theories and showed that the amount of energy did not change when different energy forms were converted to each other. This shattered the idea of energy as a substance. Instead, heat began to be associated with motion. Today, energy is often related to an ability, or capacity, to change or perform work, where also motion is included. However, there is no unambiguous and common definition of the concept of energy. Instead, different definitions are used for different energy forms, such as kinetic energy and chemical energy. For the different forms of energy, it is possible to specify which factors that influence. It is also possible to make calculations regarding different energy forms based on quantities such as mass, speed, and temperature.

Energy is used as a designation or collective name for

- Energy forms, for example, potential energy and kinetic energy
- Energy carriers, for example, electricity and district heating
- Energy sources, for example, oil and wind power.

Subsequent to a brief introduction of the concepts energy and power, an exposition of energy forms, energy carriers, and energy sources follows.

2.3 Energy and power

Energy may be regarded as the capacity to perform a work. Power is the rate at which this work is performed, or otherwise expressed; the rate at which energy is converted from one form to another. The relationship is as follows:

$$P = E/t \tag{2.1}$$

where P is the power, measured in joules per second (J/s), Watt (W), (1 J/s = 1 W) and horsepower (hp); E is the energy, measured in different units where joule (J) is the base unit, another unit is watt hour (Wh), (3,600 J = 1 Wh); and t is the time (s).

Some literature indicates electrical energy in watt hours, other literature uses joule. However, by consistently using the same unit comparisons between different types of energy are facilitated. Watt hours (Wh) are therefore used throughout this book.

An example to clarify the difference between energy and power

Two similar balloons are inflated equally, which means that the same amount of energy is stored in both balloons. Two approaches are used to "release" energy from the balloons: in balloon 1, a hole is pierced with a needle, which means that all energy is discharged over a very short period of time. Balloon 2 is released and flies around for some time while the energy is released from the balloon. The power when the amount of energy is released is higher for balloon 1 than for balloon 2, because the energy is released during a shorter period of time.

Some examples of energy and power

- The power of a hotplate of a stove is approximately 1 kW (1000 W). If the hotplate is turned on for 1 hour, 1 kWh of energy (electricity) is used.
- A larger car consumes about 0.9 L of petrol per 10 km. The energy content of 1 L of petrol is approximately 11 kWh. The energy requirement for driving a car 1000 km will then be almost 1 MWh (1000 kWh). The power during these 1000 km varies depending on how the car is driven and whether you are accelerating or braking. Maximum power for a car varies but is in the order of 100 kW (100,000 W) (100 kW = 135 hp).
- A relatively normal-sized house in a relatively cold climate (e.g., in Sweden) has an energy requirement of 25 MWh (25,000 kWh) for 1 year, but the need naturally varies depending on how the house is used and how well insulated it is. The power varies with outdoor temperature, the use of the house and how well insulated it is. The peak power (winter time) may amount to about 12 kW (12,000 W).

2.4 Energy forms

Energy may be divided into different energy forms for which different types of calculations can be performed. The energy forms are as follows:

- Potential energy
- Kinetic energy
- Thermal energy
- Electrical energy
- Chemical energy
- Electromagnetic energy
- Nuclear energy

This division into seven energy forms is the most common, but other divisions can also occur. Sometimes mechanical energy is regarded as its own form of energy. Mechanical energy is the sum of potential energy and kinetic energy. A pendulum is a good example for illustrating mechanical energy because potential energy and kinetic energy vary depending on the position of the pendulum.

2.4.1 Potential energy

Potential energy is present in an object that is affected by a force field. One example is water in a hydropower plant. Calculations of potential energy are made according to the following formula

$$E = m \cdot g \cdot h \tag{2.2}$$

where E is the energy (J); m is the mass (kg); g is the gravity acceleration (m/s^2); and h is the height (m).

The potential energy unit is kg m/s$^2 \cdot$ m $=$ kg $\cdot$ m^2/s^2.

2.4.2 Kinetic energy

Kinetic energy equals force multiplied by distance

$$E = F \cdot ds \tag{2.3}$$

where E is the energy (J); F is the force (N); and s is the distance (m).

Kinetic energy can also be calculated according to the following formula

$$E = \left(m \cdot v^2\right)/2 \tag{2.4}$$

where E is the energy (J); m is the mass (kg); and v is the speed (m/s).

The kinetic energy unit is kg $\cdot$ m^2/s^2, that is, kinetic energy has the same unit as potential energy. In comparison with Einstein's well-known formula, kinetic energy is calculated: $E = m \cdot c^2$, where c is the speed of light (s). Here too, the unit is kg $\cdot$ m^2/s^2.

2.4.2.1 The relationship between the two kinetic energy equations

The relationship between the two equations for kinetic energy $E = F \cdot ds$ and $E = (m \cdot v^2)/2$ is derived as follows:

Power is defined, according to Newton's second law, as the change of a system's momentum ($m \cdot v$) over time (t). The force is thus defined as the time derivative of the momentum, that is

$$F = m \cdot dv/dt \tag{2.5}$$

where F is the force (N); m is the mass (kg); v is the speed (m/s); and t is the time (s).

By using the above equation in the equation for kinetic energy and integrating, the following is given

$$E = \int F \cdot ds \tag{2.6}$$

$$E = \int m \cdot dv/dt \tag{2.7}$$

$$E = \int m \cdot dv/dt \cdot ds = \int m \cdot dv \cdot ds/dt = \int m \cdot v \cdot dv = (m \cdot v^2)/2 \tag{2.8}$$

From $E = (m \cdot v^2)/2$ it is also possible to construct an equation for the energy content in a flow over a cross-sectional area as follows:

$$P = (m \cdot v^2)/(2 \cdot t) \tag{2.9}$$

$$m \cdot = m/t = \rho \cdot A \cdot v \tag{2.10}$$

$$P = (\rho \cdot A \cdot v^3)/2 \tag{2.11}$$

where P is the power (W); m is the mass flow (kg/s); ρ is the density (kg/m^3); A is the area (m^2); and v = speed (m/s).

It is thus possible to calculate the power that passes through a cross-sectional area based on the kinetic energy. The power is thus a function of the speed in cubic. By multiplying over time, energy is obtained as

$$E = \rho \cdot A \cdot v^3 \cdot t/2.$$

2.4.3 Thermal energy

When calculating thermal energy, one must know the temperature difference, over, for example, a heat exchanger. Thermal energy is calculated according to the following formula:

$$E = m \cdot c_p \cdot dT \tag{2.12}$$

where E is the energy (J); m is the mass (kg); c_p is the specific heat capacity [J/(kg · K)]; and T is the temperature (K).

Different materials have different specific heat capacity. Specific heat capacity thus depends on which material that, for example, flows through a pipe where a temperature change occurs. The specific heat capacity of some materials (at normal temperature and pressure) is

- Water: 4.18 kJ/(kg · K)
- Air: 1.01 kJ/(kg · K)
- Ethanol: 2.44 kJ/(kg · K).

Ethanol is usually not associated with thermal energy, but rather with chemical energy as it is incinerated. But when the ethanol is burned, the stored chemical energy is converted into thermal energy. However, ethanol has a specific heat capacity just like any other material.

Thermal energy can be sorted under the category of kinetic energy. The reason is that depending on temperature molecules move differently quickly. Thermal energy may also be sorted under the category of potential energy since there is potential energy between the molecules. This is not due to the force of gravity but to the fact that atoms and molecules are surrounded by electron envelopes that take hold of each other, which leads to the molecules being affected by forces from each other. These forces are the source of potential energy and depend on the position of the molecules in relation to each other.

2.4.4 Electrical energy

To develop an equation for calculating electrical energy, one can start from the base units in the SI system. There are seven base units, namely

- Time (second, s)
- Mass (kilograms, kg)
- Length (meters, m)
- Electric current (amps, A)
- Thermodynamic temperature (kelvin, K)
- Amount of substance (mole, mol)
- Luminous intensity (candela, cd).

Energy and power (as previously reported) can be derived from the base units for mass, time, and length. The following equation for electrical energy can be derived by using current (which is a measure of the number of electrons passing per second) and energy as a starting point.

$$E = U \cdot I \cdot t \tag{2.13}$$

where E is the energy (J); U is the voltage (V); I is the current (A); and t is the time (s).

Voltage is a measure of the energy per electron and is not a base unit according to the SI system. It is defined as the voltage needed over a resistance so that the current 1 A creates the power 1 W, according to the equation

$$P = U \cdot I \tag{2.14}$$

where P is the power (W); U is the voltage (V); and I is the current (A).

2.4.5 Chemical energy

Through photosynthesis, solar energy (electromagnetic energy) is bound as chemical energy in coal, oil, natural gas, and bioenergy. Chemical energy is stored in the bonds between atoms and when a fuel is burned, the links are broken and new links with a lower energy content are created. The excess of chemical energy is emitted as light and thermal energy. For example, when methane is incinerated, the following chemical reaction occurs

$$CH_4 + 2O_2 \rightarrow CO_2 + 2H_2O + \text{thermal energy}$$

The thermal energy emitted amounts to just under 10 kWh per normal cubic meter of methane burned. This is called the heating value and is specific to each energy source. Some examples of approximate heating values (MWh/ton) are

- Firewood (humid) 2 MWh/ton
- Firewood (air-dried) 4 MWh/ton
- Coal 8 MWh/ton
- Oil 12 MWh/ton
- Natural gas 15 MWh/ton

Chemical energy can be sorted under the category of electrical energy. The reason is that the molecules' bonds with electrons are broken up during combustion. In a molecule, the electrons of the atoms can sense the electric force field from its own atomic nucleus and its electrons, and partly from neighboring atoms. In a chemical reaction, the atoms are redistributed, and the electrical potential energy is changed.

2.4.6 Electromagnetic energy

An example of electromagnetic energy is the solar energy that hits the earth. This takes place in the form of photons, which can have both particle and wave properties. The photon sometimes shows particle properties, for example, in the solar cell, and sometimes wave properties, for example, in the color changes in a soap bubble. The energy content of a photon is

$$E = h \cdot f = h \cdot c/\lambda \tag{2.15}$$

where E is the energy (J); h is the Planck constant ($6.63 \cdot 10^{-34}$ J · s); f is the frequency (derived SI unit from time); c is the light speed in vacuum ($3 \cdot 10^8$ m/s); and λ is the wavelength (m).

Because a photon has a low energy content, the energy unit usually used is electronvolt (1 eV = $1.6 \cdot 10^{-19}$ J).

2.4.7 Nuclear energy

Nuclear energy is obtained when atomic nuclei are split, and the parts repel each other with enormous force. Most of the energy is converted into kinetic energy. When the particles hit the surrounding material, thermal energy is created. One common reaction is

$$^{235}U + 1 \text{ neutron} \rightarrow {}^{92}Kr + {}^{141}Ba + 3 \text{ neutrons} + \text{energy} \tag{2.16}$$

where U is the uranium; Kr is the krypton; and Ba is the barium.

2.5 Energy carriers

Before energy can be used for the purpose of creating benefits, it usually needs refining and then transported—an energy carrier is needed. When the concept of energy carrier is used, it is thus not the energy that is to create the benefit that is intended, but the matter that "carries" the energy. The following are typical energy carriers:

- Electricity
  - Electricity can be produced in many ways (e.g., from wind power, solar cells or via combustion of fuels). Electricity is an unpreserved product that must be consumed at the same time as produced.
- Bioenergy
  - Bioenergy is usually refined before it is used. When refined, the following energy carriers are for example available: chips, pellets, briquettes, rapeseed oil, biogas, and methanol.
- Fossil fuels
  - Fossil fuels, such as crude oil, are refined first to extract, for example, petrol, diesel, and gas.
- District heating
  - District heating is produced centrally in large heat production facilities (e.g., by burning fuel or by utilizing excess heat from other operations) and heat is then delivered via hot water pipes (in rare cases with steam) to customers.
- Compressed air
  - Compressed air is produced in air compressors and delivered via the compressed air system to users.

2.6 Energy sources

Energy sources of different origins can be used for different types of energy services. The energy sources may be divided into different categories and one possible subdivision is primary and secondary energy sources. However, in order to avoid misunderstandings, it is important to be aware that the categorizations can be made in different ways. Two possible categories are presented below.

Primary and Secondary energy sources—Category 1:

- Primary energy sources
  - Solar energy
  - Nuclear energy
  - Earth's rotation (creates ebb and flow)

- Secondary energy sources (whose origin is primary energy sources)
  - Oil
  - Coal
  - Natural gas
  - Bioenergy
  - Hydropower
  - Wind power
  - Tidal energy

Among the primary energy sources, solar energy constitutes about 97% and nuclear energy about 3%. Thus, the Earth's rotation is negligible in comparison.

Primary and Secondary energy sources—Category 2:

- Primary energy sources
  - Oil
  - Coal
  - Natural gas
  - Bioenergy
  - Hydropower
  - Wind power
  - Tidal energy
  - Solar energy (not the sun)
  - Nuclear power (not nuclear energy)
  - Geothermal energy
- Secondary energy sources (whose origin is primary energy sources)
  - Electricity
  - District heating
  - Fuels

The seven first primary energy sources according to category 2 correspond to the secondary energy sources according to category 1. Furthermore, the secondary energy sources according to category 2 are energy carriers.

2.7 Concepts describing energy supply and energy utilization

By merging the above information, the picture presented in Fig. 2.1 is obtained. In the figure, category 1 is used with respect to primary energy sources.

2.8 The laws of thermodynamics

Below is the first and second laws of thermodynamics presented. There is also a zeroth law of thermodynamics that refers to the relationship between three subsystems (if A is in thermal equilibrium with both B and C, then B is also in thermal equilibrium with C) and a third law of thermodynamics that gives a connection

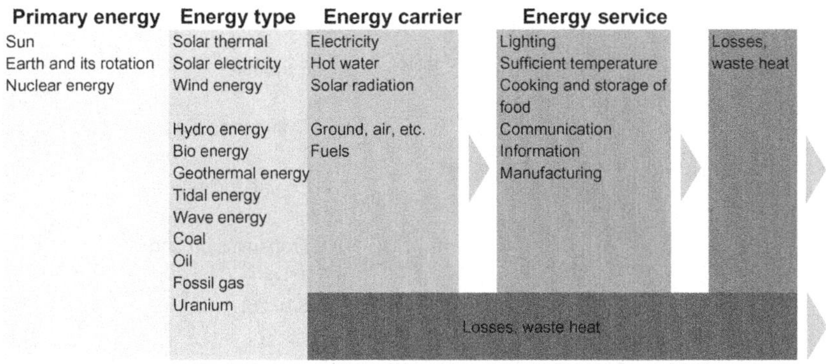

Figure 2.1 Relations between different concepts, from primary energy sources to losses. *Source*: Based on Areskoug, M., Eliasson, P., 2011. Energi för hållbar utveckling – ett historiskt och naturvetenskapligt perspektiv. Studentlitteratur, Lund. (Areskoug and Eliasson, 2011).

between entropy and the absolute zero (temperature). However, these are not dealt with in the book.

2.8.1 The first law of thermodynamics

The first law of thermodynamics clarifies that energy cannot be created or destroyed, but only converted. An energy form can be converted to another, but the total amount of energy is always the same, as James Prescott Joule demonstrated in his experiments during the 19th century. An automobile engine can serve as an example in the following exercise. The energy in the petrol is not consumed when burned in the engine. During acceleration, the energy is converted into kinetic energy (the engine and tires provide heat). When braking, friction heat is created. An overview of the energy conversion of energy supplied to a car is shown in Fig. 2.2.

One consequence of the first law of thermodynamics is that concepts such as "energy production," and "energy consumption" are not correct. However, these concepts are found in many texts, including industry-specific journals aimed at the energy industry, and the concepts are also used by companies that work with energy issues. The concepts also occur in the academic world and its scientific journals.

2.8.2 The second law of thermodynamics

The first law of thermodynamics equals a state before and after a reaction or event. However, the first law of thermodynamics says nothing about the direction in which a reaction or event occurs. Instead, it does the second law of thermodynamics. The second law of thermodynamics clarifies that heat cannot, by itself, be transported from a colder to a warmer body. A thought exercise regarding the second law of

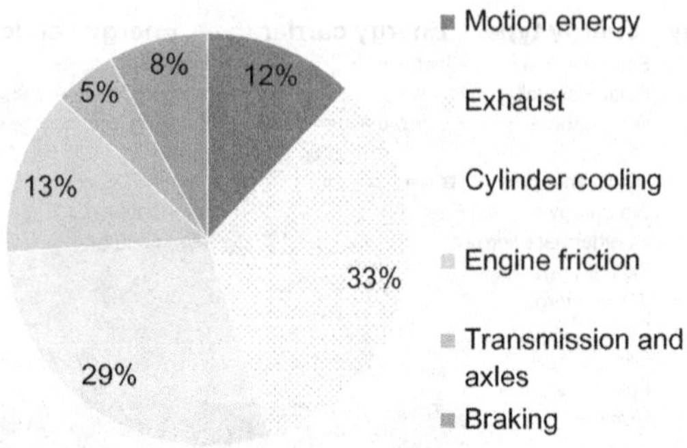

Figure 2.2 Typical energy use for a car (Rodrigue, 2010).

thermodynamics is to imagine a film showing a smoking chimney from coal combustion. It is unreasonable to imagine the reverse, that is, that the flue gases is sucked into the chimney (as if the film was played backward) and that the ash that together with the flue gases and the heat would turn into coal.

Some energy forms can be converted into other forms of energy, as in the example above where coal (chemical energy) is converted into thermal energy, but where a reverse reaction is not reasonable. This is evident from the second law of thermodynamics, which clarifies that different energy forms have different quality.

The importance of the second law of thermodynamics is great and it has certain consequences, such as that differences in temperatures always level out. For example, a house (which is externally cooled or heated) adopt the ambient temperature after a period of time when external cooling or heating ends. Another consequence is that all energy supplied will eventually turn into heat, that is, approaching the ambient temperature, for example, the electricity supplied to a refrigerator is converted into heat.

2.9 Energy quality

The second law of thermodynamics clarifies that different energy forms have different qualities. Energy quality is determined as the amount of energy of a particular form of energy that can be converted into mechanical energy. Table 2.1 shows such energy quality factors.

Entropy is a concept that considers changes in energy conversions and is a measure of how much of the thermal energy in a system can *not* be converted into work. In all energy conversions, losses always occur, and entropy is a measure of these. This means that entropy always increases at energy conversions. An engine operating between two given temperatures has the greatest efficiency when the minimum total entropy increase is achieved. The fact that entropy is always increasing at energy conversions is an alternative definition of the second law of thermodynamics. Entropy in the universe thus tends to increase.

Table 2.1 Energy quality factors.

Energy type	Quality factor
Mechanical energy (kinetic and potential)	1.0
Electrical energy	1.0
Nuclear energy[a]	1.0
Electromagnetic energy	0.95
Chemical energy	0.90
Thermal energy, 300°C	0.52
Thermal energy, 80°C	0.23
Thermal energy, 20°C	0.07
Thermal energy, 0°C	0.0

[a]Nuclear energy in reality contains a lower quality than what is displayed in this table. The quality factor means that the temperature of nuclear energy should be 5600°C, while its melting point is 2300°C. A factor of 1.0 would mean an accident, a military application or emanating from the fact that a nuclear particle, from splitting an atom, gains pure motion energy.

Another concept relates to energy quality, namely exergy. Exergy is the maximum available work when a system reaches equilibrium with the environment or the part of the energy that can be converted into work. Exergy is thus another way of expressing order but is formulated inversely against entropy,[2] that is, exergy refers to available work while entropy refers to losses. Energy, on the other hand, refers to the potential to be converted into work.

Exergy must be related to a reference temperature. Table 2.1 uses the reference temperature 0°C. However, there is no definite reference temperature that is always used, which means that the concept of exergy can be difficult to apply. When using the concept "energy" as the basis for calculations, one must therefore consider the temperature (quality). An example that illustrates this is that 1 MWh of water with the temperature 80°C does not correspond to water with the energy content 1 MWh and the temperature 40°C. The pulp and paper industry is an example of an industry with large excess heat flows. Just specifying the amount of MWh would be deceiving, because it is the temperature levels of any users of the excess heat who decides whether it is useful or not.

Exergy analysis

An energy carrier[3] with 100% exergy corresponds to a 100% possibility of converting the energy carrier into mechanical work. Exergy yield per mass unit can be obtained from Eq. (2.17).

[2] Entropy is a measure of the disorder of a system/substance, with the unit kJ/(kg · K).

[3] Energy carriers are something that transfers energy. It can be compressed air, electricity, a hot air stream, and so on.

$$e = (h - T_0 \cdot s) - (h_0 - T_0 \cdot s_0) \tag{2.17}$$

where e is the exergy per mass unit (kJ/kg); h is the enthalpy[4] per mass unit (kJ/kg); h_0 is the original enthalpy per mass unit (kJ/kg); s is the entropy per unit mass (kJ/kg · K); s_0 is the original entropy per unit mass (kJ/kg · K); and T_0 is the original temperature (K).

Different energy carriers have different exergy values, that is, various possibilities to convert energy into mechanical work. Electricity has close to 100% exergy value because the energy is of high degree of order (low entropy), while 70-degrees hot water has an exergy value of 4%.

The exergy value for a boiler is about 4%, direct electric heating about 5%, a heat pump about 15%, and a cogeneration plant about 40%. Based on its energy quality, that is, exergy value, heating of water with electricity is waste of exergy (energy quality). Instead, electricity with its high quality (100% exergy content) should rather be used for electricity specific processes.

2.10 Energy conversion

Energy forms with different energy quality can be converted in a certain direction according to the second law of thermodynamics and the energy quality concept. At energy conversions, losses occur. The efficiency of the conversion between different energy forms is expressed according to the following formula:

$$\text{Efficiency} = \text{useful energy}/\text{energy supplied} \tag{2.18}$$

The useful energy is never as large as the energy supplied, which means that the conversion efficiency is always lower than 100%. Some examples of conversion efficiencies and energy forms for some different techniques are presented below.

- Water turbine: potential→kinetic→electric ~90%
- Electric motor: electric→kinetic > 90%
- Condensation power plant: chemical→thermal→kinetic→electrical ~40%
- Internal combustion engine: chemical→thermal→kinetic ~20%
- Thermal solar electricity: electromagnetic→thermal→kinetic→electric ~20%

2.11 Managing energy quality and energy quantity

Energy quality or energy quantity management can be done in different ways, as described in the following text.

[4] Entalphy is a measure of the energy content of a system/substance measured in kJ, and is often related to the mass (kJ/kg).

2.11.1 Managing energy quality

Using electricity for heating is waste of quality, as electricity can be used for so much else, for example, to operate engines. Heating of houses and hot water can be accomplished with energy forms of lower energy quality, such as hot water (60°C−70°C). In countries where electricity is a scarcity, electricity is not used for heating (as it is in, for example, Sweden). When it comes to the use of electricity, other factors also must be considered, such as how the electricity is produced, but this discussion is not included in this book.

It is possible to take advantage of that different energy forms have different energy qualities. For example, if steam is produced at high pressure and high temperature for some purpose in an industry, the steam can later be used in other processes at lower pressures and at lower temperatures. If the pressure is high enough, electricity can be produced in a back-pressure turbine or in a cogeneration plant. Finally, the steam is condensed and converted from gas to liquid phase. Then, the hot water that is formed can be used for heating.

2.11.2 Managing energy quantity

Managing energy quantity differs from managing energy quality. Some examples of managing energy quantity are

- Lowered indoor temperature (when there is a heating requirement)
- Fewer and shorter freights, for example, optimize cargo volumes and select subcontractors closer to production facility
- Increased energy efficiency in a production line

Of course, these examples also have other dimensions and values, which are discussed more under nonenergy benefits in this book.

2.11.3 Reasons why some non-electricity-specific processes should be converted

To illustrate why a non-electricity specific process driven by electricity should be converted, an electric boiler and a fuel boiler are compared here. The boilers are compared from two aspects, namely fuel consumption and energy quality. An electric boiler is assumed to be replaced by a local fuel boiler fired with wood, oil, and so on or with district heating where the heat is produced in an externally placed boiler.

The most common way of producing electricity in Europe is through so-called condensing power plants. The efficiency of a condensing power plant depends partly on its age. An older plant has an efficiency of about 30% while a newer plant has an efficiency of at best 40%. The losses consist mainly of condenser losses. Reaching higher efficiency in a condensing power plant is theoretically impossible.

Producing electricity in a condensing power plant thus entails losses in the order of 60%−70%. This means that the efficiency of obtaining warm water (for domestic hot water and heating purposes) with a fuel boiler is higher than for an electric boiler since considerable losses have already occurred at an earlier stage when the electricity was produced. An example illustrates this.

$$\eta_{\mathrm{fb}}/\text{electric heating} = \eta_{\mathrm{fb}}/(\eta_{\mathrm{cpv}} \times \eta_{\mathrm{wh}}) \tag{2.19}$$

where η_{fb} is the 0.95, efficiency, fuel boiler; η_{cpv} is the 0.35, efficiency, condensing power plant; and η_{wh} is the 1.0, efficiency, water heater.

Eq. (2.19) gives

$$\eta_{\mathrm{fb}}/\text{electric heating} = \eta_{\mathrm{fb}}/(\eta_{\mathrm{cpv}} \times \eta_{\mathrm{wh}}) = 0.95/(0.35 \times 1.0) \approx 2.7$$

By converting the operation of the heating from direct electric heating to fuel boiler heating, the efficiency is increased by almost the factor three. The calculation example is applicable to all fuel boilers. What, on the other hand, distinguishes the fuel boilers is among other things the amount of carbon dioxide emissions, which depends on the fuel the boiler is fired with. The efficiencies between different boilers also differ depending on the system design, for example, whether the system has an accumulator tank or not, and the boiler's performance.

References

Areskoug, M., Eliasson, P., 2011. Energi för hållbar utveckling – ett historiskt och naturvetenskapligt perspektiv. Studentlitteratur, Lund.

Rodrigue, J.P., 2010. The Geography of Transport Systems. Routledge, New York.

Part I

Energy efficiency gap

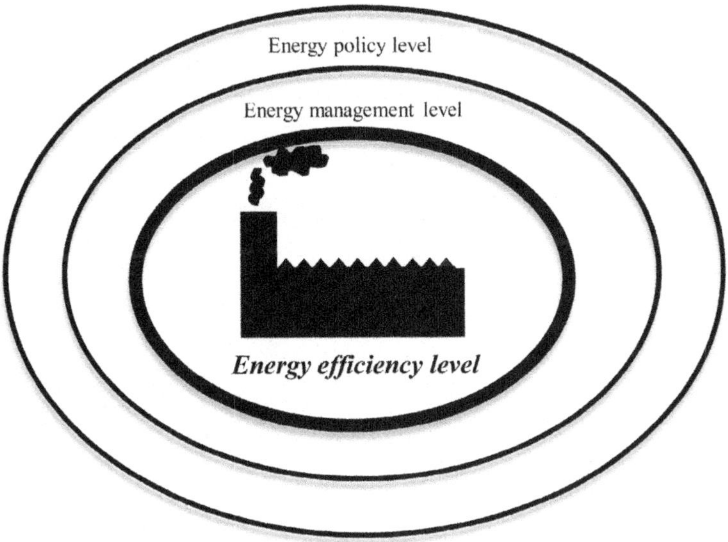

The difference between the potential for energy efficiency and what is de facto implemented is entitled the energy efficiency gap. Part I focuses on energy auditing and improving energy efficiency at companies from a primarily technical perspective and is based on the so-called energy efficiency gap.

Industrial energy systems

3

Chapter Outline

3.1 Industries - energy use

Industry accounts for a large share of global energy end-use. Depending on how the term *industry* is defined in different reports, the industry accounts for between 25% and 50% of global energy end-use. In 2014 the industrial sector used 43,300 TWh/year or 36% of the world's total energy end-use (IEA, 2017). The world's most energy-intensive industries are according to the IEA (2017) chemicals and petrochemicals, iron and steel, cement, pulp and paper, and aluminum representing about 69% of the global energy end-use (IEA, 2017). Chemicals and petrochemicals and iron and steel account for about 10% and more than 8% of the world's energy end-use, respectively (IEA, 2017).

About half of the energy end-use in the chemical and petrochemical industry comes from oil, and most of the remaining energy end-use emanates from natural gas. The iron and steel industry mainly uses coal in its processes. Both industries also use a lot of electricity. More than half of the energy end-use in the cement industry comes from coal, while the remainder is relatively evenly distributed between electricity, natural gas, and oil, and a very small part of biomass. For the pulp and paper industry, almost half of energy end-use comes from fossil fuels, while the remaining share is biofuels. Electricity is the dominant energy form for the aluminum industry, while almost half of the energy end-use is fossil fuels.

Introduction to Industrial Energy Efficiency. DOI: https://doi.org/10.1016/B978-0-12-817247-6.00003-1

3.2 Classification of industries according to Nace rev. standard

Nace rev. 2 codes are used in order to classify companies and based on the activities they are undertaking. Initially, activities are formed into groups from A to U, see Table 3.1.

Industries are classified according to the activity being conducted and each industry can have several activities (Nace Rev. 2 codes). Nace Rev. 2 codes for industrial sectors (code C, see Table 3.1) are shown in Table 3.2.

Each Nace Rev. 2 code in Table 3.2 can be divided into additional "subcodes." For example, code 17 can be divided into 17.1 (Manufacture of pulp, paper, and paperboard) and 17.2 (Manufacture of articles of paper and paperboard). Furthermore, the codes 17.1 and 17.2 can be further divided into more specific. Under code 17.2 there are, for example, 17.21 (Manufacture of corrugated paper and paperboard and of containers of paper and paperboard).

In the following section the world's most energy-intensive industries, according to the IEA (2017), are presented, that is, chemicals and petrochemicals, iron and steel, cement, pulp and paper, and aluminum. These industries represent about 69% of the global energy end-use (IEA, 2017).

Table 3.1 Nace Rev. 2 categorization.

A	Agriculture, forestry, and fishing
B	Mining and quarrying
C	*Manufacturing*
D	Electricity, gas, steam, and air-conditioning supply
E	Water supply, sewerage, waste management, and remediation activities
F	Construction
G	Wholesale and retail trade; repair of motor vehicles and motorcycles
H	Transportation and storage
I	Accommodation and food service activities
J	Information and communication
K	Financial and insurance activities
L	Real estate activities
M	Professional, scientific, and technical activities
N	Administrative and support service activities
O	Public administration and defense; compulsory social security
P	Education
Q	Human health and social work activities
R	Arts, entertainment, and recreation
S	Other service activities
T	Activities of households as employers; undifferentiated goods- and services-producing activities of households for own use
U	Activities of extraterritorial organization and bodies

Table 3.2 Nace Rev. 2, for the industrial sector (SCB, 2014).

C	Manufacturing
10	Manufacture of food products
11	Manufacture of beverages
12	Manufacture of tobacco products
13	Manufacture of textiles
14	Manufacture of wearing apparel
15	Manufacture of leather and related products
16	Manufacture of wood and of products of wood and cork, except furniture; manufacture of articles of straw and plaiting materials
17	Manufacture of paper and paper products
18	Printing and reproduction of recorded media
19	Manufacture of coke and refined petroleum products
20	Manufacture of chemicals and chemical products
21	Manufacture of basic pharmaceutical products and pharmaceutical preparations
22	Manufacture of rubber and plastic products
23	Manufacture of other nonmetallic mineral products
24	Manufacture of basic metals
25	Manufacture of fabricated metal products, except machinery and equipment
26	Manufacture of computer, electronic, and optical products
27	Manufacture of electrical equipment
28	Manufacture of machinery and equipment n.e.c.
29	Manufacture of motor vehicles, trailers, and semitrailers
30	Manufacture of other transport equipment
31	Manufacture of furniture
32	Other manufacturing
33	Repair and installation of machinery and equipment

3.3 Chemical and petrochemical industry

The chemical and petrochemical industry produces thousands of products, in quantities varying from thousands of tons to just a few kilograms, but only a few products account for the bulk of energy use in the industrial sector (IEA, 2007). Energy constitutes for up to 60% of the production costs for bulk chemical manufacturing, and for ammonia it may be as high as 80% (IEA, 2007).

The processes are uniform worldwide and according to IEA (2007) nine processes account for 65% of the energy use (including feedstock) within the industrial sector, namely:

- Petrochemicals
  - steam cracking of naphtha, ethane, and other feedstocks to produce ethylene, propylene, butadiene, and aromatics
  - aromatics processing
  - methanol production
  - olefins and aromatics processing

- Inorganic chemicals
  - chlorine and sodium hydroxide production
  - carbon black
  - soda ash
  - industrial gases
- Fertilizers
  - ammonia production

More than half of the total energy used are locked into the feedstock that is used to produce the product (IEA, 2007). That is, the carbon from the feedstock (e.g., oil and natural gas) is included in the final products, such as plastics, solvents, urea, and methanol. When, or if, the products are incinerated as waste treatment, this energy is used. However, energy used for feedstock is not possible to reduce through energy efficiency measures (IEA, 2007).

Fossil fuels are the main raw materials used in the industry, primarily crude oil and natural gas, but also coal (IEA, 2007). Around 75% of all feedstock is oil. Oil is used to produce intermediate chemical products, such as olefins (ethylene and propylene) and aromatics (benzene, toluene, and xylenes). These intermediate chemical products are thereafter further processed. The final products are plastics, rubbers, resins, and solvents as well as other petrochemical products. The second largest feedstock is natural gas, and products using natural gas as feedstock is, for example, ammonia and methanol. Ammonia is mainly used to produce fertilizer. The natural gas components ethane, propane, and butane are used for olefins production (IEA, 2007).

Olefins are, for example, produced in oil refineries by steam cracking and fluidized catalytic cracking (FCC) (IEA, 2007). In steam cracking the gaseous or liquid hydrocarbon feed, for example, naphtha produced from raw oil in the oil refinery, is broken down into smaller hydrocarbons, by diluting the hydrocarbon feed with steam and then heat in a furnace for a short time, without the presence of oxygen (Posch, 2011). For FCC a catalyst is used instead of steam. One olefin, the relatively inexpensive ethylene, is the most used petrochemical intermediate and generates several derivatives using processes, such as oxidation, hydration, oligomerization, alkylation, and chlorination (IEA, 2007). Aromatics are also, for example, produced in oil refineries using steam cracking of catalytic reforming and naphtha extraction (IEA, 2007).

In a report by IEA (2009) the 57 most important chemical and petrochemical processes, producing the 66 largest chemicals in terms of physical production volumes, were analyzed. Electricity use and potential savings in the industry were shown in the report and are presented in Table 3.3, where "Others" comprises heating, ventilation, HVAC (air-conditioning), and other support processes (e.g., office equipment).

As presented earlier, nine processes account for 65% of the energy use (including feedstock) within the industry (IEA, 2007). In addition, another study showed that 49 products stand for more than 95% of all energy used in the industry (IEA, 2007). Energy use (heat and steam use, excluding electricity use), when best

Table 3.3 Electricity use and potential savings in the global chemical and petrochemical industry (IEA, 2009).

Demand category	Process electricity use (TWh/year)	Share of total electricity use (%)	Process electricity savings (TWh/year)	Process electricity savings compared to total final electricity use (%)
Electrolysis	140	13	9	1
Motors	670	65	139	14
Lighting	70	7	14	1
Others	150	15	33	3
Total	1030	100	195	19

Table 3.4 Examples of energy efficiency measures from the database NEAD regarding production processes from manufacturing of rubber and plastic products (NEAD, 2018).

Energy efficiency measures	Savings (MWh/year)	Cost (thousand SEK)
Utilize free cooling from nearby creek to cool production	43	70
Machine park – shut down machines	29	0
Production, eliminate idle loads	52	0
Include water cooling for the forming machines	5	20
Shut down mixing machine night time and weekends	37	0
Install timers	45	27
Optimize operation times for production	8	5
Install timers for the forming machines	15	30

practice is used for these 49 products, is presented in IEA (2007), showing that energy use for production of ethylene is by far the largest energy user, amounting to 1700 TWh annually followed by production of ammonia, 1070 TWh/year. One of the reasons is that the produced amount of both ethylene and ammonia is large. Per ton production benzene has the largest figure, closely followed by toulene and zylene, amounting to almost 19 MWh/ton, when using steam cracking as production method. Using naphtha extraction as production method reduces the figures by approximately 20% to 30% (IEA, 2007).

Table 3.4 shows some examples of energy efficiency measures from manufacturing of rubber and plastic products for production processes. These have been implemented within the framework of the Swedish policy instruments PFE and the energy audit support between 2010 and 2014.

3.4 Iron and steel industry

Of the total energy use in Swedish industry, the iron and steel industry accounts for 15%, and this level has remained constant in recent decades. The same proportion applies to the total electricity consumption. The use of fossil fuels has been relatively constant over the past few decades (although the proportion of different energy carriers has changed over time), but now accounts for more than 55% of the total use of fossil fuels in Swedish industry, compared with 35% a few decades ago. This is because other industrial sectors have reduced their use of fossil fuels.

Iron is produced from iron ore in a blast furnace, where iron ore is added together with mainly lime and coke (which is made from coal in a coke oven, where coke oven gas is produced that may be used for various energy purposes), or via direct reduction of iron ore in solid form using natural gas. The sponge iron formed in the direct reduction process is then melted in an electric arc furnace. In these processes, the iron oxides in the ore are reduced to metallic iron. In the next step, the raw iron is usually dropped into a torpedo ladle car where reagent is added to purify the iron from sulfur. Then the steel production begins, which in principle is a matter of lowering the carbon content in the iron by the carbon reacting with oxygen and forming carbon dioxide. The reaction is performed in a converter where high speed oxygen is blown onto the surface of the liquid iron. External energy supply is not necessary because the carbon in the liquid iron is the fuel for the reaction. Here a gas (consisting of, among other things, carbon monoxide) is formed, which can be used for various energy purposes. Then the crude steel is poured into a ladle and alloying elements, for example, chrome and nickel, are added to get the steel that is in demand. Here, a certain amount of energy is required to keep the melt at the right temperature (coke oven gas is used for iron production in blast furnaces, otherwise other gases, oil, or electricity is used). The steel is thereafter casted by continuous casting or ingot casting, which requires electrical energy.

It is also possible to produce steel from recycled iron and steel scrap by melting in an electric arc furnace. Arc furnaces use large amounts of electricity to produce liquid crude steel. Temperature and alloy contents are then adjusted in a ladle before the steel is casted to desired dimensions by continuous casting or ingot casting.

When producing iron in a blast furnace, large amounts of coal are used, both as fuel and reducing agent. Coal is supplied to the blast furnace in the form of coke produced from coal in a coke oven. A small proportion of coal is also supplied in powdered form. In coke production coke oven gas is formed, which is an energy-rich gas. It can be used internally as fuel in heating furnaces, for production of process steam or for production of electricity in a steam turbine with subsequent generator. At the same time as the blast furnace produces crude iron, blast furnace gas is also produced. This gas is used internally in the plant, but the excess can also be used for the production of electricity and steam.

In Sweden (2018) there are two iron ore-based iron and steel plants (blast furnace), one iron ore-based sponge iron plant (direct reduction of iron ore and further processing in an electric arc furnace) and 10 scrap-based steelworks. In addition, there are 15 steel processing plants, for example, rolling (hot or cold), wire drawing, and extrusion (profiles and pipes). Since the 1970s, the production of crude steel has varied between 4000 and 6000 ton per year. During the same period, electricity consumption has remained relatively constant at 4 TWh/year, while LPG and natural gas use have increased from just over zero to about 2 TWh/year. Oil use dropped drastically between 1970 and 1990, from 10 TWh/year to about 2 TWh/year, while the use of coke (and coke oven gas) is at the same level, 12 TWh/year, as 50 years ago. On the other hand, the use of coke (and coke oven gas) has changed over this time period, initially a decline, then a rise, and then another decline.

Table 3.5 illustrates some examples of energy efficiency measures in the iron and steel industry regarding production processes. These have been implemented within the framework of the Swedish policy instruments PFE and the energy audit support between 2010 and 2014.

Table 3.5 Examples of energy efficiency measures from the database NEAD regarding production processes in the iron and steel industry (NEAD, 2018).

Energy efficiency measures	Savings (MWh/year)	Cost (thousand SEK)
Ny steering system, from pneumatic to fluid-based	0	100
Installation of pool coverage	12	60
Change of drilling machines	100	1800
Inverter	87	15
New filter system	500	0
Reduced idle load (lid, middle pipe, etc.)	70	0
Temperature control of circulation pump in cooling tank	10	24
Reduce oven temperature when not in use	288	50
Reduced grinding in the grinding machine	480	60

3.5 Cement industry

China is by far the world leading producer of cement producing about half of the demand in the world (Madlool et al., 2011). Cement is most commonly used to make concrete to be used in, for example, buildings. Besides cement, concrete also contains gravel and sand (aggregates) as well as water.

To produce cement materials like limestone, marl or chalk is needed, providing calcium carbonate ($CaCO_3$). These materials are usually extracted close to cement plants. Very small amounts of other materials (e.g., iron oxide and alumina) are also needed to produce the cement. This material is then transported to crushers and grounded into a fine powder. Before reaching the cement kiln the powder is preheated. The preheated powder enters the kiln at temperature of around 1000°C. In the kiln the powder is heated to about 1450°C using fuels such as coal, petroleum coke, gas, oil, and alternative fuels. In the kiln, which is a tube between 3 and 5 m wide, from 30 to 80 m long, and rotating about 3 to 5 times per minute, the powder is converted into an intermediate called clinker. The clinker is thereafter cooled down using air and the clinker may be transported to other grinding plants but are usually used on site. Four to five percent of gypsum is added to clinker and the mixture is grounded into a gray powder called ordinary Portland cement (OPC). Adding other minerals while grinding produces other types of cements, however, OPC is the most commonly used cement in the world.

There are also older technologies to produce cement using a wet slurry instead of power, but today over 90% of European cement clinker is produced in dry kilns using powder.

The cement industry is the third largest energy using industry in the world and emits large amounts of CO_2. Emissions from the cement industry arise partly from the combustion of the fuels required for the production, and from the calcination process where the carbon dioxide that is bound in the limestone are "liberated" when heated. About 60%–65% of the total carbon dioxide emissions derive from the calcination process and the remainder from the use of fuels.

The energy used in the cement industry emanates from mining, transportation to the plant, crushing, heating in the kiln, fans for cooling and grinding the clinker (Madlool et al., 2011). The largest electricity consumer is cement grinding and transport together with kiln feed, kiln, and cooler. The specific electricity consumption amounts to 65–140 kWh/ton, while specific thermal consumption amounts to 10–17 MWh/ton (Madlool et al., 2011).

3.6 Pulp and paper industry

The pulp and paper industry accounts for about 50% of the Swedish industry's total energy use during the recent decades. This industry uses a large amount of biomass to support its processes with heat (about 90% of bioenergy use in

Swedish industry), but electricity consumption is also large (about 40% of electricity use in Swedish industry). At the same time, excess heat is produced, which in some cases is used for district heating production and electricity production (SEA, 2017).

The type of energy used in the pulp and paper industry depends to a great extent on the type of production method for the pulp being used. Pulp is produced chemically (wood chips are boiled with chemicals and the cellulose fibers are thus separated from the binder (lignin)), mechanically (wood chips are grounded to pulp, where cellulose fibers and lignin are included) and by using recycled fibers (used paper is grounded and decolorized). In the production of chemical pulp, steam is preferably required, while mechanical pulp primarily requires electricity. The properties of chemical and mechanical pulp differ, which means that they have different application areas, for example, paper with low opacity (transparency), paper that is allowed to age, or a strong paper.

The production of *chemical pulp* follows some typical process steps. The log of wood first passes a barking drum that removes the bark from the log. The bark can then be burned in a bark boiler. Then a wood chipper is used to make wood chips from the log. Sometimes chips can also be purchased from other manufacturers. The wood chips are then impregnated with white liquor, which comes from the chemical recycling (see the following text), and thereafter cooked. After cooking, washing and screening of the fibers follow a separation of the chemicals from the fibers and removal of uncooked fibers and other debris is carried out (usually these steps are integrated). Finally, the pulp, as it is denominated after these steps, is bleached. When the fibers are washed, the washing water is usually taken care of for further delivery to the chemical recycling process. In this chemical recovery, the washing water, which is called black liquor at this stage, is first evaporated in an evaporator. The black liquor is converted to thick liquor in the evaporator and then combusted in the recovery boiler. In the recovery boiler, green liquor is formed, which in turn is converted to white liquor using burnt lime. The white liquor is then used for impregnating the wood fiber. The burnt lime is produced in a lime kiln. The cooking requires large amounts of steam and the washing takes place with hot water taken from the secondary heating system. The evaporation of the black liquor also requires large amounts of steam and is usually the process in the pulp mill which requires the most heat energy. In the combustion of the thick liquor in the recovery boiler, such large amounts of steam are produced that the needs of the plant are covered. Oil or natural gas is most often used in the lime kiln, but other fuels can also be used, such as tall oil pitch and wood powder.

In the production of mechanical pulp, wood chips are used, which are usually produced at the mill using barking drum and wood chipper. The bark can, as in the production of chemical pulp, be burned in a bark boiler. In the wood chipper logs are turned into chips. Chips may also be purchased from other manufacturers. Then the wood chips are grounded, which requires large amounts of electricity. There are different types of techniques, some of which add a smaller amount of chemicals and also preheat the chips. In the production of mechanical pulp, large amounts of steam are formed which are utilized in other processes.

In papermaking, drying (drying section) is by far the largest energy user. Steam is the main source of energy used in papermaking, but smaller amounts of electricity are required, for example, to power pumps. A paper machine consists of three different parts; the wire section, the press section, and the drying section. The stock (fiber slurry or suspension of pulp in water) is distributed on a wire (an endless cloth) and in the wire section the water is drained off and the moisture content is lowered by about 20% (from about 99%). In the subsequent press section, the moisture content is lowered to about 50%–60% when the sheet (formerly the stock) passing system of nips formed by rolls pressing against each other aided by press felts, where the water is pressed out. Next, the drying section follows, which consists of steam-heated cylinders. Most paper machines have many cylinders (sometimes up to 100 pieces), but there are also those that have a single large cylinder (and also combinations of these). Subsequently, when the paper is rolled up on large rolls (reel drum) at the end of the paper machine, the paper holds a moisture content of about 6%.

The pulp is manufactured in integrated paper mills or nonintegrated pulp mills (market pulp). A nonintegrated pulp mill sells the pulp to paper mills, while integrated paper mills has paper production where the pulp is produced.

The number of pulp and paper mills in Sweden has decreased in recent years, in 2018 there were 51, at the same time as production has increased, see Fig. 3.1 and Table 3.6. Over the past 20 years, the electricity demand for the pulp and paper industry has been around 20 TWh/year, with a tendency toward a downward curve, of which about 5 TWh/year has been covered by self-produced electricity. The use of biomass has increased from about 35 TWh/year (1990) to about 50 TWh/year (2017) and has largely replaced oil use, which in 2017 amounted to approximately 2 TWh/year (Skogsindustrierna, 2019).

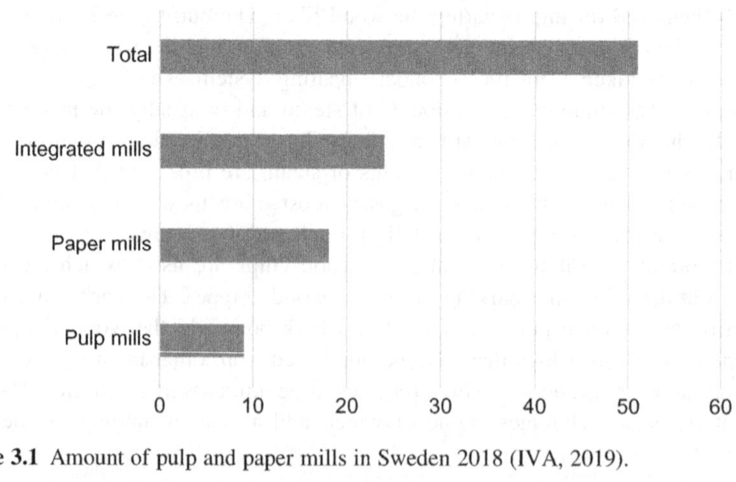

Figure 3.1 Amount of pulp and paper mills in Sweden 2018 (IVA, 2019).

Table 3.6 Structural changes in the Swedish pulp and paper industry (SFI, 2018).

Pulp	**1980**	**1990**	**2000**	**2010**	**2016**
Number of mills	72	48	45	41	40
Total capacity, million ton	10,5	10,9	11,7	13,1	12,7
Capacity per mill, 1000 ton	145	225	253	320	318
Paper					
Number of mills	62	51	48	40	38
Total capacity, million ton	7,2	9,5	11,1	12,1	11,2
Capacity per mill, 1000 ton	115	185	232	295	295

A relatively new concept is "biorefineries" in which the pulp and paper industry plays a central role (already existing in some mills). At biorefineries, not only pulp and paper are produced, but also other products, for example, textiles and chemicals, but also biogas and the basic raw material for other fuels such as hydrogenated vegetable oil.

Table 3.7 illustrates some examples of energy efficiency measures in the pulp and paper industry regarding the process of pumping. These were implemented within the framework of the Swedish instrument PFE[1] between 2005 and 2014.

3.7 Aluminum industry

Aluminum may be produced from primary resources, that is, aluminum oxide (alumina), so-called primary aluminum, or through recycling, so-called secondary aluminum. The major process to produce primary aluminum is the electrolytic process called the Hall-Héroult process (Tabereaux and Peterson, 2014). By connecting several electrolysis cells in series an electrical reduction line is formed (EIPPCB, 2014). In the cell, alumina is reduced into metallic aluminum (EIPPCB, 2014). There are two main types of electrolytic cells called Søderberg and Prebaked.

To store the liquid aluminum from the electrolysis holding furnaces are used, and these furnaces are also used for adding alloying metals or master alloys. Metallic impurities, nonmetallic particles, and hydrogen are removed and finally the metal is filtered. Scrap from internal sources and external deliveries is also melted at primary aluminum sites, and is either added to liquid metal or is melted separately before adding liquid metal (EIPPCB, 2014).

[1] For more information about PFE, see Part III in the book.

Table 3.7 Examples of measures from NEAD regarding pumping in pulp and paper mills (NEAD, 2018).

Energy efficiency measures	Savings (MWh/year)	Cost (thousand SEK)
Changed operation of hot water pump	250	25
Rebuild fire pump for continuous operation	70	10
Intermittent operations of backwater pump to resolver	50	5
Intermittent operations of backwater pump for the mixer	50	5
Change of pulp pump	257	251
Inverters for pulp pump	100	117
Inverters for condensate pump	159	40
Intermittent operations of cleaning water pump	76	78
Control of pulp line 2, pressure after valve opening	132	0

In the process of secondary aluminum scrap is melted, and as there is a variety of raw materials a variety of furnaces are used (EIPPCB, 2014). First, the raw materials must be sorted into wrought alloys (e.g., rolled plates, foils, and extrusions) and cast alloys (aluminum poured into a mold) and different furnaces are used for remelting the scrap dependent on the alloys. The metal is refined, by removing other materials and gases in, for example, a holding furnace and alloys may, for example, be added directly into the holding furnace.

The world average energy use for primary aluminum is 13.4 kWh/kg aluminum produced, and if auxiliaries (support processes) are added the average rises to 14.2 kWh/kg (Haraldsson and Johansson, 2018). In contrast to the large energy use for producing primary aluminum production of secondary aluminum requires only about 5% of the primary aluminum use and amounts to 0.56–2.5 kWh/kg, and the lower the quality of the scrap the more energy is usually needed (EIPPCB, 2014).

After producing the molten aluminum, and alloys are added, casting is performed or there is a direct delivery of molten aluminum to a customer. Different kinds of casting methods and machines are used in the casting of ingots, slabs, and billets in both primary and secondary production. Thereafter the ingots, slabs, and billets are processed either by extrusion, rolling, or shape casting.

Before extrusion, the billet is preheated to between 450°C and 550°C, depending on the product design, desired mechanical characteristics, and alloy. While extruding, a billet is forced through a steel die by hydraulic pressure, and an elongated shape with a consistent cross-section is formed (BCS, 2007). When rolling, the

Table 3.8 Examples of measures from NEAD regarding production processes in manufacturing of metal goods except machines and apparatus (NEAD, 2018).

Energy efficiency measures	Savings (MWh/year)	Cost (thousand SEK)
Remove holding furnace	1000	5
Reduce idle loads	30	0
Change cooling tower to closed cooling system	0	0
Inform staff for changed behavior	33	5
Install free cooling for plastic profiles	257	251
Improve insulation for pot	100	50
Exhaust fan improvements	80	150
Change in surface treatment line	172	300

thickness of the slab is reduced by forcing the slab to pass between counter-rotating steel rolls. Usually the slab passes through both a hot rolling step and a cold rolling step. The slab is preheated in a furnace to a temperature between 400°C and 500°C in the hot rolling step prior to rolling. The cold rolling step takes place at room temperature and after the hot rolling step.

Finally, the product may be treated by heat to change physical and mechanical properties of the metal. The reason is to prepare the metal for certain applications or for further processing (Rajkumar et al., 2014). The surface of the product may also be treated by anodic oxidation (anodizing), normally followed by a sealing process (hot or boiling deionized water), to enhance the properties of the metal surface.

In a review by Haraldsson and Johansson (2018), 52 production-related process measures for improved energy efficiency in the aluminum industry were identified. The majority of the identified energy efficiency measures address electrolysis, recycling, and general measures.

3.8 Mechanical engineering industry

The mechanical engineering industry is characterized by a relatively low use of energy, where a large part of the energy use stems from the so-called support processes. According to ECVG (2012), the primary production processes in the mechanical engineering industry are plastic forming, chip-cutting machining, piece-cutting machining, finishing, joining, degreasing, and other cleaning, painting, rust protection, pickling, heat treatment, chemical and electrolytic surface transformation, metal coating, and demetallization.

Table 3.8 shows some examples of energy efficiency measures from the manufacture of metal products, except machines and apparatus, regarding production processes, implemented within the framework of the Swedish policy instrument energy audit support between 2010 and 2014.

3.9 Other industries

Other industries not listed in the previous sections of this chapter, include, for example, ceramic industry, textile industry, wood manufacturing industry, and food manufacturing industry. These other industrial sectors may account for large shares of the energy use in specific countries. In Sweden, for example, other industrial sectors account for about 20% of the Swedish industry's total energy use (besides, pulp and paper industry, iron and steel industry, chemical industry, and manufacturing industry), where the wood manufacturing industry are the largest followed by the food manufacturing industry.

References

BCS (Brown Construction Services Inc.), 2007. U.S. Energy Requirements for Aluminium Production: Historical Perspective, Theoretical Limits and Current Practices. U.S. Department of Energy, Washington DC.

ECVG (Environmental Cooperation Västra Götaland), 2012. Tillsynshandledning Verkstadsindustri (Environmental audit manual). <http://extra.lansstyrelsen.se/miljosamverkan/SiteCollectionDocuments/Publikationer/2012/2012-tillsynshandledning-verkstadsindustri.pdf> (accessed 27.05.19.).

EIPPCB, European Integrated Pollution Prevention and Control Bureau, 2014. Best Available Technique (BAT) Reference Document for the Non-Ferrous Metals Industries. European Commission's Joint Research Centre, Seville.

Haraldsson, J., Johansson, M.T., 2018. Review of measures for improved energy efficiency in production-related processes in the aluminium industry – from electrolysis to recycling. Renew. Sustain. Energy Rev. 93, 525–548.

IEA (International Energy Agency), 2007. Tracking Industrial Energy Efficiency and CO_2 Emissions – Energy Indicators. OECD/IEA, ISBN 978-92-64-03016–9, IEA Publications.

IEA (International Energy Agency). 2009. Chemical and Petrochemical Sector – Potential of Best Practice Technology and Other Measures for Improving Energy Efficiency, OECD/IEA, IEA Information Paper.

IEA (International Energy Agency), 2017. Tracking Clean Energy Progress 2017, Energy Technology Perspectives 2017, Excerpt Informing Energy Sector Transformations. IEA.

IEA (International Energy Agency), 2018. Technology Roadmap – Low-Carbon Transition in the Cement Industry. IEA.

IVA, 2019. Så klarar svensk industri klimatmålen. En delrapport från IVA-projektet Vägval för klimatet. IVA, Kungl. IngenjörsVetenskaps Akademien, ISBN 978-91-7082-981-9, EO Grafiska, Sweden (in Swedish).

Madlool, N.A., Saidur, R., Hossain, M.S., Rahim, N.A., 2011. A critical review on energy use and savings in the cement industries. Renew. Sustain. Energy Rev 15, 2042–2060.

Nordic Energy Audit, 2018. NEAD (Nordic Energy Audit Database). <www.nordicenergyaudit.com> (accessed 02.01.19.).

Posch, W., 2011. Applied Plastics Engineering Handbook. William Andrew, Applied Science Publishers.

Rajkumar, K., Rajan, P., Charles, J.M.A., 2014. Microwave heat treatment on aluminium 6061 alloy–boron carbide composites. Proc. Eng. 86, 34–41.

SCB (Swedish Statistical Bureau), 2014. Uppgång i industriproduktionen (Rise in industrial production). <http://www.scb.se/sv_/Hitta-statistik/Statistik-efter-amne/Naringsverksamhet/Industriproduktionens-utveckling/Industriproduktionsindex-IPI/11307/11314/Behallare-for-Press/376113/> (accessed 28.02.18.) (in Swedish).

SEA (Swedish Energy Agency), 2017. Energy in Sweden 2017. Swedish Energy Agency Publication Department, Eskilstuna.

SFI (Swedish Forest Industry), 2018. Massa & pappersindustrin (Pulp and paper industry). <http://www.skogsindustrierna.se/skogsindustrin/branschstatistik/massa--pappersindustrin/> (accessed 24.04.18.).

Skogsindustrierna, 2019. Massa & pappersindustrin. <https://www.skogsindustrierna.se/skogsindustrin/branschstatistik/massa--pappersindustrin/> (accessed 27.05.19.) (in Swedish).

Tabereaux, A.T., Peterson, R.D., 2014. Aluminum production. In: Seetharaman, S., McLean, A., Guthrie, R., Sridhar, S. (Eds.), Treatise on Process Metallurgy: Industrial Processes, 3. Elsevier Ltd., Oxford Waltham, pp. 839–917.

Further reading

SEA (Swedish Energy Agency), 2018. Energy in Sweden 2017. Swedish Energy Agency Publication Department, Eskilstuna.

Industrial energy use

4

Chapter Outline

4.1 Benchmarking of energy use

The concept of "industry" includes a variety of industries. Industrial companies can be different in size, included in different sectors (e.g., textile industry, iron and steel industry, and food industry) and have different types of processes. Comparing these different types of industrial companies with each other and obtaining key figures regarding, for example, energy use is usually complicated. Key performance indicators (KPIs) that can be used include the following when comparing energy usage:

- energy use/area [kWh/(m^2, year)],
- energy use/person [kWh/(person, year)],
- energy use/air volume [kWh/(m^3, year)] alternatively [kWh/(m^3/s)] alternatively [kW/(m^3/s)],
- energy use/product (kWh/kg) alternatively (kWh/m^3) alternatively (kWh/quantity), and
- energy use/production hours (kWh/production hours).

Introduction to Industrial Energy Efficiency. DOI: https://doi.org/10.1016/B978-0-12-817247-6.00004-3

One may also use KPIs which relate to energy costs, such as:

- energy cost/m^2,
- energy cost/person,
- energy cost/turnover, and
- energy cost/value added.

All KPIs have both advantages and disadvantages. For example, two companies may have the same type of production but operate in two buildings with considerably different areas, which make it difficult to compare companies in terms of energy use per square meter. Another, and common way of categorizing energy use in order to enable comparisons is the so-called unit process concept. When applying the unit process concept, energy use is divided into different smaller entities or parts, where the overall breakdown is made between support processes and production processes. Support processes are the processes that support production, such as ventilation, lighting, and compressed air. Production processes are those processes that directly contribute to the production of a product, for example, melting and jointing. In some cases, industry-specific divisions are made where energy use is categorized in industry-specific processes. Creating KPIs for benchmarking can thus be done at different levels (see Fig. 4.1), where Level 4 corresponds to for example a breakdown into unit processes for both individual production processes as well as for individual support processes.

At Level 1, the factory is viewed as one entity, while an increasingly sophisticated and advanced form of division is made on Level 2 to Level 4.

In Andersson et al. (2018) a new method for benchmarking is presented using an energy efficiency index (EEI). Using the EEI benchmarking of the energy performance may be accomplished on both site level and process level. Energy managers at companies may use the EEI, but also organizations with an auditing role. In Johnsson et al. (2019), a taxonomy for energy end-use and greenhouse gas (GHG) emissions on a process and energy carrier level is presented for the wood industry, as well as potentially viable energy key performance indicators (KPIs), and suggested new energy KPIs within the industry.

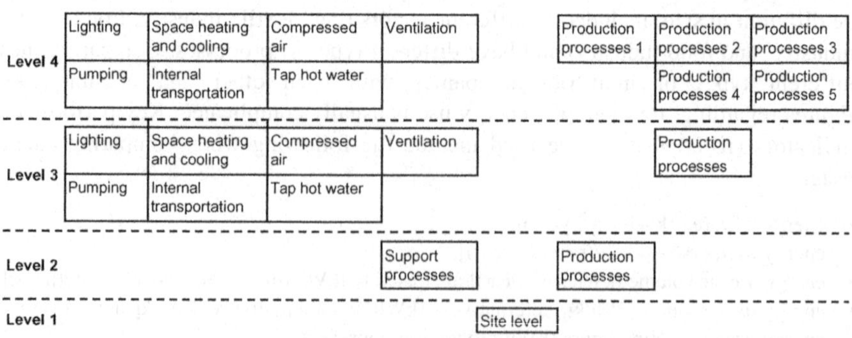

Figure 4.1 Example of four different benchmarking levels (Andersson and Arfwidsson, 2016).

4.2 The unit process concept

According to the unit process concept, energy use in an industry is divided into smaller parts or units (Söderström, 1996). A unit process is based on the purpose of a given industrial process, for example, to mix materials, to cool or dry products, to produce compressed air, or to move materials. Unit processes thus constitute the smallest parts of an industrial production system and its related energy use. Unit processes are therefore general to all industries and therefore enable comparisons between industries in terms of energy efficiency.

Unit processes are defined by the energy service they provide and there are two main categories: production processes and support processes. These unit processes constitute the "building blocks" for energy use, which enables comparisons between different companies within the same industry, but also between industries.

One concept with clear connection to the concept unit processes is unit operations. The concept of unit operation is used primarily in chemical technology, but also in other related areas such as the process industry. Examples of unit operations are separation, distillation, milling, enrichment, humidification, evaporation, and fluidization. These are similar in some respects to the division into unit processes. But while the number of unit processes is limited (there are 11 production processes and 10 support processes), a large quantity of unit operations occurs. The introduction of the unit process concept has entailed that several similar unit operations, which basically relate to the same thing, have been merged into a unit process. The following unit processes occur in the industry:

Production processes:

- coating,
- molding,
- heating,
- melting,
- drying/concentrating,
- packing,
- disintegrating,
- mixing,
- disjointing,
- jointing, and
- cooling/freezing.

Support processes:

- steam,
- pumping,
- compressed air,
- ventilation,
- lighting,
- space heating,
- space cooling,
- domestic hot water,

Table 4.1 Examples of coating in different industries (based on Söderström, 1996).

Industry	Coating
Textile industry	Dyeing, printing
Wood industry	Varnishing
Pulp and paper industry	Coating
Plastic industry	Foiling
Foundry industry	Painting
Mechanical engineering industry	Painting, thermal spraying, electrolytic surface treatment

- internal transport, and
- administration.

For example, in Andersson et al. (2018) the unit process concept is being used, where data from the Swedish Energy Audit Programme have been used. Energy use has been allocated into both production processes and support processes in wood, food, and metal industry, showing large discrepancies of the use of energy between the industries.

The production processes are reported in the next section and the support processes are presented in Chapters 8–11, with a short notification directly after the section regarding the production processes.

4.3 Production processes[1]

4.3.1 Coating

Coating is a process whose purpose is to give a product a new surface by applying a material layer of some thickness. Usually, the coating is not very energy-intensive. Instead it is the subsequent drying and the treatment of ventilation air for the environment (the premises) which are the major energy users. For coating, both physical and chemical processes are used (see Table 4.1).

4.3.2 Molding

Molding is a process whose purpose is to give shape to a product. Some molding processes require high pressure, which means a great power requirement. The energy is supplied directly via electricity, compressed air, or hydraulics. Table 4.2 presents some molding processes.

[1] This text is based on Söderström (1996).

Table 4.2 Examples of molding in different industries (based on Söderström, 1996).

Industry	Molding
Food industry	Physical processes
Bakery industry	Dough shaping
Glass industry	Hand-shaping, pressing
Foundry industry	Sand casting
Mechanical engineering industry	Plastic forming

Table 4.3 Examples of heating in different industries (based on Söderström, 1996).

Industry	Heating
Food industry	Blanching, boiling, frying
Textile industry	Dye bath heating
Pulp and paper industry	Boiling, preheating
Rubber industry	Vulcanization
Brick industry	Brick burning
Foundry industry	Heat retention
Mechanical engineering industry	Heat treating forging heating

4.3.3 Heating

Heating is a process whose purpose is to raise the temperature and, in some cases, cause changes in the product's material. In principle, any type of energy can be used for the supply of energy during heating. The power requirement varies depending on the heat transfer principle applied and the required temperature. Table 4.3 presents some heating processes.

4.3.4 Melting

Melting is a process whose purpose is to convert the material from solid to liquid form. In principle, any type of energy can be used for the supply of energy during melting. The power requirement varies depending on the heat transfer principle applied, but also depends on the melting point of the treated material. Table 4.4 presents some melting processes.

4.3.5 Drying/concentrating

Drying/concentrating is a process whose purpose is to remove solvents from a material. In principle, any type of energy can be used for supplying energy during

Table 4.4 Examples of melting in different industries (based on Söderström, 1996).

Industry	Melting
Food industry	Fat melting
Wood industry	Glue melting
Glass industry	Glass melting
Iron and steel industry	Metal melting
Foundry industry	Metal melting

Table 4.5 Examples of drying and concentration in different industries (based on Söderström, 1996).

Industry	Drying and concentration
Food industry	Drying of milk, drying of cereals
Textile industry	Drying of textile
Wood industry	Drying of wood, drying of varnish
Pulp and paper industry	Drying of pulp
Chemical industry	Evaporation
Mechanical engineering industry	Drying of surface treatment

Table 4.6 Examples of packing in different industries (based on Söderström, 1996).

Industry	Packing
Food industry	Filling of tubes and cans
Mechanical engineering industry	Shrink wrapping of pallets

drying/concentrating. The power requirement varies depending on the heat transfer principle applied, but also depends on the initial solvent content and the final content after drying/concentrating. Table 4.5 presents some drying/concentrating processes.

4.3.6 Packing

Packing is a process whose purpose is to protect or facilitate the handling of a product or set of products from the environment, or alternatively to protect and facilitate the handling of the product. Packing can be carried out in different ways, which means that the energy demand for this production process is very variable. Energy can be supplied in several ways; with electricity, compressed air, or other. Table 4.6 presents some packing processes.

Table 4.7 Examples of disintegrating in different industries (based on Söderström, 1996).

Industry	Disintegration
Food industry	Physical processes (e.g., milling)
Slaughter industry	Cleaving, cutting
Bakery industry	Crushing for rusk flour
Tanning industry	Cleaving (cleaving of hides)
Sawmill industry	Sawing
Pulp and paper industry	Chipping, grinding
Chemical industry	Grinding
Glass industry	Crushing (of recycled glass)
Brick industry	Crushing (of clay)
Foundry industry	Shakeout
Mechanical engineering industry	Cutting

4.3.7 Disintegrating

Disintegrating is a process whose purpose is to divide raw materials, materials in production or products, into several parts. In some companies, large installed power occurs, but usually the energy use in these processes is small compared to other processes in the company. Generally, electricity is used in disintegrating, but sometimes the disintegration is handled manually. Table 4.7 presents some disintegrating processes.

4.3.8 Mixing

Mixing is a process whose purpose is to create a new homogeneous product by combining several substances. The mixing processes are mainly electricity-based. Energy use is usually minor compared to other processes, but there are exceptions, such as mixing raw materials in the rubber industry or mixing dough in the baking industry. Table 4.8 presents some mixing processes.

4.3.9 Disjointing

Disjointing is a process whose purpose is to remove parts from raw materials. The parts that have been removed cannot be used directly. Disjointing usually takes place through physical processes, which are electricity-based with relatively low electricity consumption. Sometimes chemical processes are also used. Table 4.9 presents some disjointing processes.

Table 4.8 Examples of mixing in different industries (based on Söderström, 1996).

Industry	Mixing
Food industry	Mixing of raw materials
Wood industry	Mixing (chipboards, varnish)
Plastic industry	Mixing of raw materials
Rubber industry	Mixing of raw materials
Chemical industry	Mixing of raw materials
Bakery industry	Mixing of dough
Brick industry	Preparation of clay
Brick industry	Preparation of clay (e.g., incorporation of sawdust)
Foundry industry	Preparation of sand
Mechanical engineering industry	Mixing of varnish and glue

Table 4.9 Examples of disjointing in different industries (based on Söderström, 1996).

Industry	Disjointing
Wood industry	Debarking, milling, turning
Pulp and paper industry	Debarking
Plastic industry	Cleansing
Rubber industry	Cleansing
Foundry industry	Cleansing, milling, turning
Mechanical engineering industry	Milling, turning, electrolytical processes

4.3.10 Jointing

Jointing is a process whose purpose is to create a new product or new semifinished product by jointing two or more parts. In some cases, pressure and heat are required when jointing, which means that the shape may also change. This is thus a mixed process where molding, jointing, and sometimes even heating is included. Table 4.10 presents some jointing processes.

4.3.11 Cooling/freezing

Cooling is a process where the purpose is to lower a material's temperature. When freezing, the purpose is to lower the temperature of a material so that the

Table 4.10 Examples of jointing in different industries (based on Söderström, 1996).

Industry	Jointing
Wood industry	Gluing, screwing
Rubber industry	Gluing
Plastic industry	Welding
Foundry industry	Welding, screwing
Mechanical engineering industry	Gluing, welding, screwing

Table 4.11 Examples of cooling/freezing in different industries (based on Söderström, 1996).

Industry	Cooling/freezing
Food industry	Cooling of milk and freezing of ice cream
Rubber industry	Cooling of unvulcanized intermediate products
Plastic industry	Cooling of films, hoses, injection-molded products
Foundry industry	Cooling of casting

freezing point for water is passed, which is mainly done in the food industry. The energy supply usually is realized with the energy carrier electricity. The power requirement varies depending on both the heat transfer principle used and the properties of the treated material. Table 4.11 presents some cooling/freezing processes.

4.4 Support processes

The support processes are used to support the production. Support processes are sometimes also called generic processes, supplementary processes, and auxiliary processes. Sometimes these different names, or concepts, refer to the same type of processes, and sometimes not. Thus it is important to state the meaning of the concepts when using them. However, the concept support processes is well-defined, as it involves 10 processes:

- steam,
- pumping,
- compressed air,
- ventilation,
- lighting,

- space heating,
- space cooling,
- domestic hot water,
- internal transport, and
- administration.

The support processes are described in Chapters 8–11.

4.5 Examples of production processes and benchmarking within industries

4.5.1 Example: production processes within the engineering industry

Table 4.12 describes the connection between the engineering industry's unit processes and the included processes.

Processes 1–7 are common in the Swedish engineering industry, numbers 8–10 are less frequent, while numbers 11 and 12 occur mainly in specialized companies that perform surface treatments. For a more detailed description of the various processes, see ECVG (2012).

Table 4.12 The connection between the unit processes and the in-depth processes for the engineering industry.

Real process	Unit process
1. Plastic molding	Molding
2. Machining	Disjointing
3. Cutting	Disjointing
4. Finishing	Disjointing
5. Grouting	Jointing
6. Degreasing and other types of cleaning	Disjointing
7. Varnishing	Coating/drying
8. Grazing	Disjointing
9. Heat treatment	Heating
10. Chemical/electrolytical/resurfacing	Coating
11. Metal coating	Coating
12. Demetallization	Disjointing

4.5.2 Example: benchmarking of energy use in a corporate group or industry

For an individual company, it is usually not important to categorize energy use in detail. The benefit arises when comparisons with other companies are to be realized. When reporting energy end-use in a Swedish policy instrument (the energy audit support) a breakdown into essential support processes was realized but only one category for production, namely, production processes. The reason is that the support processes account for the largest share of the energy use of most Swedish small and medium-sized enterprises.

Before data collection begins, whether it relates to a region, a corporate group, a country, the EU or the international energy agency, it is important to determine which categories or processes are to be used. In an international project within the framework of the International Energy Agency research program, data were collected from several countries. The study showed that data for energy end-use, to the available extent, had been categorized in different ways. Thus some comparisons were not possible between the countries (Thollander et al., 2015).

If data collection is carried out in a uniform manner, comparisons are made possible between companies as well as within a region or corporate group. Furthermore, longitudinal comparisons are possible, that is, comparisons over time, both between individual companies and within the corporate group, regarding the processes that are included in the category. When selecting processes, it is recommended to use expertise, as it may be difficult, or in some cases impossible, to restructure the categorization afterward.

4.6 Energy and digitalization

The development in information technology enables very large data sets to be collected. Concepts that are usually used are digitalization and big data. The concept of industry 4.0 originates from Germany, which includes digitalization of the entire manufacturing industry. Energy use linked to digitization opens exceptional opportunities for detailed management and monitoring of energy use, as well as energy efficiency and monitoring of key figures. However, in order to enable efficient utilization of the digitalization potential, not only does data need to be collected, it must also be properly visualized and structured, otherwise the large amount of data become unmanageable. Real-time measurements of companies' energy use are already in operation regarding energy supplied. In the same way as for benchmarking the industry's energy use, certain typical data categories must also be determined when digitalizing energy use. For example, unit processes can advantageously be used for categorization. The support process division can be used for the company's support processes and then a selection of the essential energy-using production processes can be used.

If a company decides to start measuring a sample of the processes' energy use in real time, it is recommended that the company in advance prepares a plan for how to follow up energy data and how to integrate the data into the ongoing internal energy efficiency effort. In this case too, it is recommended that expertise be used.

References

Andersson E., Arfwidsson O., 2016. *Benchmarking av industriella små och medelstora företags energiprestanda. Presentation av en metod för beräkning av energieffektiviseringsindex.* [Benchmarking of the energy performance of industrial SMEs. Presentation of a method for calculating energy efficiency index]. LIU-IEI-TEK-A-16/02432-SE.

Andersson, E., Arfwidsson, O., Thollander, P., 2018. Benchmarking energy performance of industrial small and medium-sized enterprises using an energy efficiency index: results based on an energy audit policy program. J. Cleaner Prod. 182 (1), 883–895.

ECVG (Environmental Cooperation Västra Götaland), 2012. Tillsynshandledning Verkstadsindustri (Environmental audit manual). <http://extra.lansstyrelsen.se/miljo-samverkan/SiteCollectionDocuments/Publikationer/2012/2012-tillsynshandledning-verkstadsindustri.pdf> (accessed 27.05.19.).

Johnsson, S., Andersson, E., Thollander, P., Karlsson, M., 2019. Energy savings and greenhouse gas mitigation potential in the Swedish wood industry. Energy 187 (15).

Söderström, M., 1996, Industrial electricity use characterized by unit processes – a tool for analysis and forecasting. In: Proceedings of the 13th International Congress on Electrical Materials and Applications, June 16–20. Birmingham, UK.

Thollander, P., Paramonova, S., Cornelis, E., Kimura, O., Trianni, A., Karlsson, M., et al., 2015. International study on energy end-use data among industrial SMEs and energy end-use efficiency improvement opportunities. J. Cleaner Prod. 104, 282–296.

Energy auditing[1]

5

Chapter Outline

5.1 Introduction to energy auditing

Industrial energy efficiency is an important part of the economic transition toward improved sustainability. For an industrial company, there are four principal means of reducing energy costs, namely

- Operation changes, for example, more efficient operating and maintenance routines and changed behavior.
- Investments in energy-efficient technologies.
- Time control, for example, load management.
- Conversion from one energy carrier to another, for example, from electricity to district heating.

[1] Part of this text was published in an earlier version in Rosenqvist et al. (2012).

Introduction to Industrial Energy Efficiency. DOI: https://doi.org/10.1016/B978-0-12-817247-6.00005-5

The EU Energy Efficiency Directive promotes, among other things, the elimination of existing market barriers and inadequacies which may impede effective energy end-use. Energy auditing is an important tool for reducing barriers to energy efficiency. In addition, a well-structured energy audit is a first important step for successful energy management practices in most industrial companies.

The European Energy Audit Standard (EN 16247-1) defines energy auditing as a "*systematic inspection and analysis of energy use and energy consumption of a site, building, system or organisation with the objective of identifying energy flows and the potential for energy efficiency improvements and reporting them.*"

The importance of energy auditing varies depending on the type of industrial activity and in which country the company operates. The way in which the energy audit can be used as a means of promoting energy efficiency also varies for different industries.

The energy audit method described in this chapter has been developed in Sweden with the Swedish manufacturing industry as a base. The development of the method has mainly taken place within the academy from 1980s onwards. The Swedish market for energy audits in industry began to pick up around 2005, a few years after the first major electricity price increases in the Nordic electricity market. Due to the earlier low electricity prices, the energy issue was not particularly prioritized. Energy auditing was given further focus as Sweden in 2014 introduced the Act on Energy Audits for Large Enterprises. The Act states that large companies must carry out an energy audit every fourth year, either with the help of an external certified energy auditor or with internal competence within the framework of an existing management system, for example, ISO 14001 (environmental management) or ISO 50001 (energy management).

Today there are two international standards in the field of energy auditing; EN 16247 and ISO 50002. The standards are general in nature and wherefore a more detailed description of the energy auditing method is necessary. Evaluations of implemented energy audits in small and medium-sized manufacturing companies shows that the calculated (mainly technical) energy efficiency potential for improving energy efficiency varies between 16% and 40% of the total energy end-use, for electricity the estimated saving potential is approximately 60%. If the proposed measures are implemented and if the potential is achieved or not depends on the barriers to and driving forces for energy efficiency improvement.

The method described here is primarily intended for a technical energy audit in order to make energy end-use more efficient.

5.2 History of energy auditing

Before 2005, energy audits were carried out mainly within the academy and within the framework of various public policy programs. Energy auditing program is by far the most commonly used means of policy internationally for improved energy efficiency in companies. The world's largest energy audit program, the American IAC, has carried out more than 18,000 energy audits since its inception in 1976.

In 2006, EU's first directive on energy efficiency was published and stated that energy auditing was the most important policy for small and medium-sized enterprises. Based on the Directive, EU member states launched various energy efficiency policy programs, e.g. the Swedish Energy Agency introduced the first Swedish national energy audit policy program in 2010, which meant that companies could apply for grants corresponding to up to 50% of the cost of the audit.

In EU's second energy efficiency directive from 2012, there are requirements for large companies to carry out energy audits every 4 years. The energy audits can either be carried out with internal competence within the framework of existing certified energy or environmental management systems or with the help of an external certified energy auditor. EU member states have chosen to apply the directive's requirements for energy audits for large companies every 4 years in different ways. In Sweden, the energy audits must be carried out by a certified energy auditor and the energy auditor must comply with one of the current standards for energy audits. As regards requirements for becoming a certified energy auditor, one has to pass a certification test. The criteria the test is based upon are very similar to the European energy audit standard EN 16247-5. The requirement for certified energy auditors so far only concerns the Energy Audit Act for large companies.

5.3 Energy audit standards

5.3.1 EN 16247

The European standard for energy auditing, EN 16247, was published in 2012 and consists of five parts:

- Part 1: General requirements
- Part 2: Buildings
- Part 3: Processes
- Part 4: Transport
- Part 5: Competence for energy auditors

EN 16247 Part 1: General requirements

Chapters 1–3 of the Energy Audit Standard contain scope, normative references, and definitions of important concepts. Chapter 4, Industrial energy use, contains quality requirements and is broken down as follows:

- 4.1 Energy auditor
  - 4.1.1 Competence
  - 4.1.2 Confidentiality
  - 4.1.3 Objectivity
  - 4.1.4 Transparency
- 4.2 Energy audit process

Chapter 5 in Part 1 contains the energy audit process' parts:

- 5.1 Preliminary contact
- 5.2 Start-up meeting
- 5.3 Collecting data
- 5.4 Field work
- 5.5 Analysis
- 5.6 Report
- 5.7 Final meeting

EN 16247 Part 2–4: Building, processes, and transport

Chapters 1–3, in Part 2-4, contains scope, normative references, and definitions of important concepts, while Chapter 4 contains quality requirements divided into energy auditor and energy audit process. Chapter 5, in Part 2-4, gives an overview of the parts of the energy auditing process and follows the same outline as part 1: general requirements.

EN 16247 Part 5: Competence for energy auditors

Chapters 1–3, in Part 5, contains scope, normative references, and definitions of important concepts, just like Part 1-4. Chapter 4, in Part 5, reviews personal attributes divided into general, professional skills, and ethical principles. This chapter gives an overview of the requirements that are placed on the general and specific knowledge and skills of the auditor. Chapter 6, in Part 5, provides an overview of the acquisition of competence, how it can be maintained or improved, and is divided into general requirements, initial education, work experience, training and maintenance, and improvement of competence.

5.3.2 ISO 50002

The international standard for energy auditing, ISO 50002, was published in 2014 and consists of five parts:

1. Scope
2. Normative references
3. Terms and definitions
4. Principles
5. Performing an energy audit

5.3.3 Differences between energy audit standards

The international standard differs to some extent from the European. Among other things, a distinction is made regarding the three levels of energy audits provided by ASHRAE.[2]

[2] American Society of Heating, Refrigerating, and Air-Conditioning Engineers.

- Level I: Walk through assessment—invoices, easily accessible information, transit, focus on "low hanging fruits."
- Level II: Energy analysis—simpler measurements of interesting processes, continuation at level 1. Energy use is divided into different subprocesses.
- Level III: Detailed analysis of capital-intensive modifications—a detailed analysis of capital-intensive measures. This level is often characterized by advanced simulation, optimization, pinch analysis, and so on, which requires detailed measurements.

Furthermore, the European standard is divided into five substantially different parts. Another significant difference is that the ISO 50002 standard requires nonenergy benefits to be included in investment assessments of energy efficiency measures.

5.4 Classification of energy audits

Energy auditing methods can be described in terms of scope, depth, and purpose of the audit. The method presented in this chapter normally covers the entire plant studied, while other auditing methods may focus on parts of a plant. The depth of a study varies, from overall estimates to detailed studies, and often the depth varies between different processes in the same energy audit. In the method described in this chapter, a top-down strategy is used, and the energy audit always begins with a broad systems approach. As the audit progresses and the most important areas are identified, both data collection and analysis become more detailed. The purpose of an energy audit can be either to identify general energy-saving areas or to propose specific energy-saving measures. The method described here shows general energy-saving areas, which in turn can be used to identify and prioritize specific measures.

Energy auditing can also be described based on the classification of energy audits provided by ASHRAE. The three levels are as as shown above: level I (walk through assessment), level II (energy auditing and analysis), and level III (detailed analysis of capital-intensive changes).

Level I involve an assessment of energy costs and energy efficiency through analysis of energy invoices and possibly a short visit to the plant. Energy auditing at this level is looking for measures that have low or no cost and offer a list of improvements that need further study or evaluation. Level II contains a more detailed audit and analysis of the studied plant. Usually, some form of a detailed breakdown of energy use is made, either in activities and energy carriers or, as in this chapter, in unit processes, that is, ventilation, compressed air, hot tap water, and so on. Level III often involves an in-depth study, for example, a building simulation or energy system optimization where data from the level II audit is used. An energy audit using the method described here corresponds to levels I and II.

An obvious risk with the level I audits is that the division of the unit processes may be lacking or misleading. Research shows that the energy audit can then give a misleading picture of the actual end-use of energy, which in turn risks affecting the ongoing work on energy efficiency negatively.

5.5 The main parts of the energy auditing method

An energy audit is expected to provide information regarding end-use of energy in the processes. In addition, the results from the audit may be used to identify measures that increase energy efficiency, for example, switching from non-renewable to renewable energy sources and reduction of energy use in the company or in the energy system as a whole. The result is normally presented in a report.

This section describes a method for auditing and analyzing energy use in an industry. The three main parts of the industrial energy audit are as follows:

- Auditing of energy use
- Analysis
- Suggested measures

5.5.1 The three main parts of the energy audit method have different purposes

Auditing of energy use—what needs to be done?

- Define the system boundary
- Identify unit processes
- Quantify energy supply
- Distribute the energy use to the various unit processes

Analysis—how "bad" is the utilization of the energy supplied, what is the reason, and how good can it be?

- Identify both system errors and specific problems
- Identify the size of idling (overall and at process level)
- Identify technology that does not meet the requirements of BAT (Best Available Technology)
- Identify the need for energy quality for different processes in order to avoid that high-quality energy is used in processes with lower energy quality requirements, for example, electricity for heating
- Identify obstacles and driving forces
- Sum up the potential of the system as a whole

Suggested measures—how can the analyzed energy system be improved?

- Identify possible solutions
- Calculate the energy demand reduction of the possible solutions and analyze them
- Calculate the financial consequences of possible solutions.

5.5.2 Top-down strategy and iterative method

The energy audit is basically a process whose purpose is to obtain a picture of in which processes the energy end-use takes place, and to propose energy-efficient measures and possible changes of energy carriers. The parts of the energy audit method have been presented above, but the actual workflow does not follow a straight line from energy

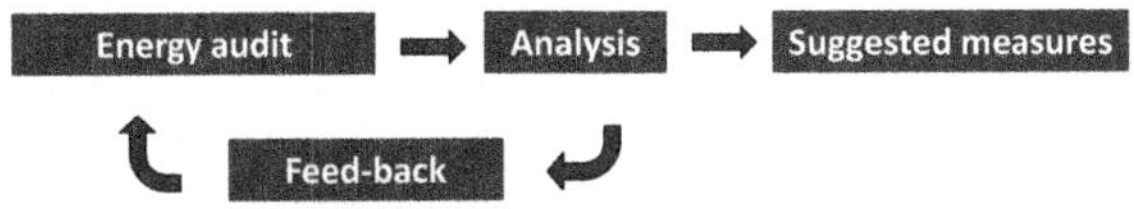

Figure 5.1 Schematic workflow for an industrial energy audit.

auditing to proposed measures. In practice, the energy analysis must be started, and measures are to be identified as soon as the collection of data for the energy audit starts. The energy analysis and ideas for measures also influence the need for data collection. Ideally, the method can be described as an iterative process, see Fig. 5.1. Iterations are terminated when a sufficient amount of data has been collected in order to propose relevant measures. This is also influenced by the difficulties to gauge data collection.

The iterative process is combined with a top-down strategy. In this case, this means that the studied object (organization, company, or plant) is initially regarded as a unit with its own streams of energy and materials, see Fig. 5.2. The total use of purchased energy is quantified. The use of materials and the products can be described but must not necessarily be quantified.

The next step is to describe the unit processes and analyze them in terms of energy use, see Fig. 5.3. This represents the second level of the top-down strategy and provides an initial overview of energy use, which enables a relatively rapid identification of key areas where additional resources for measurements and analyses are required (Franzén, 2005).

On the second level, data should be identified that enables distribution of energy use to relevant unit processes. Which unit processes are relevant depends on the purpose of the audit and which type of company that is audited. The top-down strategy may, if necessary, continue further down in the processes, for example, with focus on individual components. The amount of detailed information needed depends, among other things, on the purpose of the audit, the complexity of the proposed measures, and the timeframe for the energy audit. The ability to collect data is another issue regarding the information needed and data for thermal energy (steam, hot water, etc.) are usually more difficult and more expensive to collect compared to electricity data.

The purpose of working from the top and down in the system is that detailed information only needs to be produced if the energy auditor considers it necessary. The level of detail will vary between different unit processes. Sometimes the energy use in all unit processes in a production line is summarized under the heading production processes. This can be done for practical reasons, for example, when a machine with only one energy carrier handles several device processes, or because of a delimitation in the energy auditing, if, for example, only support processes are in focus.

5.6 The energy audit activities

In practice, the schedule for the activities during an energy audit may resemble as follows (where each point in the activity schedule corresponds to about one week, but the variation is great between different energy audits and activities):

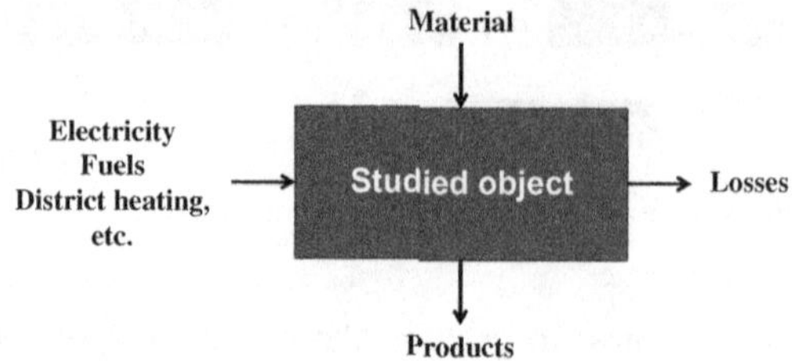

Figure 5.2 Energy audit with top-down strategy in the initial step (first level) where the studied object is considered as one unit.

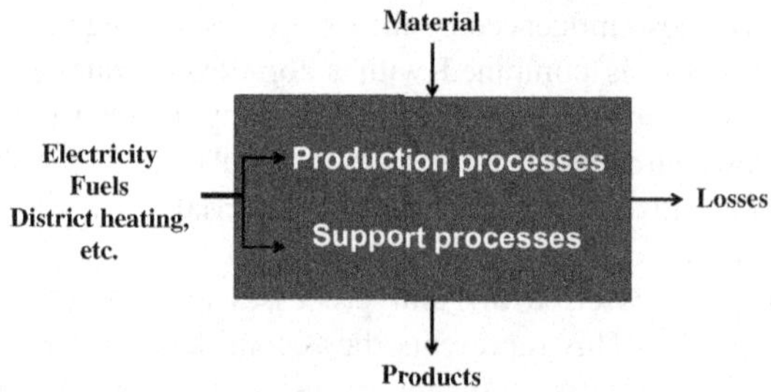

Figure 5.3 Second level in an energy audit with top-down strategy, which means that unit processes are studied in more detail compared to the first level.

1. Start-up meeting with the company in order to start collecting information from company responsible. Waiting for the company responsible to deliver data. Collection of basic data, such as annual energy statistics (from energy supplier), available data (if possible) from existing measurements at the plant and technical drawings. Description of unit processes. Finally, analysis of collected material to identify, for example, the largest energy efficiency potentials and where supplementary measurements may be needed.
2. First visit on site—auditing of energy use in different unit processes, collecting data, measuring, establish energy loggers, and establish calculations regarding the equipment. Analysis of data from the visit. At this stage, detailed information is available, but a top-down strategy is still applied. Analysis to seek answers to the following questions: Where are the greatest energy efficiency potentials? Which unit processes are most promising from an energy efficiency perspective? Which areas (system, processes, and so on) require more information or knowledge?
3. Second visit on site—removal of energy loggers and additional instantaneous measurements. Analysis of data from energy loggers. Complete the energy balance.[3]
4. Calculation of the energy efficiency potential of the measures. Identification of suppliers who can provide the equipment and system solutions required for the measures to be implemented. Waiting for data from suppliers to arrive. Completion of the report.

[3] Energy balance means that all energy supplied is assigned to the energy used in the company and a balance between them must prevail. For more information, see section 5.6.9 Data collection, step 4: Create power balance and energy balance and Figs. 5.7 and 5.9.

5. Third visit on site—the result is presented. Validation of the proposed measures with the company's staff. Some energy efficiency measures, although the calculations show that they are optimal, can be down-prioritized to second or even third best choices after discussion with the company's staff. In rare cases, a proposed measure can be removed completely if the company considers it to be uninteresting.

Note that the activity schedule is only an example of how an energy audit may be carried out. The same procedure can be applied to different scales. An energy audit according to Level I at a small company (about five employees and 500 MWh/year) can be completed within two weeks and requires fewer than 30 manhours. On the other hand, it would be quite possible to spend several months on a larger and more complex plant (e.g., with more than 5 buildings, many different types of energy, and several production lines) that require a more basic energy audit and detailed measure proposals. The depth and extent of the audit is often governed, given that the energy auditor possesses the necessary skills, the company's willingness to allocate resources in the form of both money and time. An insecure client who wants to keep costs down often places tight demands on the maximum number of hours for the audit, which in turn can force the energy auditor to prioritize down parts of the energy audit.

5.6.1 Example: planning an energy audit for a large company based on EN 16247

Needs and objective

The company must carry out an energy audit according to EKL (the Swedish Energy Auditing Act for large companies). The objective is thus to carry out an energy audit that meets the requirements of EKL.

Expectations, scope, and delimitations

The company CaseCompany Inc. expects the project responsible, in this case EnergyAudit Inc., to carry out an energy audit based on current legislation and regulations and to propose profitable energy efficiency measures.

Project responsible party (EnergyAudit Inc.) expects active participation by the company (CaseCompany Inc.) regarding data collection and, if necessary, advice and suggestions on directions for establishing energy balance and energy efficiency measures.

The energy audit covers the entire plant. All energy-using systems in the company are divided into support and production processes.

Schedule, GANTT scheme

Fig. 5.4 shows an example of the GANTT scheme.

A first sub-report of the support process space heating including the building envelope will be presented in March 20YY. Afterward, the energy audit focuses on

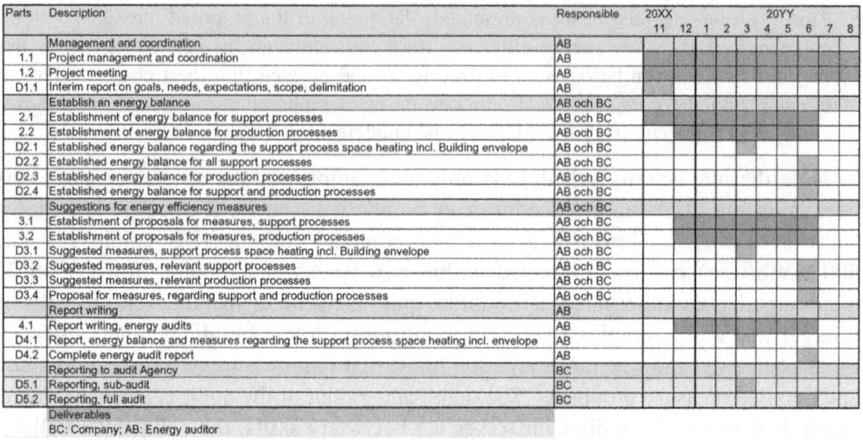

Figure 5.4 Example on GANTT-scheme, time plan, for an energy audit.

lighting and compressed air and then on other energy-using support and production processes.

Given that the auditing of the process space heating, including building envelope, takes time, an energy audit of either the process lighting or the process compressed air is completed, in order to comply with current legislation with requirements for subreporting in quarter 1 (q1), 20YY.

Criteria for assessing measures

Unless otherwise stated, the auditor and the company representative agree that payback criteria shall be used for profitability calculations.

Time and resource consumption

Expected time spent for project responsible party (EnergyAudit Inc.) and the company (CaseCompany Inc.) is negotiated in a separate process.

Requirements for data to be collected before the energy audit commences, as well as availability, validity, and format of energy and activity data

The company (CaseCompany Inc.) is expected to provide statistics on operational data for all energy carriers as well as, where this occurs, measured data regarding energy-using processes of significance for the energy audit. Part of the data collection is done by the company via the provided software for energy auditing.

Possible measurements or inspections, or both

In agreement with the company (CaseCompany Inc.), the party responsible for the project (EnergyAudit Inc.) conducts measurements of the energy-using processes that are considered to be of importance for the energy audit.

Other collection of information

The company (CaseCompany Inc.) provides information in accordance with EN 16247 regarding

- Background to the energy audit
- Legislation, regulations, or other restrictions affecting the scope or other aspects of the proposed energy audit
- Agenda in a greater strategic context (planned projects, management of facilities when outsourcing)
- Management system (environmental, quality, and energy management systems or other systems)
- Changes that may affect the energy audit and its conclusions
- Possible opinions, ideas, and limitations on potential measures for improved energy efficiency
- Expected results and requirements on the format of the report
- Whether a draft of the final report is to be submitted to the organization for commenting.

Protocol on implementation

When carrying out an energy audit, a protocol should be provided, according to Table 5.1, which describes, among other things, which standard to be followed and responsible persons.

Method description in report

The project responsible (EnergyAudit Inc.) describes the processes and tools that have been used in the energy audit and establishes timetables and identifies any need for additional measuring equipment in the method section of the audit report.

Data gathering

Data are collected in accordance with the valid standard EN 16247.

Field work, accomplishment, and site visits

Field work, accomplishment, and site visits are carried out in accordance with the valid standard EN 16247.

Analysis

Analysis follows the valid standard EN 16247.

Table 5.1 Protocol when carrying out an energy audit.

Questions	Answers	Date (if needed)
Standard to be followed	En 16247	
Responsible person at the company	Name	
Responsible auditor	Name	
Other persons	Name	
Inform the relevant staff and other stakeholders about the energy audit and any requirements imposed on them in connection with the energy audit	Company has been informed about this	20XX-12-01
Assure that regarded actors cooperate	Company assures this	20XX-12-01
Inform about abnormal conditions, maintenance work, and other activities that occur while the energy survey is being carried out	Company assures this	Ongoing based on demand
Agreement with the Energy auditor about: 1. Actions that give the energy auditor access 2. Safety and protection regulations 3. Resources and data to be made available 4. Confidentiality agreements (e.g., tenants in a building) 5. Suggested times for visits with priorities for each visit 6. Requirements for special measurements 7. Procedure that should be followed when installing measuring equipment	Company assures this	Ongoing based on demand

Firstly, the ranking of measures is based on the payback time and, secondly, the investment cost.

Report

Report follows the valid standard EN 16247.

Final meeting

At a final meeting, the energy auditor shall

1. Hand over the report from the energy audit
2. Present the results of the energy audit in a way that facilitates the decision making of the company
3. Be able to explain the results

The need for a follow-up should be discussed and decisions made.

5.6.2 Definition of the energy auditing object and system boundaries

The object of the energy audit and its system boundaries must be defined. Since it is an energy audit, the energy flows will be studied, but the delimitation can be done in different ways. For example, the study can be limited to focus on reducing the need for a certain type of energy (e.g., electricity), reducing the need for fossil fuels, or reducing the need for purchased energy in general. The delimitation can also be done, for example, by focusing on a certain part of the object (e.g., the building envelope) or only support processes. The system boundary can be the company or part of the company, a building or a physical area with several companies. The boundary can also be wider. For example, management systems and economic aspects can be considered, or consideration can be given to the energy embedded in different products, that is, the value chains of the products.

Since the method is iterative, the system boundary may need to be limited or widened depending on the limitations and possibilities that appear during the energy audit. One such example is if the company being studied owns the building where the business is conducted, but that in connection with the energy audit it become known that the company has a tenant who conducts business in a smaller part of the building. Instead of trying to figure out how much energy the tenant uses, the system limit can be redefined so that the tenant is also included. Another example is when there is an existing or possible symbiosis between two neighboring activities. In this case, the system can be extended from one to two companies. Another common example is that the company rents the building where the business is conducted. In this case, it is possible to exclude the building from the energy audit, but it is probably easier to find more and better efficiency measures by treating the building and the business as a system, with two participants: the company's owner and the property owner.

5.6.3 Specific requirements for energy auditing of large companies

For energy audits that are made within the framework of the Energy auditing act for larger companies in Europe, there are for some EU Member States strict requirements that specify what is to be included and how the choice of system boundary is to be made, for example, only the energy costs that the company can prevail over must be included in the audit. This means that, for example, air travel and transport performed by a procured forwarding agent should not be included in the audit, while company car trips by rental car, own car, or the company's vehicle should be included. For more information, see the Swedish Energy Agency's website.

5.6.4 Energy audit and energy analysis

With the aid of the energy audit method, power balance and energy balance can be created, where energy supplied is distributed among different unit processes. In order to be able to propose measures for energy efficiency or changing energy carriers, an understanding of how the processes function is required. The purpose of the energy audit is not necessarily to produce a detailed description of the energy use, but rather to identify potential for energy efficiency and possibilities for conversion to renewable energy sources.

By initially applying a top-down strategy, an overview is created that enables the identification of potential changes at the structural level. Such an example may be that the energy auditor starts with a simpler analysis of energy use in general, that is, supplied energy given in the form of invoices, and then, in general, breaks down energy supplied to different processes. Furthermore, the combination of top-down strategy and iterative method for data collection means that it is possible to study details without being burdened with unnecessary details.

Data collection can be done in different ways. However, provided that a top-down strategy is applied, it is not necessary that all available data on each device are to be collected. Thus, the data collection is initially focused on easily accessible information at the general level that the energy auditor finds appropriate. Next comes an energy analysis of the collected data. The energy analysis probably generates new issues, which are used to direct the continued data collection.

At some point, data collection must be terminated, even though all questions have not yet been answered. It is often the time reserved for the project that determines when to terminate. Other reasons may be that continued measurements are not possible, that there are practical obstacles to changing a process or that historical data are missing. It is often necessary (and possible) to use data from a more aggregated level together with calculations, assumptions, and analyses. Data collection can be performed according to the bullet list below. These steps for data collection are well in line with the activity schedule presented earlier in the beginning of Section 5.6 where steps 4 and 5 in the data collection are integrated in point 4 of the activity schedule, and step 6 in data collection is done in connection with bullet 5 of the activity schedule. Some differences between the steps in the data collection and the activities in the activity schedule may of course occur.

- Data collection, step 1: Collect statistics regarding electricity, district heating, fuels, water, production rate, and so on. Compile drawings and production plans.
- Data collection, step 2: Visit the company to collect data.
- Data collection, step 3: Visit the company during nonproduction hours.
- Data collection, step 4: Create power balance and energy balance.

To complete the energy audit, two more steps are required:

- Data collection, step 5: Create an energy balance including proposed measures (showing the effects of the proposed measures).
- Data collection, step 6: Confirm the energy audit result.

The main purpose of the data collection is to provide basic data that can be used to identify relevant unit processes as well as create power and energy balances. Another aim is that it should provide sufficient information to be able to propose suitable measures for energy efficiency and the exchange of energy carriers. The origin and quality of the information can vary considerably, therefore sources of energy data should be provided.

5.6.5 Data collection, step 1: Gather statistics, drawings, and production plans

Before the first visit to the company to be audited, statistics on electricity, district heating, fuels, and water use can be studied. Also, information about other media, such as steam and compressed air, can sometimes be available before the company visit. Statistics of (or estimate of) the production rate, during the same time period as the above-mentioned media, can also sometimes be provided before the company visit. Production data make it possible to estimate the specific energy use, for example, in the form of kWh per unit produced, per hour worked or per ton of raw material used. These data can be useful in the distribution of energy and in calculating the result of proposed measures.

Prior to the first visit, it is also possible to ask the staff to provide information about which activities are being carried out at the studied facility, in order to get an idea of which unit processes and energy flows should be investigated in connection with the visit. Before the visit it is also possible to ask the staff to provide information about a relevant period to schedule a visit depending on the production level.

5.6.5.1 Electricity supply

Companies that sell electricity and electricity grid owners can assist with information about the electricity supply to the studied company. As detailed information as possible is advantageously and for example, electricity grid owners have consumption data broken down by hour for their customers (at least in Sweden). If the company where the energy audit is carried out is small or if the company has under-measurement, that is, individual measurements of, for example, a machine or a machine group, it may be problematic to get data divided on an hourly basis. Energy data for an entire year, regardless of time resolution, is then recommended to enable analysis of the outdoor temperature's impact on the energy demand.

Detailed information on total electricity consumption during the period for which the company's unit processes are audited is also important. One option is to log all incoming electricity during the same period as the measurements are carried out, but with today's hourly measurement from the electricity grid companies this is in practice never necessary.

From invoices it is possible to get information about prices, power output, reactive power, and so on. An invoice from the month of January normally contains a compilation of the energy use for the previous year (at least in Sweden).

To some extent, the data analysis can already begin before the first company visit. For example, diagrams of the energy use for production hours during different seasons can be compared in order to identify the heating and cooling requirements that are met with electricity. It is also possible to compare idle losses during

Electricity trading	Electricity grid company
Power subscription	Power subscription
Electricity use	Electricity use
(energy tax)	Subscribed demand
(VAT)	Load tariff
	High load tariff
	(reactive power)
	(energy tax)
	(VAT)

Figure 5.5 Difference in electricity contract between grid owner and power supplier.

different seasons and analyze the total energy demand, power peaks, reactive power, and so on.

5.6.5.2 *Electricity contracts*

An electricity contract consists in principle of two parts, see Fig. 5.5.

Electricity trading companies include companies that sell electricity directly to consumers. The electricity contract has a fixed fee and a variable part, which is influenced by the amount of electricity used. The variable part may in turn consist of either a locked forward price or a spot price (from the power market; Nordpool in large parts of Europe). Larger companies usually have a portfolio of different contracts to minimize the risk of high electricity costs. In addition to the cost of energy, there may also be a cost for electricity certificates (e.g., in Sweden), which can vary from year to year.

Electricity distribution companies include those companies that have a monopoly on the local electricity grid where the consumer (the company or the private person) is located. How the electricity contract is designed differs between different electricity grid companies. Normally, the contract consists of a fixed fee which varies depending on the power of the subscription and the type of subscription. There is also a variable part which normally, in Sweden, consists of two parts where the consumer, for example, pays a lower price during the period April to October, and a higher price during weekdays between 7 and 21 during the period November to March.

For apartment and property owners, the subscription is often constituted by a so-called fuse subscription, where the size of the fuse (e.g., 16 or 25 amps) determines how large the fixed fee is. For companies, instead, power subscriptions are offered where the fixed fee is determined by how large the power output is. The power fee refers to a certain period of time, for example, SEK per kW and month or SEK per kW and year. An increased power fee is normally taken out if the power subscription is exceeded.

Reactive power arises as a result of inductive loads in the company's electricity consumption and is stated as a cost for over-charging of reactive power, for example, SEK per kvar (kilovolt-ampere reactive) and month. The cost arises if the reactive power exceeds the one that the electricity grid company has prescribed.

Normally this is priced by a maximum allowable share (%) of the active power. The fee is added if, for example, the reactive power output exceeds 50% of the active power.

In Sweden, since January 1, 2018, an energy tax on electricity (electricity tax) is levied by the electricity distribution company, and not as before by the electricity trading company. For nonmanufacturing companies, the electricity tax is around 0.35 SEK per kWh excluding VAT (varies over time). For manufacturing companies, the electricity tax is considerably lower, more precisely 0.005 SEK per kWh, excluding VAT. Companies that run so-called virgin processing, for example, metallurgical processes, manufacture of mineral products, chemical reduction, or electrolytic process, pay no electricity tax.

The relation between Sweden's electricity use, electricity production, and electricity prices

Historically, Sweden has had very low electricity prices, which contributed to a high electricity consumption per capita (partly due to the expansion of nuclear power in the 1970s and 1980s). The high electricity consumption originates from, among other things, that the property stock have direct electricity as the primary form of heating, and that Swedish companies preferably chose electricity as energy carriers. The Swedish expansion of electricity production can in turn be seen as a reaction to earlier oil crises.

Deregulation of the electric market

The Swedish electricity market was deregulated in 1996, which meant that the electricity market was exposed to competition and that the barriers that existed for electricity trading were abolished. However, electricity networks are still a regulated monopoly. As with other markets, the electricity price decreases if the supply increases and vice versa if the demand decreases.

The power market Nordpool

The deregulation of the electricity market in Sweden in 1996 entailed that electricity could be bought and sold in the same way as other commodities. Sweden and Norway first formed the power market Nordpool, which up to 2018 have included some more countries as members. The spot market price varies hour by hour. Furthermore, electricity can also be traded in futures, which means that prices are fixed for a longer period of time. Sweden is divided into four different electricity price areas, SE1 to SE4 where the electricity prices can differ between the areas at the same point of time.

5.6.5.3 District heating supply

Previously, it has been difficult to access data with hourly resolution for the district heating used. Therefore it has often been necessary to keep up with information about the amount of district heating used per month. However, with the energy efficiency directive's requirement for charging based on measurement, it is now usually possible to obtain district heating data divided by hour (at least in Sweden). Sometimes, at least in Sweden, it is also possible to use the customer number and PIN code to retrieve data from the supplier via the Internet. If the service is requested in advance, the supplier can sometimes deliver hourly resolution for the period that the energy audit refers to.

5.6.5.4 Fuel supply

Fuels such as oil, wood pellets, or wood chips are usually purchased in larger quantities during individual procurements. In order to calculate the amount of fuel that has been used for a certain period of time, information is needed regarding the stock of fuel at the beginning of the period, the purchased amount during the period in question and the stock at the end of the period. If this information is missing, the fuel consumption can be estimated by calculating the heat losses.

5.6.5.5 Other energy supply

There are also other possible energy supplies, such as steam, compressed air, liquid nitrogen, and welding gas. If the supplier charges the energy delivery, which is not always the case, it may be possible to access historical data. It is advisable to identify "unofficial" energy flows. For example, two adjacent companies may share the same district heating supply, air handling unit, or air compressor, there may be solar heating for heating of the domestic hot water, or alternatively the company may use residual products, such as wood chips from the neighboring company for heating. Energy supply that is not normally considered energy can sometimes prove to be important. Liquid nitrogen for cooling and welding gas for gas cutting (oxy-fuel) are two examples.

5.6.5.6 Various background data

Other information that may be useful in the first step of the energy audit includes:

- Water supply, in order to calculate hot water use
- Building drawings
- Drawings of the electricity supply system
- Drawings of other energy supply systems, for example, hot water pipes and compressed air systems
- Ventilation, for example, drawings of ventilation ducts, number of air handling units, and airflows
- Operating times for production processes, ventilation, and lighting
- Energy efficiency measures already carried out

Drawings can provide guidance as to where in the system or building the measurements can be realized and what information to ask for. The drawings can also provide information that can be used directly, for example, air flows may have been noted on the ventilation drawings. In addition, rated power for electrical equipment such as motors and heaters may have been documented in detailed electrical drawings. When using this type of data, it is wise to investigate whether the information is up-to-date. This is preferably done by asking the staff for historical changes or by measurements to confirm the accuracy of the information.

Since the energy audit is a time-limited project, it does not justify installation of permanent measuring devices. If sensors for, for example, measurement of power, current, air flow, pressure, or temperature are already installed and relevant data are already logged, or are possible to log, they should of course be used in the audit. Information on installed sensors may be given by staff prior to the visit.

5.6.6 Data collection, step 2: Visit the company to collect data

By visiting the company during production time, it is possible to study how the processes function, study the design and function of systems and equipment, validate descriptions, drawings and sketches that have already been obtained (e.g., for building and air handling units) and fill in any gaps where information is missing. During a company visit, it is possible to

- Note downtime for different processes, as well as when they are in operation, by registering how the processes work and by asking the staff
- Check the building envelope and heating system, possibly using a thermographic camera
- Examine service and maintenance procedures
- Systematically go through all unit processes
- Assess the function and condition of the equipment
- Inventory of the installed power, for example, for motors, light sources, and other devices
- Measure the power of a selection of equipment, such as the most energy-intensive process that is crucial for dividing the energy flow into different unit processes, or other equipment identified as crucial
- Look for control systems (e.g., for ventilation or a production line); if these occur, it should be investigated whether they are used and how they work (e.g., from control systems it is possible to get tables with operating hours or logged data concerning the production rate)
- Count the number of units (e.g., light source or computers) for later energy calculations
- Speak to key people in different areas, such as operation and maintenance

By talking to key employees at the company, it is possible to get information about, among other things, which unit processes are used, accessibility and downtime for different processes and operating hours. These conversations may also reveal energy efficiency actions that have already been identified by the staff. These key persons can be found among energy managers, production managers, caretakers, machine operators, or others, depending on the structure of the organization. For example, a caretaker or electrician at the company can help identify which processes need to be audited and show the placement of the switchgear for electricity measurements. The production manager can further inform about the different production processes and their respective operating times.

Information on the use of electricity is usually important but also data relating to air flow, water flow, pressure, temperature, and size can be useful. Often it is possible to find set values and statistics from, for example, plans, technical documentation, and monitoring systems. The main purpose of the data collection at the company is to get an overview, that is, it is not necessary to measure all parameters to create the power and energy balance. Still, sometimes measurement is the easiest solution. More precise measurements may be required, for example, to predict the results of proposed energy efficiency measures or changes of energy carriers.

Sometimes it is also possible to log the output signal directly from the control system using a logger that measures voltage or current, depending on the type of signal (usually 0–10 V or 4–20 mA). The logged information can be used directly as an indicator of the operation of various processes, such as an air handling unit. For further calculations, it is necessary to know the real state the signal represents and then adapt and adjust the logged values.

5.6.6.1 *Electricity measurements*

Data from the electricity supplier only is not enough to distribute the energy use on different unit processes. Internal electrical measurements for the studied plant occur sometimes, especially for the most important production processes. However, measurements of the electrical power are usually required for a selection of important processes in order to be able to distribute the energy use and construct the power and energy balance. Completing electricity measurements is often the main task for company visits.

In addition to measuring the electricity consumption directly with permanently installed meters, there are different ways to measure the power or current for electrical equipment in order to calculate the electricity consumption. To calculate the electricity consumption from electricity measurements, information about the active power is required. During the company visit, it is advisable to measure the active power directly with a wattmeter. Separate measurement of current and voltage enables calculations of the apparent power, which, however, cannot be used directly for energy calculations. The apparent power is only useful when the power factor is known, either from previous measurements or from technical documentation. Rated power for electrical equipment can be used in some contexts, but these data should not normally be used in the calculations. It is more appropriate to obtain information on power, or current and power factor, directly from measurements. The equations for calculating the power output are given below.

For single phase applies

$$P = U \times I \times \cos\varphi \tag{5.1}$$

For three phase applies

$$P = 3\sqrt{U} \times I \times \cos\varphi \tag{5.2}$$

where P is the power (W); U is the voltage (V); I is the current (A); and cos φ is the phase angle between voltage and current

A short company visit at the studied plant does not provide sufficient information to distribute the energy use. To see changes in power consumption over time, a combination of current transducers (AC current clamps) and data recording equipment can be used. One useful method is to allow data loggers to log the current for a week. This provides information on changes in usage patterns during both working hours and nights and weekends.

The use of electrical energy is often the focus of energy audits. Methods and materials for electricity measurements and evaluation of data are also developed for energy audits.

5.6.6.2 Active and reactive effect

Different industrial processes cause phase shifts, due to inductive and capacitive loads. When the phase angle φ is greater than zero, reactive power occurs. Reactive power (Q) has the unit voltampere reactive (var). Cos φ is the phase angle between voltage and current, while apparent power (S) is the product of voltage and current with the unit voltampere (VA). Active power (P) has the unit watts (W) (see Fig. 5.6).

5.6.6.3 Airflow measurements

Air flow measurements can partly be used to calculate heat losses in the ventilation, and partly to roughly estimate the electricity used to operate the fan motors. On the other hand, the electric power used to operate the fan motor can also give an idea of the air flow, and electricity is usually easier to measure. Sometimes it may be best to find out the operating time of a machine or working hours in a building, other times it may be more appropriate to log the air flow or the temperature in the ventilation, or to log the electricity used to operate the fan motor.

5.6.6.4 Temperature measurements

Temperature data are particularly useful for calculating the energy use of the unit processes space heating, space cooling, and ventilation. Information from

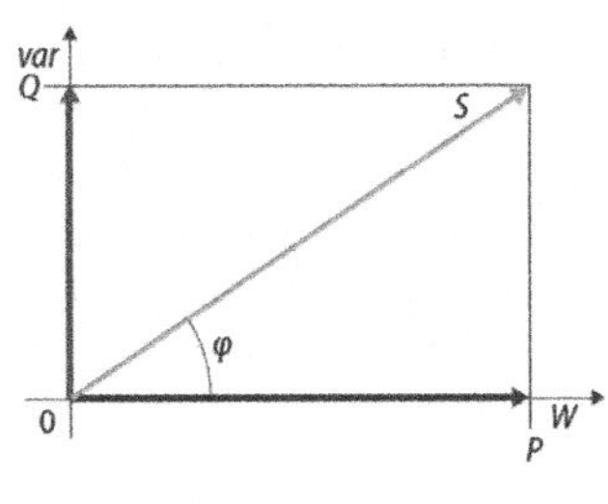

Figure 5.6 Active (P), apparent (S), and reactive power (Q).

temperature measurements can also be used to calculate the efficiency of heat exchangers. The exhaust air temperature is crucial for calculating the heat losses but is rarely recorded by the monitoring system and logged even less frequently. Using standard values for calculating the efficiency of the heat exchangers is a possible shortcut, which must sometimes be used when other available data are missing.

Temperature logging can be used to identify the operating time for different processes. A temperature probe can be attached directly to a machine or to an air flow or other medium connected to the studied process.

5.6.6.5 Size measurements

In order to analyze the heat losses from a building, information about the size of the building and the properties of the building materials included is required. Measurement of window and door sizes, wall thicknesses, and ceiling height often form part of the energy audit. When completing air flow measurements in ventilation ducts, the size of the duct must be known.

5.6.6.6 Inventory of installed power and operating time

One quick and effective way to find out installed power is to count the number of engines, light sources, computers, and other devices, and note the power output. Rated power for motors and other processes is available on nameplates or in technical documentation. Also note the rated cos φ, that is, the phase angle, for the motors, for future logging of electricity. If it is not possible to measure the actual power factor (cos φ), the cos φ in, for example, the nameplate gives an idea of the magnitude. However, it is strongly recommended that the power factor is obtained through measurements. This is due to the large variation of power factors that occur in different applications, for example, a compilation of 400 ventilation units showed no correlation with respect to the power factor cos φ (Reichenberg et al., 2009).

A reasonable estimate of the energy use of computers, copiers, servers, and printers is obtained by counting the number of units and calculating the power and energy use. Examples of useful tools are different templates and estimates for these calculations, for example, commercial software.

A suitable method for calculating power and energy requirements for light sources is to count the number of different light sources and note the rated power printed on the light source, on the packaging or in the technical documentation. The method works under the condition that the lighting is only controlled by an on/off switch. If there are discharge lamps, the power used by the ballast (control gear) must be added. It is important to know if it is a magnetic ballast (which is very common in older installations but is still used) or a more modern electronic device. Magnetic and electronic ballasts differ in power. For magnetic ballast light sources, the nominal power is multiplied by 1.2 and for electronic ballasts the nominal power is multiplied by 1.1. Even for energy auditing of lighting, commercial software is available on the market, which can make the work more efficient.

It can be difficult to get information about the operating time of the lighting. One way is to ask the staff to log the lighting or temperature of the selected light sources or to log the current used for lighting in order to reveal the actual operating time.

5.6.7 Data collection, step 3: Visit the company during nonproduction time

This step is performed to get information on which equipment is running during nonproduction time and to verify whether the equipment is necessary to be in operation or not. During the visit, it is possible to assess which support processes are in operation, such as lighting, ventilation, and compressed air, even though no operations are required that need these support processes. It is also possible to find production processes in idle or in standby mode.

By using eyesight and hearing, equipment running during nonproduction time can be identified. The measurement methods used during nonproduction hours are the same as during production hours, but the focus may be on other processes.

5.6.8 Data collection, step 4: Create power balance and energy balance

When the energy balance is created, data with different quality will be available, such as detailed production data, calculated operating hours and rated data. Furthermore, data relating to different time periods and time resolutions, such as statistical data for a full year, measurements over a week and instantaneous measurements, will be available. Data will also relate to different parts of the energy flow; sometimes energy supply, other times energy use or energy losses. Fig. 5.7 shows an energy balance, where the energy supplied of different kinds is distributed among different processes.

Another way of illustrating the energy use is using a so-called Sankey diagram, see Fig. 5.8 where the width of the arrows (flows) is related to the size of the energy use for the different energy carriers and the unit processes.

5.6.8.1 Some common cases to illustrate how energy use in a company may be distributed

Case 1: All energy supplied is assigned to a specific unit process. Sometimes this phenomenon occurs for support processes, such as fuels for internal transport or district heating for heating. Parts of the supplied energy may also be very specific to a certain manufacturing process, such as gas for welding or liquid nitrogen for cooling.

Case 2: Almost all energy supplied is assigned to a specific unit process. This is the case for the unit processes of space heating and hot tap water, see Fig. 5.7, which are the only processes that receive energy from the oil boiler. Moreover, this may occur if a hot water system is used for production, heating, and cooling. One example is whether district heating is used for both a heated cleaning bath for surface treatment and for heating of

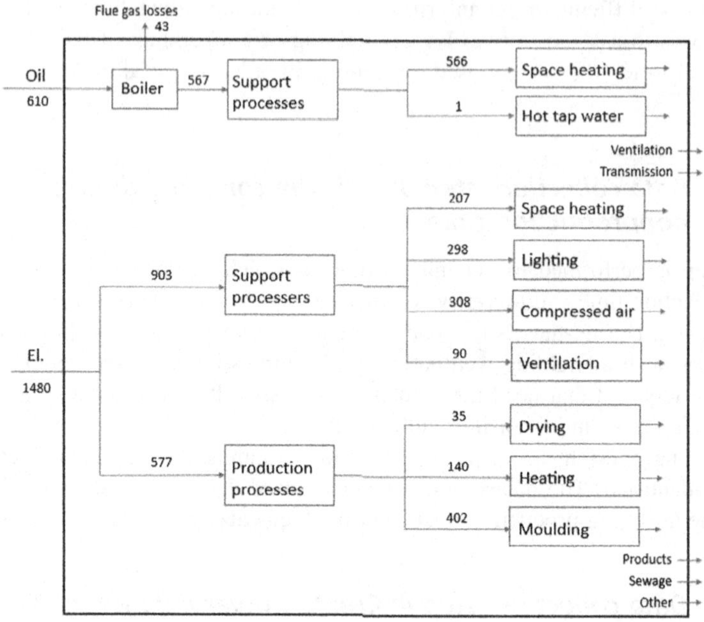

Figure 5.7 Energy balance with the help of an energy audit of a mechanical engineering company (Based on Trygg and Karlsson, 2005).

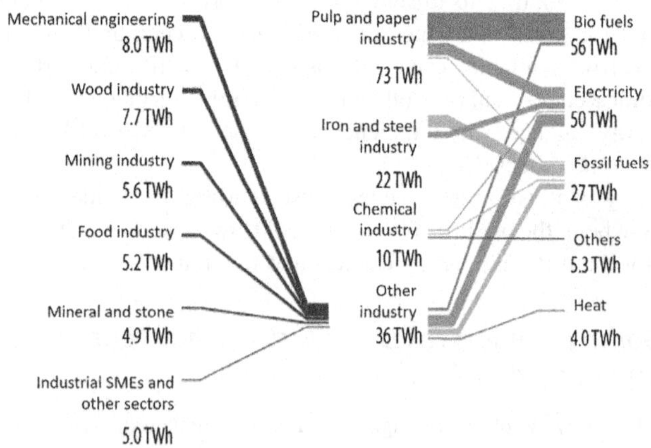

Figure 5.8 Sankey diagram for Sweden's industrial energy use (SEA, 2014).

premises. Space heating is often weather dependent and varies with the seasons. The energy use of the cleaning bath depends entirely on the production speed. If the monthly data from the district heating supplier is available and if the production rate during the summer is known (i.e., when the heating system is mostly switched off), the energy use of the cleaning bath can be estimated. Subsequently, the energy use during the summer and winter can be compared, whereon the energy use for the washing machine can be subtracted. The remaining energy demand then corresponds to space heating.

Case 3: Different unit processes may have the same delivery point, as is often the case with electricity. From the switchgear or the distribution board, where there is often practical opportunity to measure, electricity is supplied to many different processes. Usually, only one delivery point exists for all processes in one part of a building. In order to be able to distinguish the processes, at least some of them must be operated in a predictable manner. If, for example, a distribution board is completely dominated by a manufacturing process (e.g., a laser cutting machine), support processes for the manufacturing process (e.g., a compressor and a cooler) can also be connected to the same distribution board. The energy use for the laser cutter will vary, but if the cooling unit is operating continuously as soon as the laser cutter is in operation, and the operating time of the compressor is already measured for maintenance reasons, it is possible to distribute the energy use by logging the total amount of power to the distribution board. Thus, if there is only a common supply for several different processes, constant power output processes, or constant power consumption and programmed operating time (such as constant air volume ventilation units connected to an autotimer) should be identified.

Case 4: A unit process has more than one energy supply point. This is of course a common case. In a building with many air handling units, all units contribute to the same benefit, that is, they serve the ventilation process. Although there is a central system for remote control of all units, the energy use is usually not logged. The solution is to add all the energy use of all units and accept that the quality of the input data will vary, from less accurate data, such as observed rated power and estimated operating hours, to detailed measurements.

Case 5: Processes can have different energy carriers, for example, a drying plant can consist of an electric fan and a heater with hot water. If it is not possible to measure everything desired, it might be possible to use data from one of the energy carriers to calculate the operating hours for the other. For example, the temperature of the heat exchanger may reveal whether the fan is running or not.

Case 6: For reasons of safety, in certain cases there are two separate electricity supplies to the same unit process, for instance two separate air compressors can be connected to two different switchgears to prevent interruptions. Of course, there may be other reasons for having more than one delivery point to a unit process, for example, if the electricity distribution system has been supplemented because the process needs more electricity than the original switchgear was designed for.

5.6.9 Data collection, step 5: Create energy balance after adjustment (showing the effects of the proposed measures)

Depending on the type of problem the energy auditor has identified, various solutions are proposed. As previously mentioned, there are four main ways to reduce the industry's energy costs:

- Operation changes, for example, more efficient operating and maintenance routines and changed behavior
- Investments in energy-efficient technologies

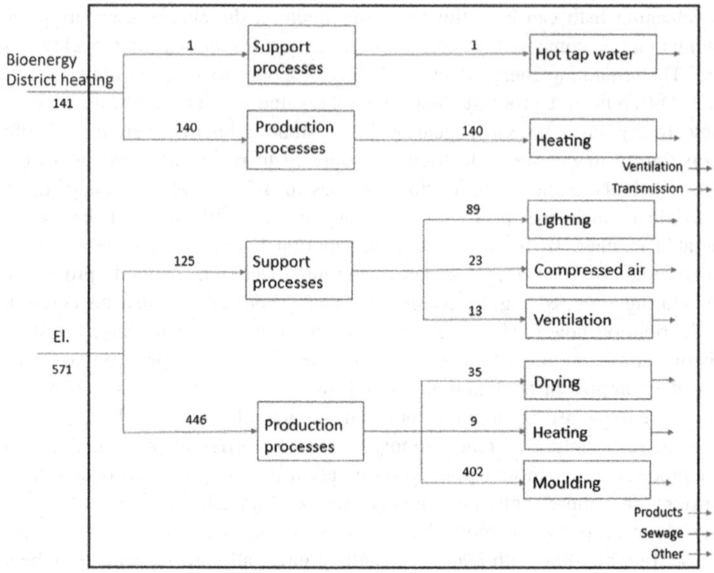

Figure 5.9 Energy balance after proposed energy efficiency measures (to be compared with Fig. 5.7) before measures have been proposed (Based on Trygg and Karlsson, 2005).

- Time control, for example, load management
- Conversion from one energy carrier to another, for example, from electricity to district heating

It should be noted that the list refers to reduced energy costs, which is not necessarily the same as more efficient energy use. Time control is not an energy efficiency measure, but a measure that only shifts energy use to another time. Furthermore, changing energy carriers can be an excellent way to reduce the use of nonrenewable resources or reduce energy costs. However, this does not automatically lead to more efficient energy use.

When the effects of the proposed measures are calculated, a new energy balance is presented, see Fig. 5.9. The new balance may have two different purposes: partly to show the potential for changes in energy use and partly to report the estimated effect of the measures. The potential for energy efficiency can be estimated with standard values for best available technology. If the effects of the proposed measures are to be presented, more detailed technical and financial calculations must be undertaken.

5.6.10 Data collection, step 6: Confirm energy audit results

This part is crucial, not least when it comes to whether the energy audit results will be accepted and used by the company. Normally, representatives from the company that has been analyzed and the energy auditors meet to discuss the proposed energy efficiency measures and the analysis of the energy balance. At this stage, the focus is primarily on the proposed measures. For some processes, there may be different alternatives. At the meeting, the energy auditors and representatives of the company are given the opportunity to discuss which measures are most suitable. Normally, no senior managers attend this meeting, instead the company is usually represented by a maintenance manager or energy manager.

References

Franzén, T., 2005. EnergiSystemAnalysMetod, EnSAM, för industriella energisystem [Energy System Analysis Method, EnSAM, for industrial energy systems]. Linköping Institute of Technology, Linköping [in Swedish].

Reichenberg, L., Borg, O., Franck, P.-Å., Nyström, I., Åsblad, Å., Thollander, P., et al., 2009. STIND—Energistatistik för Industrin, Etapp 2 (STIND—Energy statistics for industry). Swedish Energy Agency. Available at: <http://www.energimyndigheten.se/globalassets/statistik/festis/industrins-energianvandning-stind/stind-energistatistik-for-industrin-etapp-23.pdf> (accessed 04.04.19.) (in Swedish).

Rosenqvist, J., Thollander, P., Rohdin, P., 2012. Industrial energy auditing for increased sustainability—methodology and measurements. Sustainable Energy-Recent Research. InTech Publisher, ISBN 980-953-307-587-2. Available at: <http://www.intechopen.com/books/sustainable-energy-recent-studies/industrial-energy-auditing-for-increased-sustainability-methodology-and-measurements>.

SEA (Swedish Energy Agency), 2014. Sankey-diagram, Sweden. Eskilstuna. Available at: <https://www.energimyndigheten.se/globalassets/statistik/energibalans/3mb_sankey-sveriges-energisystem-2015.pdf> [in Swedish].

Trygg, L., Karlsson, B., 2005. Industrial DSM in a deregulated European electricity market—a case study of 11 plants in Sweden. Energy Policy 33 (11), 1445–1459.

Energy auditing and measurements in practice

6

Chapter Outline

Introduction to Industrial Energy Efficiency. DOI: https://doi.org/10.1016/B978-0-12-817247-6.00006-7

6.1 Introduction

One thing you for sure are eager to measure when aiming to improve energy efficiency is energy end-use. But if you are not the energy manager, and visit the company as an external energy auditor, there will hardly be energy meters exactly where you want them to be. Of course, it is excellent if you have the opportunity to work so long in advance that you can set up permanent meters on important objects. You will nevertheless still need to measure energy end-using machines and devices, in order to find the important objects to monitor permanently.

Normally, you will not measure energy directly, but it is other parameters—current, voltage, temperature, air velocity, and so on—that will form the basis for your calculations. What you will measure depends on the purpose of your calculations, for example, to create an energy balance for the entire plant or process, to calculate heat losses for a building or building stock, or to motivate or follow-up your suggested energy efficiency measures.

6.2 Why measure?

6.2.1 To understand

When you conduct an energy audit, there are many reasons why one wants to undertake measurements. One important purpose of measuring is, of course, that you as an auditor should understand what is happening and when. You want to be able to allocate energy use to different processes, you want to be able to analyze and make diagnosis on the whole plant, machines, or processes and be able to adress if they work properly or if something is incorrectly installed or operate inefficiently. Once you have done this, you want to be able to find possible energy efficiency measures by analyzing the measurements. You can also measure to follow up, by repeating previous measurements.

6.2.2 To explain

Another purpose of making measurements may be to be able to explain to others, for example, how the plant's energy use is functioning. Measurements visualized in clear graphs showing what you have found during the audit is a good basis for the plant's decision-makers. You can also benefit from using graphs and detailed measurement data when you need to ask follow-up questions to production personnel or equipment suppliers.

6.2.3 To calculate

A third, important purpose of the measurements are that these form the basis for calculations of energy efficiency measures. In its simplest form, a calculation of the improved energy efficiency resulting from changed operating hours for lighting, for example, or in more advanced form as a basis for the energy-related consequences of a rebuilding or for modeling of changes in a production process. In most cases there is no room for full-scale experiments—calculations and modeling are thus crucial.

6.2.4 Other reasons to measure

There may also be other reasons to measure. Others already measure to monitor, to regulate, to alarm, or to charge. As an energy auditor, you can sometimes benefit from these measurements. Another reason to measure can simply be to learn, to gain experience and skills, and to test different possibilities.

But you will be reminded at every new energy audit that there is one very good reason to measure: There is simply no data available from other sources. If there are other sources—use them. If not—go out and measure!

6.3 Data sources

There are a number of data sources one can use such as:

- Instantaneous measurements—real-time measurements
- Logged measurements—measurements repeated at a specified time interval and saved as a series of measured values
- Accumulated values—continuous measurement where many instantaneous values in a time series are summed to one figure

Instantaneous measurements can be made on-site, with temporary meters for field-measurements or by permanently installed meters. Process and building monitoring systems can also be used—you can overview and read several meters at a distance.

The instantaneous measurements have the advantage of giving you some control over the circumstances. Seeing (and understanding) what is happening is often possible by following changes in, for example, power, power factor or temperature. Sometimes, instantaneous measurements allow for greater accuracy. The downside is that only a limited number of values can be handled manually and saved when undertaking instantaneous measurements.

Logs with measured data are useful for energy audits, to the extent that logging deserves a section on its own (see Section 6.6). The use of accumulated data is limited, but if there is a meter accumulating data, you can create your own log.

6.3.1 Billing meters

Meters for billing provide very reliable figures. Sometimes there are log files, sometimes only accumulated values. Billing meters for energy, such as electricity, gas, and steam, are obvious to use, but other meters can also provide valuable information. The cold-water meter can, for example, provide a basis for calculations of the hot water use.

6.3.2 Monitoring and controlling systems

In larger, newer properties with ambitious property owners, there may be a building monitoring system [BACS (Building Automation and Control System) and SCADA (Supervisory Control And Data Acquisition)] where you can see, for example, operating times for ventilation systems, speed of pumps, temperatures in different rooms or how much a valve is open. Fig. 6.1 displays a screenshot from a process monitoring system.

The information available varies widely from plant to plant, sometimes even within the same company group. How much that is logged and how much of the log files that are saved also varies. Sometimes it is easy to start logging a parameter that you want to follow. Here you can also find nonenergy-related information that you can benefit from. If there is information about when people pass through the front doors or about when the intruder alarm system is activated, you will get exact information about the working hours.

In the same way as there are building monitoring systems for the buildings, there are sometimes monitoring systems (SCADA) for production processes. The parameters that are logged and saved are those important for the process—temperatures, currents, speed/frequency for variable speed drives, pressure drop for filters, number of products, and so on. Sometimes there are also support processes that are particularly important for the production included in the monitoring systems.

By accessing the monitoring/control system you can gain a lot of information about the design of the technical systems and the relations between them, especially if you can also access a process engineer or technician on site who can guide you. Apart from the data you want to gain, you can also look for possibilities to present real-time energy-related data in the monitoring system. If energy use is presented in

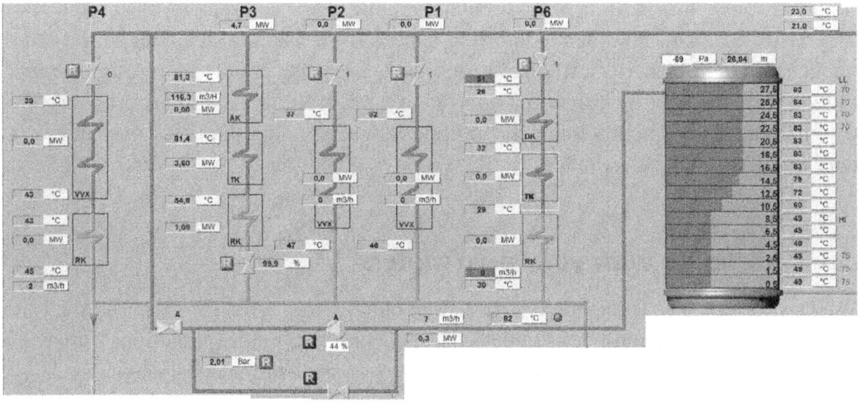

Figure 6.1 Example of process monitoring. Graphic interface. The picture shows an overview of the heat production in a district heating plant. P1 to P6 are the boilers. As the outdoor temperature is high, heat is only needed for hot tap-water and industrial processes. One boiler is running: P3 is delivering 4.7 MW to the district heating network. As only one boiler is used, only one regulating valve is active, and it is 99.9 % open. The main circulation pump is running. The number 44% indicates that is has a variable speed drive. The box with the bars to the right represents the heat accumulator.

the monitoring/control system, it can also be a monitoring/control tool for the energy management.

In manufacturing plants, there may be systems for production planning or quality monitoring with useful, stored data. In addition to direct energy-related data, you may find data that directly or indirectly gives you an idea of the rate of production, both during the time of the energy audit as well as historical data.

6.3.3 Controller device

For devices controlled or regulated electronically, there is a controller device. Either a programmable logic controller (PLC) or a controller device specially adapted for the purpose. If there is a good user interface, it is possible to retrieve some data from the controller device: measurement values from sensors, level of control signals, setpoint values, and control principles. For a ventilation unit, there may be temperatures, pressures, flows, valves and throttle control, speed for motors, and time schedules. Sometimes the control center does not have a local graphic interface, then you must access the BACS/SCADA to get the data you want. Another option is to find the analog control signal (typically 0–10 V). If there is one, it can easily be logged with a voltage logger (Fig. 6.14).

6.3.4 Data from other sources

There are also data from sources outside the studied plant that you can benefit from. An obvious example is historical weather data from various meteorological organizations. Information from regulatory authorities and certifying organizations may also be useful. Another possibility, especially if you are to calculate the consequences of major energy efficiency measures, is to use results from other people's models and calculations, as input to your own calculations. Researchers, suppliers of machines, equipment suppliers, buildings owners, and authorities may have such data that if so may be valuable.

6.3.5 Validate with your own meters

When you rely on other people's measurements, you are not sure what is measured and how. You often can only see data in one figure, without explanation. If there is a possibility to check and compare with your own measurements, instantaneously or with a shorter logging, then it is good to do so. This in order to gain more reliable data for the energy audit. For example, if you can compare ongoing temperature measurements in a process monitoring system with your own measurements or double-check the flow measurement in the ventilation unit (which is almost always calculated from differential pressure) with a separate direct measurement of the flow.

6.3.6 How carefully should you measure? How many readings should you do?

There are no exact guidelines to give as to how carefully one should measure, but the requirements for accuracy can nevertheless be said to be moderate in an energy audit. In principle, you can say that you must measure with an accuracy and sufficiency of data not to risk drawing incorrect conclusions from the measurement data you collect. A practical rule of thumb can also be that you should measure carefully enough to be able to repeat the measurement with the same result.

Whether you log or measure instantly, the goal is to make enough readings to get both a representative result and to record variations.

6.3.7 EPB standards useful for measurements

There is a series of standards aiming at international harmonization of the methodology for the assessment of the energy performance of buildings—the EPB standards. The focus of the standards is mostly production and design, but some of them also offer guidance for the assessment of existing buildings.

The standard about energy requirements for lighting (EN 15193-1:2017 Energy performance of buildings—Energy requirements for lighting—Part 1: Specifications, Module M9) contains "Simplified method for installed power Estimation" and "Calculated energy for lighting in existing buildings" with input data for a lighting installation audit. More information is provided in technical reports accompanying the standards (as CEN/TR 15193-2 for lighting). CEN/TC 169 deals with energy efficiency of lighting systems (among other subjects).

There is also a standard describing simplified methods for measurements of air flow on-site (EN 16211:2015 Ventilation for buildings—Measurement of air flows on site—Methods). Among other things, the standard guides you how to attain reliable flow values from repeated point velocity measurements in ducts.

6.4 Lack of data

There will be occasions when it is not possible to get the long, frequently sampled series of measurements that you desire. So what do you do then? One way to circumvent the problem is to measure at the normal case and at different extreme cases. It is also a useful method for validating models.

One problem is knowing what in fact is normal, for example, a normal production week. You can get an idea of the perception of "normal" by comparing your short-term measurements with longer measurement series or statistics, with poorer resolution, but based on larger data sets. Further, checking the weekly production figures with the annual one is also a good idea.

6.5 Data quality

In an energy audit, you will need to use data of varying quality, from precise measurements to oral information and sometimes even pure estimates. You can find precise measurements where someone is in need to receive a payment for the energy supplied. You can retrieve exact measurements from the energy companies' billing meters, by reading the meter yourself, reading invoices or having statistics delivered. If it is possible to obtain historical statistical data on the average power every hour (or sometimes even better resolution) for a whole year, this is very useful. Sometimes you can also log light pulses, representing the energy use, directly from the meter.

Sometimes you will have to rely upon calculations based on uncertain assumptions about, for example, operating times, power, and power factors. Assumptions are inevitable when you need to use historical energy data and when you do not know how the variations in your numbers depend on factors such as production rate or weather.

Data loggers are normally only possible to install in a limited number of places in the studied plant or process. The rest may be gained from instantaneous measurements. In such cases, annual operations of each energy end-using process may need to rely heavily on oral information gained from the in-house process engineer or technician. In the case of a workshop with varying production, this means you will have substantial data for one of the parameters you need (power) but vague information about the other (time). On the other hand, if you study a well-monitored, enclosed, and continuous process you can have detailed data about the operating

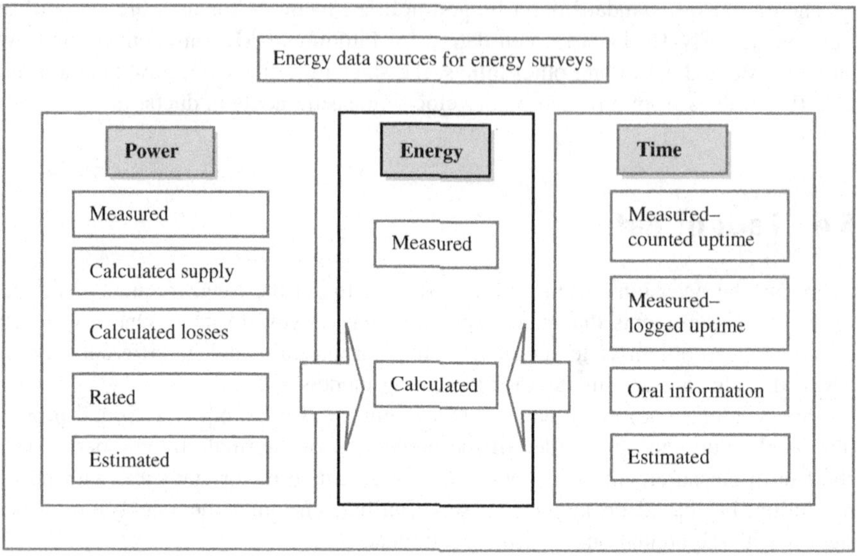

Figure 6.2 Possible data sources to calculate energy use for energy mapping purposes (Rosenqvist et al., 2012).

hours for every process, but having problems to access the machines to measure power.

Using multiple data sources to be able to validate data and confirm calculations is a good idea.

One advantage of measuring yourself is that you have control over where the numbers come from and what assumptions are made. Fig. 6.2 displays possible data sources to calculate energy use.

6.6 Logged measurements

In the case of energy audits, one can say that you only need to stay within human understandable time-spans. You rarely need longer data series than one or a few years and you rarely need more frequent logging than one sample every second.

6.6.1 Logging—for how long?

How long you should log depends on the work rhythm of the studied object. One extreme case is a continuous, constant process that is not dependent on anything else but instead runs at a steady phase, for example, a circulating pump in constant operation. The other extreme is a project that is only carried out once and never repeated, for example, the construction of a building. Most things you want to log will be somewhere in between those extremes and typically vary over the working day and week. Occasionally, practical issues in the production will cause the work rhythm to vary in other ways, for example, because something needs to be heated for 15 hours or because a machine needs to be cleaned every week. It is good to let the length of your measurements be controlled by the work rhythm. If you do not know how the production varies, expect to measure for a week.

More long-term changes in production rates due to economic cycles or seasonal variations are difficult to register with your own measurements within the framework of an energy audit. In those cases, you are dependent on others' statistics.

6.6.2 *Sampling interval*

The sampling interval you need when you log depends on which analyses you want to be able to do with your measurement data. The sampling interval is, of course, also limited by the capacity of the logger's memory.

For the data you collect in order to map the energy use of the studied object, you do not need very short sampling intervals. If you have a logger that can register a value every minute for a week, you will be able to handle most of the situations you are faced with during an energy audit.

Sometimes you do not know which analyses you want to do until afterward, when you see your logged values. Logging is also a part of understanding how the processes work. This means that at some point you will log a pump with continuous operation for a whole week, with a 10-second sampling interval, and have an abundance of data you do not need. It also means that at some point you will log, with 2-minute sampling intervals, a pump that starts every five minutes and only runs a few seconds at each start, which means that you will miss most starts and get an incomplete picture of the pump's operation.

6.6.3 *How to use different time resolutions?*

6.6.3.1 *Variations between years*

Aggregated data for a full year can be used for comparisons of current data with earlier periods, for example, before and after major changes in a building or a production process. The advantage of aggregated data for a full year is that you do not have to consider any seasonal variations in production. For outdoor temperature-dependent heating and cooling, full-year data can also be used for comparison with a statistical normal year.

6.6.3.2 *Variations between months*

Data series that reflect variations between months, for example, over a full year, can be used to analyze things that are seasonally dependent. Outdoor temperature, humidity, and light are given by nature, but it can also be seasonal variations in production. If there are historical monthly data, such as billing data from energy suppliers, it can be used for comparing energy end-use before and after a major

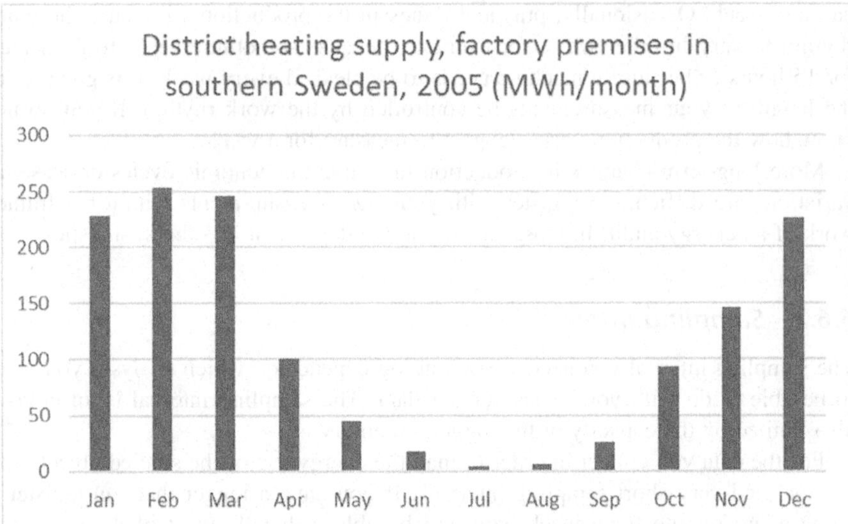

Figure 6.3 Example of variation between months. Monthly district heating supply for heating and hot tap-water at a plant in southern Sweden 2005.

change. Fig. 6.3 provides an example from a plant's annual district heating use and its variation between months.

6.6.3.3 Variations between weeks and days

To be able to capture variations over a day or a week, you often initiate your own logging measurements. Charging meters for electricity can (in some countries) record hourly values (or with even better resolution). Such hourly values are excellent because you can get reliable data series covering a long period of time, for example, the average power for every hour for a whole year.

With this time resolution, you can capture variations depending on working hours. You will be able to attain comparisons between day and night and can thus also analyze variations that are dependent on daylight or outdoor temperatures. Fig. 6.4 provides an example from a plant's warehouse and its variation over a week.

6.6.3.4 Variations between hours, minutes, or seconds

Even higher resolution can be used to see variations over an hour or minute. You need this type of data when you want to understand a sequence of events, typically in a machine or process. Perhaps you want to record production cycle times in one particular process. You can also use high-resolution data to get an idea of the size of variations, such as registering current peaks from motors or electric heaters.

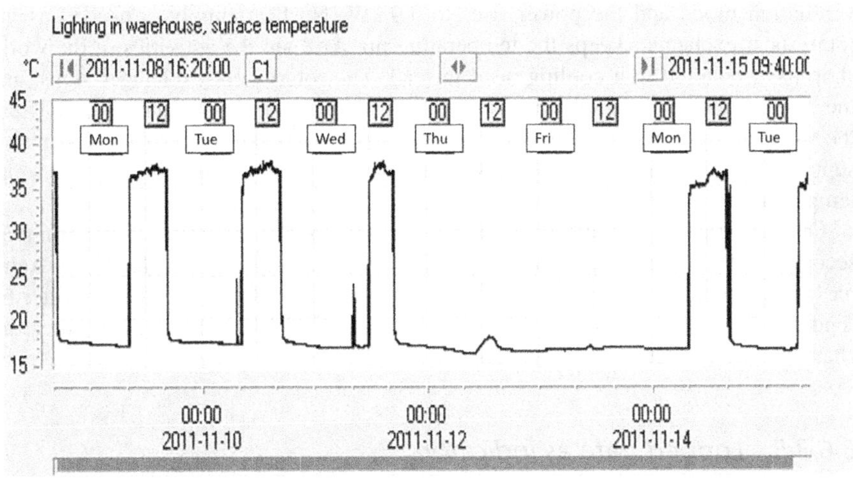

Figure 6.4 Example of variation between days. One-week logging of surface temperatures on fluorescent lighting fittings in a warehouse. The variations in temperature clearly show the working hours.

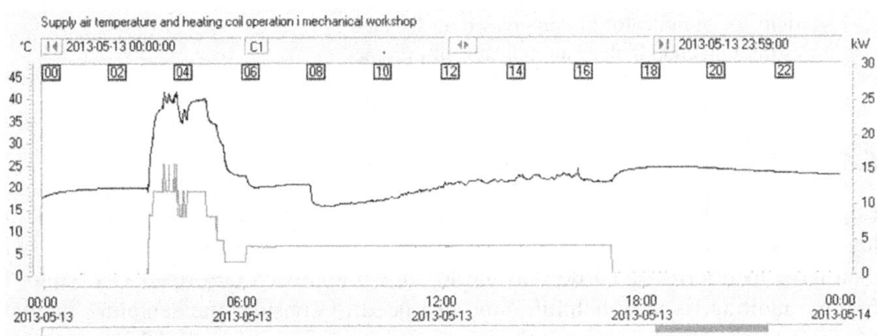

Figure 6.5 Example of variations between hours.

Fig. 6.5 provides an example of supply air temperature and heating coil operation in a mechanical workshop over 24 hours.

In Fig. 6.5, a workshop ventilation supply air temperature with an electric heating coil is logged for 24 hours. The upper line shows the supply air temperature. The lower line shows the power supplied to the air handling unit (AHU), including electric heaters. The heat in the workshop is only supplied by the air handling unit, no additional aero tempers or radiators exist. It was a cold night and at about 3 am the indoor temperature drops too low—then, the AHU starts heating. The high power (12–15 kW) and the high temperature (40°C) shows that the heaters are on. At 5:30 am the desired temperature is obtained, and the heaters stop. As the AHU is in return air mode only, the supply air fan is running, using 1.7 kW of power. At about 6 am the work starts, the exhaust fan is turned on as the AHU switches to ventilation mode and the power rises to 3.9 kW. No heat supply is needed as the rotary heat exchanger keeps the temperature up. At 8 am the activities in the workshop have resulted in a cooling need instead. The rotary heat exchanger stops, and the rising supply air temperature reflects the rising outdoor temperature. At 5 pm the workshop closes, the fans stop and the power drops to zero. As there is no air supplied, the temperature sensor in the supply air duct is rather reflecting the room temperature.

One may find it difficult to find use for logging with higher time resolution than seconds, at least in an energy auditing context. However, there is at least one example when this is needed: you can find broken filter cartridges in a flue gas filter by sending short compressed air pulses through the cartridge and record the pressure changes during fractions of a second.

6.6.3.5 Logged data as indicators

Logged data can often be used as indicators for something else:

- Surface temperature on tubes as an indicator of changes in real temperature in liquid
- Temperature, electricity, or light as indicator for operating time
- Carbon dioxide as an indicator of presence, that is, time
- Electricity as an indicator of compressed air leakage
- Entries through code lock as an indicator of presence

6.6.4 Difficulties and disadvantages of logging

Logging everything you or the manager at the plant you are auditing would like to log can be a problem. The number of loggers you have will be limited. You will then have to prioritize. To do that the top-down approach described in Chapter 5, Energy auditing, is very helpful. You also need to consider the sampling interval, the resolution of the logger, and the amount of data you can handle. Some examples are given below.

6.6.4.1 Logs with different intervals

You may not be able to select the same logging interval for all your different logs. If you use mean values from the log files for calculations only, this is often not a problem. During the analysis of data, it is often useful to combine several simultaneous measurement data series, for calculations or to be able to display them in the same diagram. If the measurements have different logging intervals or different start times, it will cause problems. One solution to this is to decide on a specific logging interval as the default setting. You can obtain the same start time if all the loggers' clocks are synchronized and if you program them to start at a set time, or you can be careful about the start time if you must start the logger manually.

6.6.4.2 Loggers with low resolution

You can have sensors with great scope and great accuracy, but loggers with rough resolution. The signal of the sensor or transducer, which is often analog, must be divided into the limited number of bits of the logger. The solution is to get loggers with more bits or to find another sensor, with exactly the measurement range you need. Sometimes it is possible to set the measuring range on the transducer, which can limit the number of sensors and loggers you need, but unfortunately also increases the possibility of making mistakes in setting.

6.6.4.3 Sampling interference

In rare cases, you will also find that the energy usage of the object you are logging varies regularly in accordance with your sampling intervals. Let's call it sampling interference because the result is reminiscent of waves interference. An example: You want to measure how much energy an air compressor uses to compensate for leakage in the distribution system. You leave the compressor on during the night and put a logging current clamp meter on the power supply. Since no production takes place during the night, the compressor will only work to cover the leakage. The leakage is equally large all the time, which means that the compressor will start and stop at a very regular interval. If that interval has a frequency that fluctuates in pace with your sampling intervals, each sampling will occur at the same location in the compressor on/off cycle. When you interpret the graph of your sampled values, it will look like the compressor is constantly on, or constantly off. A comparable case is if you would measure solar radiation at a location once a day, at 1:00 am at night. If you want to get a picture of how much the sun is shining, you will naturally get misleading, too low values.

Most likely, the sampling and the machine's variations do not have exactly the same frequency—probably there is at least a small difference. If there is a small difference in frequency, the interference will vary regularly during your logging and your graphs will have an interesting but difficult to interpret pattern, as seen in Fig. 6.6. Imagine the graph you would draw after measuring solar radiation at a location every 25 hours for a few months. Fig. 6.6 provides an example of sampling values from measurements of an air compressor for nearly 2 hours.

As can be displayed in Fig. 6.6, as the compressor is only working to cover leakages in the system, it turns on and off repeatedly, in a very regular manner as the leakage is constant. The first logging was made with a 120 seconds interval (the upper graph, the scale (A2) to the right). As the sampling interval matched the intervals of the compressor well but not perfect, the logger managed to catch almost every peak in the power use. However, you can also see from the shape of the graph that every peak is represented by only one sample—no peak last for more than 2 minutes. As the pattern seemed strange, a second logging was made with 15 seconds interval, and the real pattern of the power use became clear (the lower graph, the scale (A) to the left).

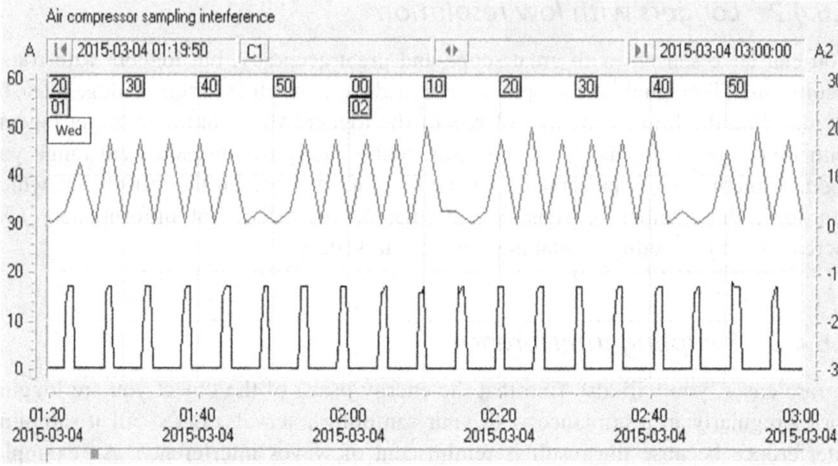

Figure 6.6 Example of sampling interference. Current used by an air compressor during nonproduction hours.

This type of extremely regular variation in energy use is not common, but the phenomenon nevertheless gives an indication that the sampling interval and the number of samples are important. You can always avoid the problem through very short sampling intervals and very long measurement series, but with the risk of getting unmanageable amounts of data and never completing your energy mapping.

So, the goal is to gain a sampling resolution high enough and measurements long enough to satisfy both your need for good average values for your energy calculations and your need to be able to analyze the changes in energy use. Some loggers can itself sample very often, for example, with second intervals, but then save only an average value, for example, a value per minute. If the logger also saves the lowest and the highest value every minute, you also get help to interpret the sequence of events during the measurement. Another variant is that the logger combines a sample at a fixed time every minute and the maximum and minimum values between the saved samples.

6.6.4.4 Do you know what you see?

One problem is that you do not always know what you see. An example is when a temperature logger, which is supposed to record the air temperature in a room or the surface temperature of a gadget, is hit by the rays of the sun once a day and registers a strange peak. If you fail to connect the peak to incoming solar radiation from a window, it is only inexplicable. If the peak is of the same magnitude as the temperature rise when the monitored machine is running, it is very easy to misinterpret the temperature peak of the solar radiation as the operating time of the machine.

Another example is machines that have a very large difference in power factor between different operating modes. If you calculate the power of a machine based on a power factor measured momentarily during operating time, together with the current you have logged for several days, you know for sure that you get a misleading power factor during idle and standby. In most cases, the proportion of idle and standby time is so small and the currents so low that the erroneous power factor does not significantly affect the result. However, sometime, the idle time will be long enough or the measured current high enough (e.g., due to high reactive current) to result in data leading you to an incorrect analysis. If you suspect that this is the case, then another visit to log the power factor or to measure the power factor during nonproduction hours is a solution.

6.6.4.5 The complication paradox

One way to solve both the problem of not always understanding what you see and the problem with the limited range of loggers and sensors is to acquire a larger number, more advanced loggers with more setting possibilities and more sensors. A gadget that fits that description is a logging three-phase wattmeter. To connect it, you use three to four current clamps and three to four voltage probes, all of which must be connected in the correct phase sequence. In some watt-meters, the setting possibilities are also extensive. In other words, you are forced to set a lot of parameters before you can even start measuring. The possibilities of making errors increase with the number of mutually dependent measurement points. The paradox is that the possibility of analyzing what went wrong also increases with the number of measured parameters. Let's call it the complication paradox.

6.6.4.6 Large amounts of data become hard to overview

Large amounts of data can cause problems. Even a single log file with tens of thousands of values can be difficult to handle. A very useful solution is to analyze graphs rather than numbers, at least if it is enhanced understanding you are looking for. Once you graphically analyze and try to understand what you see, you can more easily use the data for calculations.

6.6.4.7 Sometimes you only have one chance

Finally, there is a problem that is difficult to get around: Sometimes you only have one chance. Some measurements are neither possible to repeat nor possible to supplement afterward. If you measure energy use before and after a major change in production, you have such a case. If it turns out that the measurement before the major production change has gone wrong or that the data you collected is not sufficient for the analysis, then you will simply have to live with the fact that you missed the chance to measure.

6.7 Parameters

Since energy can be described as a product of power multiplied by time (E = P*t), time is a parameter you are very dependent on.

During an energy audit, you will be looking for operating times/durations for different operating equipment and processes. You also want to know operation times for different events. Often you want to know both.

If you find that a device has several different operating modes, you will need to know the operating time for the different modes. If you do not know if a device has different operating modes, you need to find that out too. For this you need data loggers. When you log other parameters, it will actually be the parameter's variations in the time you are looking for.

6.7.1 Data loggers

For field measurements, you will benefit greatly from small, durable, battery-powered data loggers with one or two channels. Logging current clamps and temperature loggers are very useful tools for the energy auditor.

Use measuring equipment and loggers with few setting options. It can be attractive to acquire measuring equipment with many setting possibilities, to be able to use the same equipment in many different measuring situations. But the choices allow for more mistakes in the settings and these settings will also take time and take the focus away during the often time-pressed field measurements. The fact that the measuring equipment and the logger have few setting possibilities reduces the risk of mistakes.

6.7.2 Electric power measurements

The data from the energy supply company is not enough for allocating energy use to all different processes. Internal energy measurements from the studied plant can sometimes be found, especially for well-monitored core production processes. Still you will probably have to measure the electric power used by a selection of important equipment to be able to allocate the energy use and construct the power and energy balance. To make electrical measurements is often the main activity when the studied object is visited.

Apart from measuring the energy use directly with permanently installed meters, there are different ways to measure the power or current for electric equipment to help you calculate the energy use.

To calculate the energy use from electrical measurements you will need the active power. When you do the on-site measurements, it is preferable to measure the active power directly with a wattmeter. Measuring the current and voltage separately will enable you to calculate the apparent power, which cannot be used directly for energy calculations. The apparent power is only useful when the power factor is known from previous measurements or from technical documentation.

Rated power for different electric equipment can be used to some extent, but normally it should not be used for the calculations. Rather, it is more appropriate to obtain the active power or current and power factor from measurements.

As electricity is used for many different processes, electric power measurements are mostly necessary to be able to complete an industrial energy audit. Different methods can be used to measure and analyze power and energy use for different electric equipment. The Sections 6.7.2 to 6.7.6 list some common examples.

6.7.2.1 Active, reactive, and apparent power

In an energy audit the power measurements are mostly performed in the AC-grid. That means you must deal not only with active power, P, measured in watt (W) but also with apparent power, S, with the unit voltampere (VA) and the reactive power, Q (var).

Electrical equipment with coils and capacitors causes phase difference between current and voltage. The typical example is the windings of an AC motor which induces a current back for each oscillation in the alternating voltage. When you measure the current in the conductor you will not see the active current actually used by the motor, but the result of the active and nonactive components of the AC-current. Calculating the power with that current will give you the apparent power. Electronic equipment using rectifiers also creates nonactive current. What you want in order to calculate energy use is the active power, which means you will have to measure not only current and voltage, but also the power factor, λ. Please see Section 6.9.1 regarding Electric power for further information on phase difference, power factor, and how to calculate power from electrical measurements.

6.7.2.2 Power

Power is used for:

- Calculations of energy use

With a portable wattmeter you can measure the AC-voltage and AC-current simultaneously and show the active power directly in watts (W). The power is measured for a single conductor separately, one at the time or all three conductors in a three-phase system simultaneously.

Measuring one phase at the time gives you information on how the load is distributed between the phases. If the loads are evenly distributed, typically for a three-phase motor, it is enough to measure the current in one phase to calculate the total power and energy use. Instantaneous measurements give you the real power use and the power factor, λ or PF, for different equipment or parts of the building.

If you don't have a wattmeter, current can be measured with a current clamp meter that you can simply clamp over a single conductor without breaking the electric circuit (see Fig. 6.7). Very often there is simply no room for the power meter in the switchgear or distribution board and sometimes there is no access to conductive parts to measure the voltage. The voltage is standardized and can be assumed to

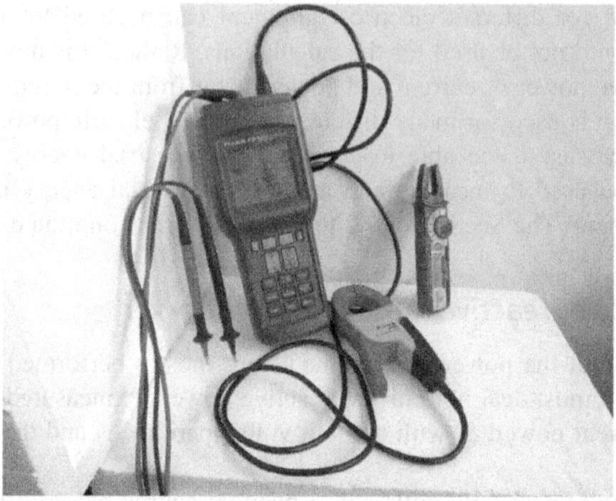

Figure 6.7 Wattmeter and clamp-on ammeter. A small, simple current clamp, or fork meter is useful to check if there is current in the conductor at all before you tear down the cabinet to access the conductors with the wattmeter.

keep the rated value. Though, the rated value must be controlled for every location. In Sweden, for example, the standard value is 230/400 V, but in some older industrial applications it is higher, for example, 500 or 690 V. The voltage can also be measured with a voltmeter, as soon as you have conductive parts available. With this information about the current and voltage, the apparent power can be calculated. The disadvantage of separate current and voltage measurements are that you can measure the apparent AC-current only. The result you will have is the apparent power in voltampere (VA), not the active power in watt. To calculate the active power, you also need to know the power factor, and that's what the wattmeter calculates for you.

6.7.2.3 In the toolbox

A wattmeter or power analyzer is a tool that can give you current, voltage, power, and power factor for one phase. A three-phase power meter is optional in the energy audit context. Fig. 6.7 displays a wattmeter and clamp-on ammeter.

6.7.3 Power factor

You can use the power factor:

- To calculate power and energy
- To see what's on the other end of the conductor
- For diagnosing equipment

To measure the power factor, you need to access conductors and live parts at the same time, and normally you measure the power factor while measuring the power, with the same instrument.

The power factor is necessary for calculating the active power if logging current clamps are used to analyze the variations in energy use over time. The clamp-on ammeters will log the apparent current, that is, the result of the active and nonactive current in the conductor you are measuring on. The current clamp detects the magnetic field around the conductor and does not distinguish whether the current is running in one direction or the other. You need to measure the power factor to find out how much of the current that is active and thus causes an energy use.

Some equipment found in the industries has a significant difference in power factor at different loads. By, if possible, measuring at different loads, by multiple instantaneous measurements or by logging, you will get an idea for which types of equipment the power factor varies. Typically, you will find different power factors at different loads for air compressors and motors for hydraulic machines but not for frequency converters, light sources, and electric heating. Sometimes the power factor is important for calculating idling energy loads correctly—in idling mode, there can be several small processes running with power factors that are different from the main process.

To measure the power factor momentarily is usually enough. There are good instruments for directly logging power and power factor as well, but for an energy audit there are many practical obstacles to do so:

- The time it takes to set the logger,
- To connect the logger's all probes,
- The lack of space for logger and probes in the distribution board,
- Purchase cost for meters,
- Lack of safety with measuring equipment connected to voltage left unattended, and
- The need for power supply to the logger.

Sometimes there is a need to log the power factor when you want to study a load with uncertain or varying power factor. A typical example is charging electric forklifts during the night, because the power factor can be suspected to vary during the charging cycle and because you probably will not be on-site to measure the power factor during the night. Another reason to log the power factor is to troubleshoot capacitors that compensate for phase differences.

6.7.3.1 Using power factor for diagnosing

Besides as a basis for energy calculations, you can also use the power factor to see what is on the other end of the cable. It is very useful when the labeling of the electrical system is unclear, or when you measure the power supply to equipment that you simply cannot access or enter. If you have an instrument that measures the PF (Power Factor, λ), you can see if it is an electric heating load ($PF = 1.0$) or something else ($PF < 1$ means that it is a load with inductive/capacitive phase differences or electronics at the other end or a mixture of several types of loads). If you have an instrument that can also show displacement power factor (DPF, equals cos ϕ at nominal frequency), you can distinguish between a (linear) inductive/capacitive load (PF = DPF) or electronics (PF $\neq$ DPF).

One example of this is when you measure the power of a fan motor. If the motor is up on the roof or hidden inside an air handling unit, you cannot easily see the type of motor or if it is connected to a variable speed drive (VSD). If, on the other hand, you can measure the PF and the DPF, you can determine whether the motor has a VSD and if the speed is regulated, or at least possible to regulate. If you during the audit find that it is a good measure to control the air flow according to need, then you know whether you should only suggest starting regulating, or you also have to propose an investment in a VSD.

Using the power factor, you can also make a diagnosis on a directly connected asynchronous motor. If the power factor for a motor is very low, for example, PF = 0.4, this means that the motor is spinning with very low resistance. The same work can then be done more energy-efficiently by a smaller motor.

6.7.3.2 Rated cos ϕ

On some devices, most commonly three-phase AC-motors, you will find a rated cos ϕ written on the nameplate of the equipment (Fig. 6.8). That cos ϕ is equivalent to the power factor at the fundamental frequency and can be used as an estimated power factor. For an AC-motor the real power factor varies with the load—a lower load gives a lower power factor—which means the rated cos ϕ can only be fully trusted if the motor is operating close to the rated power. Fig. 6.8 displays a figure of a nameplate of an electric motor.

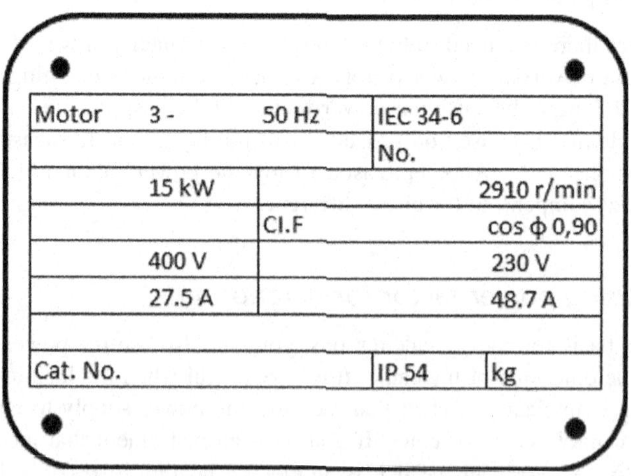

Figure 6.8 Example of a nameplate of an electric motor. For industrial energy audits, the rated power and the power factor is the most useful information.

6.7.3.3 *In the toolbox*

For power factor measurements, you need:

- Wattmeter that can measure watt, power factor, and displacement power factor
- Clamp-on ammeters in different sizes, as probes to the wattmeter
- Voltage probes of different types, to reach live parts in a safe way

6.7.4 Current

Measuring current is a common activity in energy audits. By measuring current you can:

- Calculate electrical power and energy (together with voltage and power factor)
- Calculate operating hours for electrical equipment
- Measure momentarily to prioritize which processes/machines to measure more thoroughly
- Read control signals from control centers (0–24 mA). It is more common for the control signal to be 0–10 V (see also under Section 6.7.5)

The clamp-on ammeter, the current clamp, is an invaluable tool for energy auditing. It makes it possible to measure electrical currents without breaking the current supplied to your measuring object.

The clamp-on ammeter with associated data loggers that you can leave in place to log current for a week is very useful. Since it's hardly any advantage, or even possible, to measure each individual device in the factory you are mapping, you also need to prioritize which loads to measure. To first make instantaneous measurements on the largest outgoing groups in the low voltage switchgear with a current clamp gives you an idea of where the large loads are. From this you can prioritize where to measure power and power factor and for which groups or individual machines you should log the current (see Section 6.8.1).

6.7.4.1 *About logging current*

By logging the energy use for a longer time, you can see variations in the demand, the use of energy in standby or idling mode, peak power demand, and more. Logging current clamps are good tools when you are making an energy audit (see Fig. 6.9). Standard current clamps with single-channel data loggers are useful because they are easy to handle, and no additional wiring is needed. They are placed around a conductor, in the same way as the current clamp meter for instantaneous measurements. You need to instantaneous measure or estimate the voltage and the power factor or cos ϕ to be able to calculate the energy use.

Symmetric three-phase loads are commonly occurring in industrial applications —motors, electric heaters, and forklift chargers—basically all processes with a high power demand. Often you will find a distribution board totally dominated by symmetric three-phase loads, for example, a group of cooling compressors. In case of a symmetric load, one clamp meter is enough to measure the electricity use of the equipment.

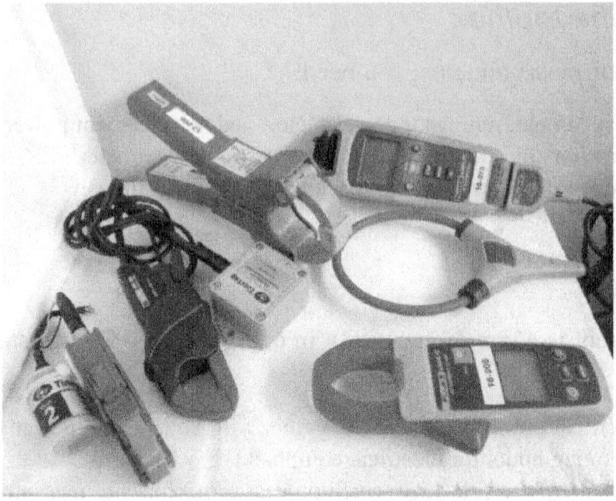

Figure 6.9 Current clamp meter loggers, covering a range from 0–20 to 0–2500 A AC. Data loggers are either external with wire connection or internal.

For switchgear with a nonsymmetric three-phase load, you need two or three clamp meters to log the different conductors, to get a good picture of the energy use. Nonsymmetric loads are typically the case in office buildings, where the power use can be high, but consisting of many one-phase devices.

Clamps of different sizes and with different metering ranges can be used to measure the current in different parts of the system. The rated current from documentation or instantaneous measurements may be used to decide the range of the current clamps needed. The access to the different parts of the electrical system and the conductors will determine what can be measured in practice. To be able to allocate the energy use to different processes, measurements in series and parallel may be needed.

How many clamp meters you need to be able to allocate the electricity for to the right process totally depends on the construction of the electrical system. A large but recently constructed sawmill can be easy to map because everything is constructed at the same time and the production is very well scheduled and dominated by three-phase loads. An older, small mechanical workshop can be harder to map because the building is rebuilt many times, there are many one-phase loads and the activities are irregular. Typically, 4–8 current clamp meters are used for a limited energy audit, level II, with wide variations depending on, for example, the above-stated parameters. Up to 30 clamp meters can easily be used for a larger plant when conducting an energy audit.

You will be able to handle most measurements with the measurement range 0–250 A, but it is good to have several sizes and types of clamp meters available, this in order to maximize metering options during the site visit.

Clamp meters with smaller measuring ranges (e.g., 0–20 or 0–100 A) provide good resolution at low loads, even with a simpler logger. Good resolution is usually not needed to be able to allocate energy use and create the energy balance in an energy audit. However, the resolution is useful when you want to analyze how several small loads vary in a switchboard or in a machine, for example, pumps and motors in a ventilation unit or in a computer numerical control (CNC) machine[1].

The measuring range 0–500 A or higher is for the larger loads, typically outgoing groups from a switchgear. Over 500 A is rarely useful. When the fuses are larger than that, there are almost always several parallel conductors, which means that you can just as easily measure on one of the conductors with a smaller clamp meter.

It is important to note that when you analyze your log files you will have to multiply your measured current by the number of parallel conductors to calculate the real current.

There are also three-phase logging power meters for electricity, but often they are found to be too clumsy and give more than necessary detailed information for an energy audit where the aim is to get an overview of the energy usage. When it comes to logging to allocate energy use to different unit processes, simultaneous information, even if it is simpler, on the relevant unit processes is more useful than detailed information on one object at the time. Simultaneous logging with one logging power meter in a central switchgear combined with logging current clamps in distribution boards or directly by a machine will deliver relevant information. In the example in Fig. 6.23, current clamp no 1 in the switchgear may be replaced by a logging three-phase power meter, especially if the load varies individually for the three phases.

6.7.4.2 In the toolbox

For current measurements, you need:

- Clamp-on ammeters for instantaneous measurements. A small fork-style ammeter for conductors up to a few mm^2, and a larger clamp for larger conductors. An ammeter with a flexible probe is sometimes the only way to be able to measure when there is a lack of space. The disadvantage is that you need two hands to connect it.
- Logging clamp-on ammeters, with the ability to log at one value per minute for one week.

Fig. 6.9 displays current clamp meter loggers, covering a range from 0–20 to 0–2500 A.

6.7.4.3 Typical measurement locations

You will typically measure current in:

- Low voltage switchgear or distribution boards, outgoing groups to machines,
- The control cabinet for a machine,
- The machine, at the electrical connection, and
- Controller, if you are going to measure control signals.

[1] A CNC machine is an automated machine that can undertake various operations such as drilling.

To be able to measure the current with a current clamp meter, you need to have access to the individual conductors in a cable. Normally, the conductors are only available where the cables are connected to something else, for example, to the machine you want to measure, to the machine's fuses or at a joint. Normally, these connections are protected to stop you from accessing live parts. The protection is there for good reasons and will sometimes prevent you from measuring. Sometimes there will also be a lack of space to clamp-on the ammeter and in some cases, the conductors will simply be too large.

6.7.5 Voltage

You use voltage measurements to:

- Calculate power,
- Register control signals (e.g., how much a damper in a ventilation system is open), and
- Measure phase sequence in three-phase power systems.

The voltage in the distribution grid is standardized and you can often expect it to be the nominal voltage, but not always. For historical or practical reasons, there may be transformers that differ from the standard, for entire plants or individual machines. Take the habit of checking the voltage with an instantaneous measurement at the location where you intend to measure power or log current.

There is rarely any reason to log the voltage in the electricity grid. An exception is if there is a lack of transmission capacity in the local grid or within the building where you map the energy use. Excessive power use will lead to overheated components in switchgears and distribution boards or voltage drops or both. To avoid being forced into expensive investments in the distribution cables and switchgear is a good reason to make energy use more efficient.

A special case of voltage measurement is when you want to register a control signal from a controller that uses 0–10 V voltage to control a device you want to map. Typically, the signal controls a damper in a ventilation system, a valve in a pipe system, a speed of a motor, or something else. One example is when you want to calculate energy use to heat or cool air that passes through an air handling unit that recirculates air. It is not enough to know the total flow of the air supplied by the AHU, you also need to know how much of the air is taken from outside and how much is recirculated through the AHU. If the mixing damper is controlled by a 0–10 V signal, it is possible to record how that signal varies and then to calculate the degree of recirculation and thus the ventilation losses. You need documentation for the ventilation unit to know the relationship between the control signal and the proportion of recirculation. You also need electrical drawings to find the outputs of the control signal in the controller device. If possible, measuring the signal directly at the actuator minimizes the risk of mistakes. In the actuator, there are sometimes outputs for the control signal where you can plug in your logger without having to break the signal to the actuator.

If you are going to measure three-phase power with a wattmeter, you need to be sure about the phase sequence between the three conductors you are going to measure. To first determine the correct phase sequence with a phase sequence indicator is recommended.

6.7.5.1 Typical measurement locations

The typical measuring location for voltage measurements is a place with warning signs that clearly tells you that you should not touch anything as long as the plant is under operation. Unfortunately, this is exactly what you need to do when measuring voltage.

It is important to note that individual countries have different laws for how and where you can measure electricity and often some form of certification to be allowed to even enter the location where you can access the switchgears and be able to undertake measurements. This chapter can merely be seen as a guide to be used when conducting an energy audit. Follow national laws and regulations when doing your energy audit! Get support from a certified electrician.

You will measure:

- Low voltage switchgear or distribution boards, outgoing groups to machines,
- The control cabinet for a machine,
- The machine, at the electrical connection, and
- Controller device, if you are going to measure control signals.

6.7.5.2 In the toolbox

For voltage measurements, you need:

- Voltmeter and live voltage detectors, preferably with phase sequence indicator. The meter must be specially designed and rated for up to 600 V. If you have a wattmeter you do not have to have a voltmeter, as the wattmeter also measures voltage.
- Voltmeter or multimeter for low voltage 0–10 V
- Logging voltmeter for 0–10 V
- Protection gear

Fig. 6.10 displays protective rubber gloves for working close to live parts.

6.7.6 To avoid measuring current and voltage

Sometimes it is a good idea to avoid measuring current, for safety reasons or due to lack of space or lack of time, for example. There are strict safety regulations on electrical installations and how to work with electrical equipment that are under operation. If it is possible to get data in some other way, it is a good reason to avoid messing around in the switchgear.

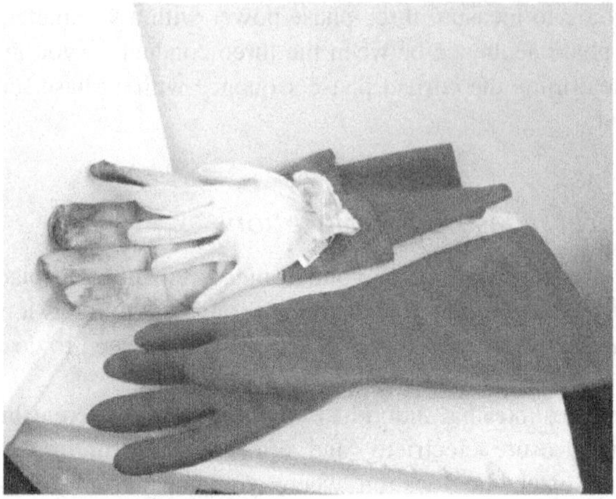

Figure 6.10 Protective rubber gloves for working close to live parts, gloves to protect the rubber from getting punctured and gloves to ensure your hands do not get stuck in the rubber gloves. A pump to ensure the rubber gloves are not punctured, safety glasses and clothing to protect you from flame in case of electric arc completes the safety equipment.

In fact, if you want operating hours you can often, instead of measuring current, measure temperature, magnetic field, light, or anything else that can indicate when a machine or process is running or not.

If there are internal current and power meters in a machine—use them. If there are internal counters or loggers, it is even better. Use these! However, you will need to make sure what is being measured. If there is an ammeter in a frequency converter, you need to know if the meter shows the current supplied to the converter or the current that the converter is supplying to the controlled equipment. If there is a power meter in a device, you need to know if it shows the power for one phase or for three.

6.7.7 Flow in air and liquids

It is often difficult, and as often dangerous, and therefore inappropriate, to connect to pressurized systems to make temporary flow measurements, especially if you cannot turn the systems off. This applies to steam, compressed air, and distribution systems for heating and cooling. There are portable ultrasonic meters to put on the outside of pipes, but for them to work, you need to know a lot about the flowing media, the material in the pipe, the thickness of the pipe walls and, in addition, any coatings on the inside of the pipe. In distribution systems for heating and cooling, there may be opportunities to measure the flow from the pressure drop over components with known properties (see Section 6.7.8).

In ventilation systems, the pressures are moderate and in most cases the medium is harmless. You can often access a duct to measure the air flow. And you often have reasons to do so in energy audits.

Measuring air flows is often useful in an energy audit in order to:

- Calculate heat losses or heat supply through the ventilation through the air flow, and
- Assess whether the air flow is adapted to the need.

Air flow measurements can be used to calculate the ventilation heat losses, and also to roughly estimate the electric power used for the fan motors. However, the electric power used for the fan motor can also give you an idea about the air flow, and the electricity is often easier to measure (see Section 6.9.8). Sometimes the optimal way to find out the operating time for a machine or the working hours in a building is to log the ventilation air flow or temperature or the electricity used by the fan motor.

A special kind of air flow measurement is the ultrasound leakage detection in compressed air systems. You do not use these measurements for an overall energy audit, but to be able to provide detailed energy efficiency measures for a compressed air system. Your current measurement loggers will probably reveal the magnitude of leakage in the compressed air system—leaving the compressor running after production time, the machine will keep working, to compensate for the leakage. A good way to find the leaks are to follow the pipes and investigate joints, connections, and hoses with a leak detector.

6.7.7.1 *In the toolbox*

For flow measurements, you need:

- Air flow meter. A hot-wire anemometer cover most needs. It is used for measurements inside ducts, but also sensitive enough to register draught in leaky buildings. The hot-wire anemometer is sensitive to humid and polluted air. A compact vane anemometer or a Prandtl tube is an option in these cases.
- Differential pressure meter combined with a Prandtl tube.
- Differential pressure transducer combined with a data logger, for long-term measurements.
- Air flow hood. You use it to measure the flow passing through an air terminal device (ATD), a supply or exhaust ATD.
- Ultrasonic leak detector for compressed air.
- Ultrasonic flow meter for pipes.

Fig. 6.11 displays an air flow meter adapted to measure air speeds and flows in ducts.

Fig. 6.12 displays a drilling machine needed to drill through the duct to place your air flow meter.

6.7.7.2 *Typical measurement locations*

You will measure flows:

- Inside ventilation ducts,
- In rooms,
- At air terminal devices, and
- In/on pipes.

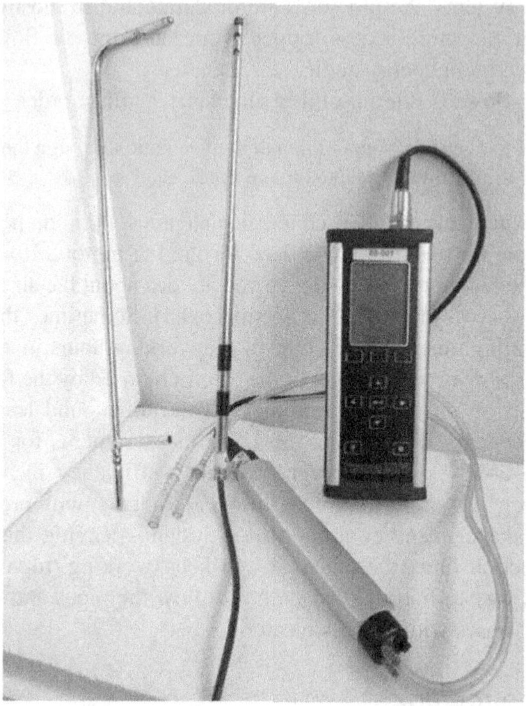

Figure 6.11 An air flow meter adapted to measure air speeds and flows in ducts. It can be used to measure pressure in the duct or difference pressure. It can measure air speed with a hot wire probe or with a difference pressure probe combined with a Prandtl tube.

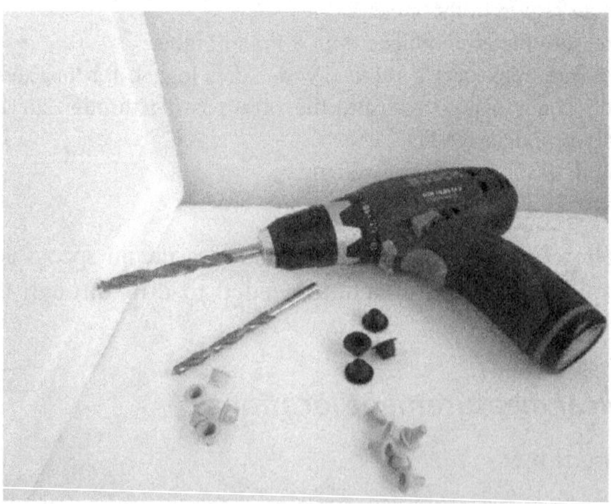

Figure 6.12 A drilling machine and drills gives you access to air flow in ducts. Plugs are used to plug the holes after measurements.

6.7.8 Pressure

Pressure can be used:

- To calculate flow and
- Together with temperatures, to calculate power and energy.

If in a pipe or duct system there are outlets for pressure measurements before and after a component with known properties, it is a good way to measure the flow past that component. Often, it is components used for adjusting flows—adjustment valves in pipe systems or adjusting dampers or adjustable supply air devices in a ventilation system. In a ventilation unit, there may be differential pressure outputs that measure the pressure before and after the fan, to get the flow for the entire unit. In all cases, in addition to your measurements, you need documentation that gives you the relationship between the differential pressure and the flow. Sometimes you will find a constant, sometimes you will find a chart. Sometimes you will find nothing at all and can only state that you cannot use your measurements to calculate the flow.

6.7.8.1 In the toolbox

For pressure measurements, you need:

- Differential pressure meter for air [typically between 0.4–2 kPa for ventilation in facilities and 8–16 kPa for fans used for transportation of materials, such as sawdust (Jernkontoret, 2018)]
- Prandtl tube for air flow measurements with differential pressure
- Differential pressure gauge for liquids (typically between 1 and 10 bar for secondary heating and cooling systems)
- Documentation of the components you measure on, to give you the k-value needed to calculate the flow from pressure

Fig. 6.13 displays a difference pressure-probe for liquids.

Fig. 6.14 displays a difference pressure transducer intended for building automation.

6.7.8.2 Typical measurement locations

Typical measurement locations for pressure are:

- Inside ventilation ducts,
- Pressure outlets on air handling units,
- Pressure outlets on valves, and
- Pressure outlets on pipes.

6.7.9 Size

To analyze the heat losses from a building the size of the building and the properties of the building materials must be known. Measuring the size of windows and doorways, thickness of walls and height of ceilings are often part of the energy

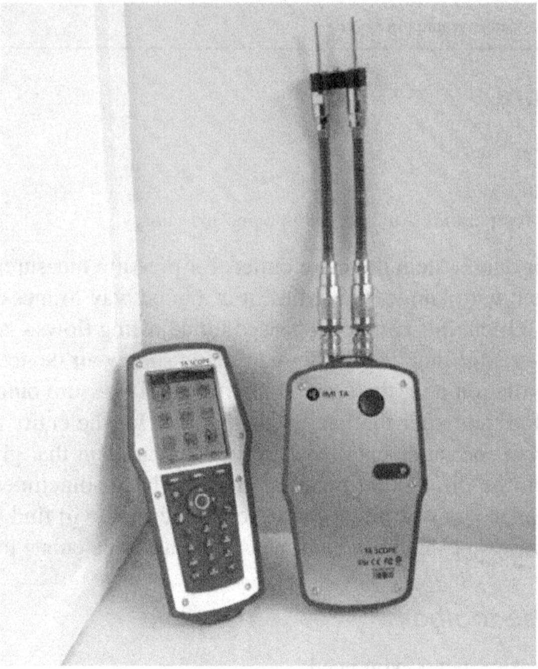

Figure 6.13 Difference pressure-probe for liquids used for heat transport, as water, propylene glycol, and ethylene glycol. This instrument is intended for adjustment of heating and cooling systems and can calculate the power for you if you add a pair of temperature sensors.

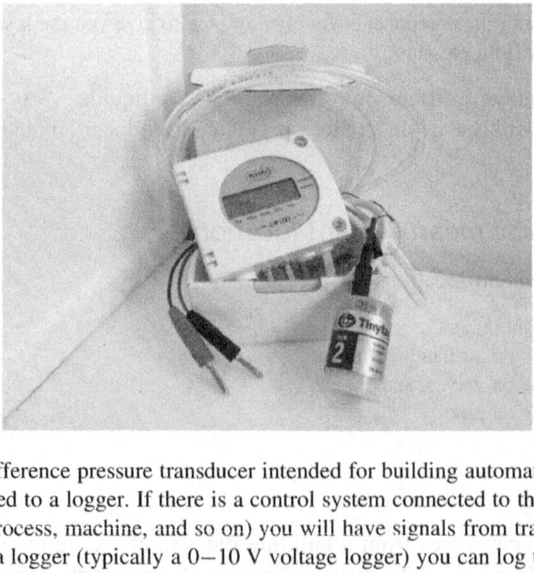

Figure 6.14 A difference pressure transducer intended for building automation, but in this case only connected to a logger. If there is a control system connected to the object you study (building, process, machine, and so on) you will have signals from transmitters and controllers. With a logger (typically a 0–10 V voltage logger) you can log the signals in the control cabinet. This also means you can build your own portable logging devices with any of the transmitters intended for process and building automation on the market. The disadvantage is that you need an electrical socket to supply the transmitter with power.

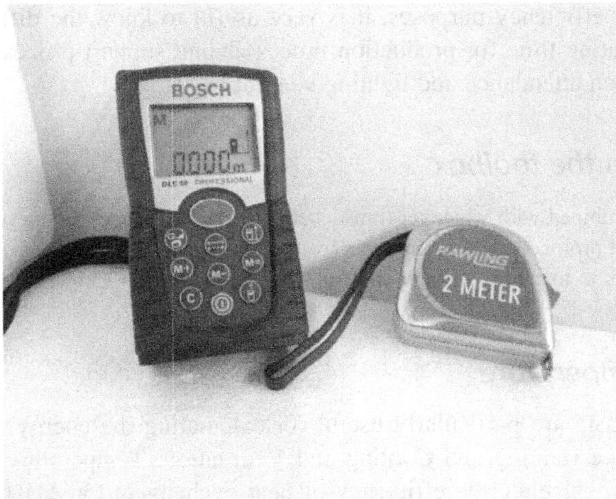

Figure 6.15 Measuring tape and distance meter to determine the size of buildings, ducts, and so on.

audit. For air flow measurements in ventilation ducts, the size of the duct must also be known. Light strips are often out of reach from the floor level, and one way to find out which type of light strips are used is to measure their length with a portable laser distance meter. And when you propose measures, you will have to know if there is room for new equipment.

6.7.9.1 In the toolbox

For size measurements, you need:

- Distance meter, laser, or ultrasound
- Tape measure

Fig. 6.15 displays a measuring tape and distance meter.

6.7.10 Time

Measuring time is central. You will want to know operating hours for machines, duration for different operating modes, times for events in processes, and so on.

One way to access this needed information is to see how different parameters vary over time. You can use current, voltage, temperature, flow, carbon dioxide concentration in indoor air, magnetic field caused by a motor, and so on. A variety of parameters can be used. To see the operating time of a machine, one can measure the variations in the electricity supplied to the machine, or the waste heat generated by the machine.

In some cases, you will find counters for operating time, typically for machines with recommended service intervals based on operating time. Noting the counted hours at two or more occasions is a useful method to see how much the machine is used.

For energy efficiency purposes, it is very useful to know the difference in time between operating time for production processes and support processes, or the difference between attendance and lighting/ventilation.

6.7.10.1 In the toolbox

- Loggers, combined with whatever transducers, sensors, or probes you need
- Standardized protocols to note numbers from counters
- A good clock to manually register short uptimes for machines

6.7.11 Temperature

Temperature data are particularly useful for calculating the energy use of the unit processes Space Heating and Cooling and Ventilation. Temperature measurements can be used to calculate the efficiency of heat exchangers for AHUs. To calculate the heat losses, the exhaust air temperature is crucial, but seldom registered by the monitoring system and even more seldom logged. To use standard values for efficiency of heat exchangers is a useful shortcut.

Temperature logging can be used to identify operating time for equipment. A temperature probe can be attached directly to the surface of a machine or in an air flow or other medium connected to the studied equipment (see Figs. 6.26 and 6.28).

Typical temperature measurements in an industrial energy analysis are:

- Outdoor temperature. To calculate heating and cooling energy demand and heat losses,
- Room temperature. To calculate heating and cooling energy demand and heat losses,
- Air temperatures in air handling units. Along with air flows to calculate heat losses, heat supply, and heat exchanger efficiency,
- Supply and return temperatures in heating or cooling systems. To make a diagnosis of the regulation in the system or together with flow to calculate the energy use,
- Surface temperatures. On pipes to get an idea of the temperature of the liquid inside the pipe. On machines to see when they are used and when they are turned off,
- Water temperatures, and
- Flue gas temperatures.

You will have use for a set of different temperature probes for air, surface temperatures, and immersion probes for liquids. A thermometer with standard thermocouple plug is good to have. This allows you to have many interchangeable probes for the same thermometer and you are not bound to any brand.

Several loggers will be needed for your temperature probes. One or two channels loggers, robust, preferably waterproof is a good choice. Use loggers with wired or wireless probes, to be able to place the equipment anywhere you want.

You will want to use several temperature loggers in many different locations simultaneously. You will not be able to install long cables for measurement data collection. You don't have much benefit from many channels—rather many loggers with one or two channels per logger.

You can have several different types of probes for your loggers, but temperature probes that are waterproof, small enough to be taped as surface sensor and at the

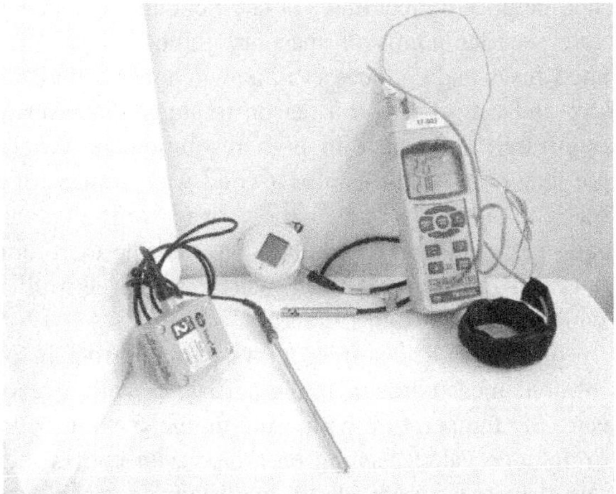

Figure 6.16 Temperature loggers and probes. To the left, a robust and waterproof logger with a stainless probe. In the middle, a temperature and relative humidity logger with a probe slim enough to fit in a small hole in a ventilation duct. It also has a display that allows you to use it for instantaneous measurements. To the right, a four-channel logger for thermo-element probes and thermistor probes. One good feature is that it can log temperature difference directly. Another good feature is that you can use it with any of the many thermo-element probes on the market.

same time quick enough to handle even measurements in air are a good compromise that can handle most situations.

The temperature range depends of course on what to measure. In most cases, you will be able to work in energy auditing contexts with accuracy as low as 0.3°C, allowing you to use temperature sensors with a large measuring range, for example, −30°C to 200°C.

If you have a thermometer for thermocouples, a temperature logger with standard contact for thermocouple sensors is valuable. The combination gives you many choices and will be useful when the probes to your standard loggers for some reason cannot handle the task. There may be special requirements, for example, on size, corrosion resistance, or temperature range.

You will probably want to log with minute intervals rather than second intervals, so the probes don't have to react faster than that.

Fig. 6.16 displays loggers and probes for measuring temperatures.

In Fig. 6.16 you see a thin standard probe and a probe with a hook-and-loop fastener strap intended for pipes, but also suitable for fluorescent tubes.

6.7.11.1 Infrared cameras and thermometers

An infrared camera is a good way of showing temperature differences and thereby, for example, illustrating leaks of hot or cold air or variations in heat transport through building constructions, poorly insulated pipes, and so on. If the difference between indoor and outdoor temperatures is large enough, one can also get a picture of where there are wooden joints or masonry joints inside the wall behind the facade. Superheated fuses and conductors in a switchgear can also be detected with an infrared camera and can give you a reason to suggest measures that reduce the power need. An infrared camera, with high resolution, or infrared thermometer, with narrow-angle lens (e.g., 12:1), is also a good way to measure temperatures on surfaces out of reach, as a basis for calculations or to see if a machine is working at all. The advantage of the camera, in comparison with the thermometer, is the visibility of the result both for yourself as an energy auditor as well as for those you want to inform with the results from the audit.

The emissivity of the studied surface plays a major role in obtaining a good result from an infrared measurement. If the surface is within reach, which allows you to also measure the temperature with a thermometer with a contact sensor, you get a very good reference value. Pasting on a tape with known, high emissivity on the surface you want to measure is also a good way to make sure you get useful values. The tape tip is also a good way to obtain reliable temperatures from surfaces with very low emissivity. Since these surfaces have high reflectivity, they will reliably reflect the ambient temperatures, rather than showing the surface temperature. The tape needs to cover at least one pixel in the thermal camera sensor for the reference to become useful.

To use an infrared camera or thermometer to measure the temperature of surfaces out of reach, you need to know something about the surface emissivity and set it in the camera/thermometer. There are emissivity tables for many materials, but it may be worth knowing that the emissivity may vary with the temperature of the material and with your viewing angle.

6.7.11.2 Typical measurement locations

You will measure temperatures:

- Inside ventilation ducts,
- In rooms,
- Outdoors,
- On pipes and ducts, and
- On operating equipment and machines.

6.7.11.3 In the toolbox

For temperature measurements you need:

- Thermometer for instantaneous measurements. With different probes for different purposes.
- Loggers with associated probes.
- Infrared camera or infrared thermometer for noncontact measurement.

Fig. 6.17 displays an IR thermometer and portable IR camera.

Figure 6.17 To the left, an IR thermometer that is lightweight and easy to use. To the right a very competent but still portable IR camera.

6.7.12 Light

A lux-meter is used to measure illumination levels in facilities.

There are recommended light levels (lumen/m^2) for different task and different types of areas. If the auditor finds areas too generously or too meagerly illuminated, both cases can give an opportunity to replace old, inefficient lighting. The standard EN 12464-1. Light and lighting—Lighting of work places—Part 1: Indoor work places, is a very good guide when planning for new lighting installations. Fig. 6.19 displays a lux-meter.

6.7.13 Carbon dioxide

Carbon dioxide measurements can be used to:

- Register presence and
- Calculate ventilation efficiency.

A transducer for carbon dioxide and an associated logger is a very good way to register the presence of breathing creatures, in the energy audit context mostly humans. If the purpose of the ventilation is to supply fresh air to the people working in the ventilated area, carbon dioxide measurements are a good way to see if the desired air circulation is achieved. With two meters you can use carbon dioxide as a tracer gas and calculate the specific air flow rate with the concentration decay method. A standardized method is described in Thermal performance of buildings and materials—Determination of specific air flow rate in buildings—Tracer gas dilution method (ISO 12569:2012).

Fig. 6.18 displays CO_2-measurement for an apartment.

Fig. 6.19 displays miscellaneous meters.

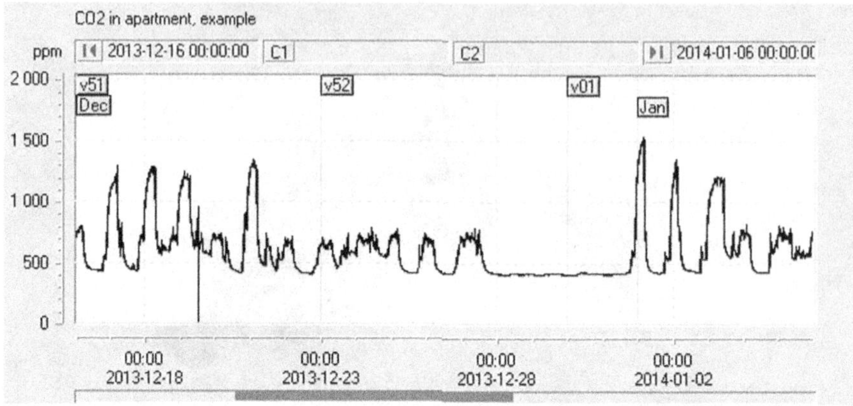

Figure 6.18 CO_2-variations in an apartment. The variations in CO_2-concentration shows the occupancy in the apartment. From the end of week 52 to the beginning of week 01, nobody is at home. During the Christmas holiday, December 24, there is at least someone at home for several days. In the beginning of January, you can see a normal working day pattern appear again, as the CO_2-concentration decreases during the day and increases during the night. In the middle of week 51 the concentration drops to zero, as the power supply to the CO_2-sensor is disconnected.

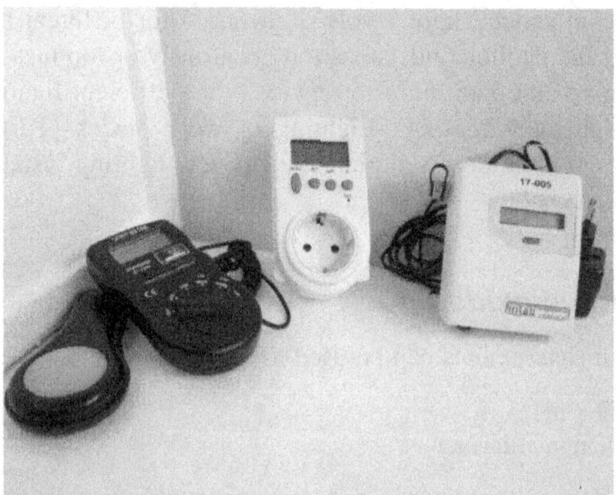

Figure 6.19 Miscellaneous meters. To the left a lux meter, in the middle a simple single-phase power, power factor, and energy meter and to the right a carbon dioxide meter.

6.7.13.1 Typical measurement locations

You will typically measure carbon dioxide concentration:

- Inside ventilation ducts
- In rooms
- Outdoors (for reference values)

6.7.14 Oxygen and carbon oxide

A combustion analyzer can be used to estimate the combustion efficiency in boilers and furnaces.

6.7.15 Relative humidity

Relative humidity can, especially together with temperature measurements, to some extent be used to register presence in the same way as carbon dioxide, but as the humidity can vary for many reasons, a reference measurement outdoor is always needed.

6.7.15.1 Typical measurement locations

Relative humidity will be measured:

- Inside ventilation ducts,
- In rooms, and
- Outdoors (for reference values).

6.7.16 Pulses

Sometimes you can record pulses from an energy meter or flow meter.

On more advanced energy or flow meters, typically a charging meter, there is often a diode that flashes in step with energy end-use. You can read these pulses with a LED pulse sensor and record them with a logger. Is there also information on the meter about the number of pulses per kilowatt hour or liter, you can use your logged pulses to analyze the energy usage.

In some applications you can record a pulse by wire. You will find outputs in electronic equipment, such as frequency converters, controllers, or meter. The output signal varies between different application.

6.7.16.1 In the toolbox

To measure pulses, you need:

- LED pulse sensor combined with data logger
- Pulse logger for wired connection

6.7.17 Production rate, weather data, and other variables affecting energy use

You need to consider which external factors that affect energy use: season, weather, production rate, and so on. Once you know that, you can also use it when analyzing historical energy data. If you have hourly or daily values for energy use during an unplanned or planned production stop (e.g., a weekend, an audit, a refurbishment, or just a shorter stop in a production line), then you can get an idea of idle energy or energy for support processes, or both.

6.7.18 Miscellaneous tools

For electrical measurements you will need tools intended to be used in switchgears, to help you avoid short circuits and to protect yourself. You will also need spare batteries for all your measuring equipment.

Fig. 6.20 displays some tools to access points of measurement.

6.7.19 Documentation and standardized protocols

It is very important to take photos of circuit diagrams, rating plates and where you placed you loggers, and to carefully take notes. You will forget what you saw, you will be unsure of what you have written, your papers will be in a state of disorder.

To create a standardized protocol for where you made your measurements and where loggers are located is also a very important part of the energy audit. Fig. 6.21 displays some tools for documentation.

6.8 Electric measurement case description

In a typical measurement case, one wants to separate the electricity use for the unit processes such as ventilation, lighting, compressed air, internal transports, and packing. Instantaneous power measurements, current logging, and rated power for equipment are used to allocate the electricity use.

Figure 6.20 Tools to access points of measurement. The tool in the lower right corner looks like a straightened paper clip, which it is, but it is also a tool for unlocking a lever switch in the switchgear. You will need to access the conductors without turning machines off.

Figure 6.21 Tools for documentation. Notebook, flashlight, small mirror, and a camera. Sometimes the camera in your phone will do, but a good zoom lens is better.

Based on information from the staff at the plant, one knows that the distribution board in Central picking stock is used for ventilation, lighting, and electric tools for packing. The Central garage distribution board is used for ventilation, lighting, and a forklift charger. From the board, Central Large Hall electricity for lighting, ventilation, and packing equipment is distributed. Based on that information, and the drawings of the electricity distribution system in Fig. 6.22, the plan for measurements may look like in Table 6.1.

The logging period is the same for all the current clamp meters, in this case 7 days, to cover the variations in energy use in production hours as well as during the weekend. The range setting of the current clamp meters is to be appropriate for the load but also depends on the type of clamps and loggers available. In this case we have 500, 200, and 20 A current clamp meters. The sampling interval for the current clamps can be changed depending on what we want to know. For the air compressor we may log more frequent, for example, every 5 seconds, if we find that the compressor is switching mode very frequently, otherwise logging with 60-second intervals may be enough. The purpose of the continuous measurement will also affect the logging interval. If the first visit gives us the impression that the compressed air system is an important part of the energy use, we may measure more carefully to have enough information to suggest detailed measures. The drawing of the electric distribution tells us that the fuse for the compressor is 250 A, but based on the information from the instantaneous measurements we plan to carry out, we might be able to replace the 500 A clamp with a 200 A clamp. Table 6.1 displays a protocol for measurement documentation.

In the case displayed in Table 6.1 we will have a measurement from a switchgear cabinet that covers several unit processes, for example, lighting and forklift charging for a garage and picking stock, as shown in Fig. 6.23. The data from clamp no. 1 in the switchgear must be complemented by simultaneous

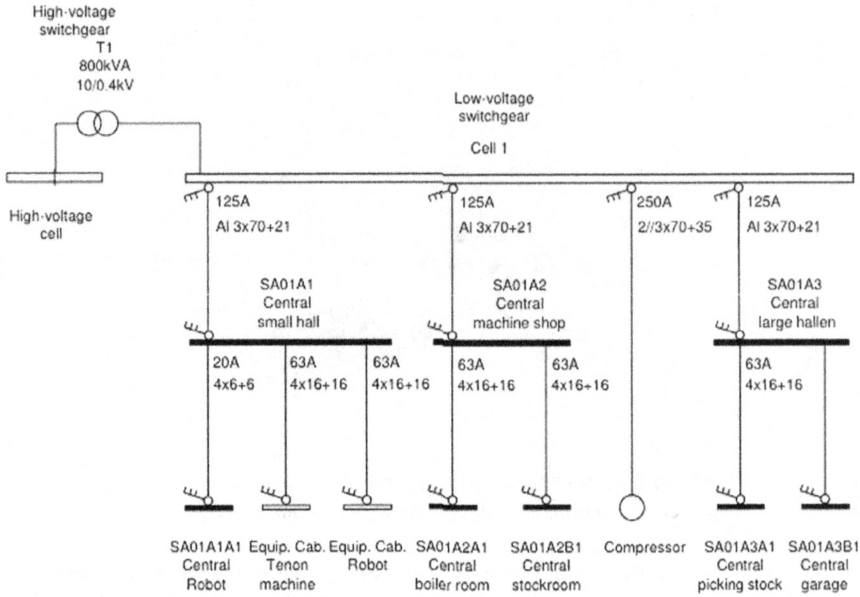

Figure 6.22 Example of a schematic drawing of an electricity distribution system. The cabinets, their location, and fuse sizes are written. The fuse sizes in the drawing indicate where there are loads big enough to measure. The compressor has its own fuse in the low voltage switchgear.

measurements from clamps 2 and 3 in the distribution board, where clamp 2 measures the total incoming current to the Central Garage board and clamp 3 measures the outgoing current to the forklift chargers.

Later, the data from clamp 2 in the Central Garage distribution board can be subtracted from clamp 1 in the switchgear to get a picture of the energy use in Central Picking Stock and Central Large Hall. At the same time, you have parallel measurements of the air compressor from clamp no. 4. The clamp meter is placed in the switchgear, where the compressor has directly connected conductors, or in the compressor, depending on where you have the best access. Fig. 6.23 displays the drawings of the electricity distribution system together with locations of the clamps.

6.8.1 Methodical measurements in a low voltage switchgear

If you start your audit without information about the electric distribution system and energy use, you must be even more systematic when you measure. This is a general method for power measurements using a portable watt- and power factor meter:

1. Measure the current in at least one of the conductors in all the large outgoing groups in the low voltage switchgear. Note the currents in a standardized protocol.
2. Check which of the groups have the largest currents and decide which of them to measure more thoroughly. Often all the groups deserve a closer investigation.

Table 6.1 Example of plan for energy measurements to allocate electricity use to unit processes.

Unit process	Power data source	Time data source	Comments
Compressed air	Instantaneous measurements	Logged data from current clamp 4	Clamp 4 settings: range 200 or 500 A, sampling interval 5 s to 60 s
Lighting	Rated power. Verified by instantaneous measurements	Oral info on working hours. Verify with data from current clamps 1 and 2	Clamp 1 settings: range 200 A, sampling interval 60 s Clamp 2 settings: range 200 A, sampling interval 60 s
Ventilation	Instantaneous measurements	Air handling unit control system. Verify with data from current clamps 1 and 2	
Internal transport	Rated power for forklift chargers combined with logged data from current clamp 3	Logged data from current clamp 3	Clamp 3 settings: range 20 A, sampling interval 60 s
Packing	Rated power, instantaneous measurements	Logged data from current clamp 1	The residual energy is logged data from current clamp 1, minus clamp 2, and the energy for lighting and ventilation. The residual is equal to the energy for packing

3. Measure current, power, and power factor for all three phases in the prioritized groups and fill in your protocol.
4. Use the drawing for the electric system or labeling related to the switchgear, to identify the loads connected to the groups where you measured the highest power demand. Information on the size of fuses and areas of the cables (cross sections) are helpful (see Fig. 6.22).
5. Take a decision to log (or not log) a complete group. Take a decision to log (or not log) the largest loads found for the various groups.
6. If a decision is taken to log one group, make sure to also momentarily meter power factor and loads for all three phases to check that loads are symmetric.
7. Take a decision for which major loads further downstream in the electric system that you want to further measure momentarily and perhaps also log.

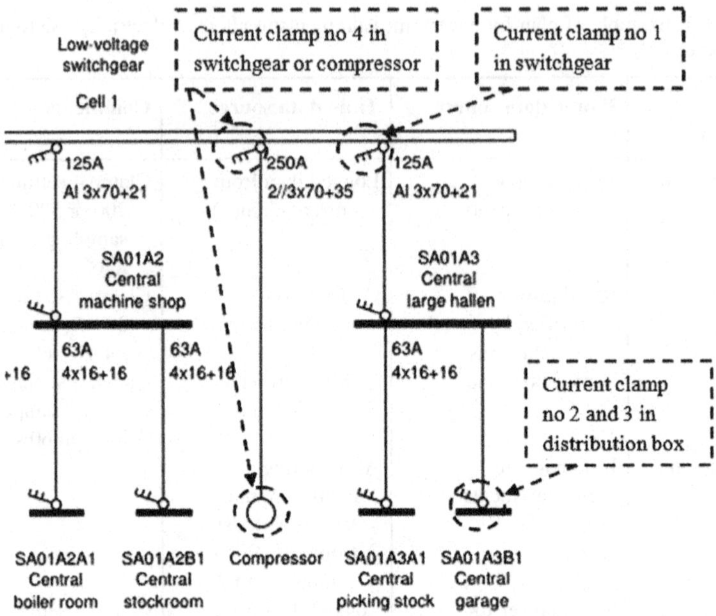

Figure 6.23 Part of a schematic drawing of an electricity distribution system, with locations for current clamps for simultaneous current measurement, to make it possible to subtract a part of the load from the total and separate electricity used for different unit processes.

8. When step 1–7 are accomplished, it may be a sound idea to check if you have found all major loads by conducting a power balance, that is, check if all major loads are somewhat close to the maximum power load the company has subscription for.
9. If step 8 fails, you need to initiate a deeper investigation to be able to spot all major loads. If step 8 fails, the perhaps most common reason for this is discontinuous power loads such as electric heating for premises or production loads (e.g., surface treatment), nocturnal forklift charging, and so on.

6.9 Analysis—electric energy use—basic calculations

This section shows a few practical examples of analysis for energy audit purposes, focusing on electricity. The building properties and heating and cooling systems are often important to examine in an energy audit. Air flow and temperature measurements are also often needed. As there are standardized methods for these tasks, these will not be further discussed in this chapter.

You can analyze your measured data to find:

- System errors and detailed problems,
- Idle and standby energy use,
- Solutions far from best available technologies,

- Incorrect energy quality for a certain use (e.g., electricity for low-temperature heat),
- Drivers and barriers for energy efficiency, and
- The potential for energy efficiency and conversion.

Sections 6.9.1 to 6.9.9 covers some basic calculations and some possible way to analyze graphs.

6.9.1 Electric power

The basic formula used for calculating the energy use is E = P*t (J). Under the pressure of energy market praxis, for practical reasons, the energy use will be described in kWh and MWh, not in the SI-unit Joule. The basic formula used for calculating the electric power is P = U*I (W). Some special versions of the formula that recurs frequently in the energy audits will be highlighted in the Sections 6.9.1.1 and 6.9.2.2.

6.9.1.1 Power factor

For the energy use the active current and active power is the most interesting part. To calculate the active power, you must take the power factor, λ, into account:

$$P = U * I * \lambda \text{ (W)}$$

This formula is valid for a one-phase device, for example, a luminaire with three 35-watt fluorescent tubes and a semiconductor ballast, located in the author's office. Current, voltage, and power factor are measured:

$$P = U * I * \lambda \text{ (W)}$$

where $U = 235$ V, $I = 0.5$ A, and $\lambda = 0.96$

$$P = 235 * 0,5 * 0,96 = 113 \text{ W}$$

The difference between the rated power, $3 * 35 = 105$ W, and the measured, 113 W, is because of the energy use of the ballast (see additional information on energy use for ballasts in Chapter 9: Energy efficiency in compressed air, ventilation, and lighting, concerning lighting).

If there is a linear load, for example, a heater or an asynchronous motor, the conditions are sinusoidal which allows you to treat the power factor as equal to the cos ϕ:

$$P = U * I * \cos\phi \text{ (W) for sinusoidal conditions}$$

where cos ϕ is the phase angle between the current and voltage.

In the distribution board where the measurements often are carried out, you can find all three phases having different current and different power factor. For such case one simply calculates the power for each of the phases and sum them to have the three-phase power distributed via the board:

$$P_{tot} = P_{\theta 1} + P_{\theta 2} + P_{\theta 3} \text{ (W) for nonsymmetric three-phase loads}$$

In industrial applications you often find symmetric three-phase loads, like three-phase motors or heaters, with the same current and power factor for all three phases. The calculation can be simplified to

$$P_{tot} = 3 * U * I * \lambda \text{ (W) for symmetrical three-phase loads}$$

An example from a studied electric motor:

$$P_{tot} = 3 * U * I * \lambda$$

where $U = 238$, $I = 3.8$ and $\lambda = 0.62 = > P_{tot} = 1.7$ (kW).

The actual motor in this case is a three-phase asynchronous motor for a process ventilation fan. The motor ratings are $P = 3$ kW, $I = 6.5$ A and $\cos \phi = 0.81$. Note that the voltage used for this calculation is always line-to-neutral voltage, the voltage between a line conductor and the neutral. If the line-to-line voltage is used, the formula $P = \sqrt{3} * U * I * \lambda$ (W) can be used.

6.9.2 Calculating the energy use from rated power

For some equipment the easiest way to gain the power is to calculate it from the rated power that you find on the nameplate or in technical documentation. It is particularly suitable for light sources with on/off-control. To use the rated power is also the only opportunity if it is not possible to measure the real power use while the equipment is operating.

6.9.2.1 Light sources

Most commonly, lighting is controlled with an on/off-switch. In that case the rated power for the light source can be used to calculate the input power.

For some light sources the input power is the same as rated power: $P_{in} = P_{rated}$. This is valid for old light bulbs and LED luminaires with integrated control gear.

For discharge lamps, like fluorescent tubes, mercury vapor lamps, and sodium lamps, the input power is the rated power plus the ballast (control gear). That means that there is some information about the ballast and some calculations needed.

Table 6.2 displays the relationship between rated power and actual power for some common light sources.

If the light source has a varying light output, the input power will also vary, and the rated power is not useful, and the energy use must be determined with measurements or estimations instead. A general guideline is that electronic ballast use approximately 10% of the light source' electricity demand and magnetic ballasts approximately 20%.

Table 6.2 Examples of relationship between rated power and actual power for some common light sources.

Light source		Power, *P* (W) at 230 V
LED	Complete luminaires	$P_{in} = P_{rated}{}^{*}1.0$
Fluorescent tube		
	T8, magnetic ballast (58 or 36 W)	$P_{in} = P_{rated}{}^{*}1.2$
	T8, electronic ballast (all sizes)	$P_{in} = P_{rated}{}^{*}1.1$
	T5, electronic ballast (from 21 to 49 W)	$P_{in} = P_{rated}{}^{*}1.1$
Mercury vapor lamp		
	50 W	$P_{in} = P_{rated}{}^{*}1.2$
	400 W	$P_{in} = P_{rated}{}^{*}1.06$
High-pressure sodium lamp		
	50 W	$P_{in} = P_{rated}{}^{*}1.2$
	400 W	$P_{in} = P_{rated}{}^{*}1.1$

6.9.2.2 Motors

The rated power for a motor is the power delivered to the shaft of the motor. The supplied electric power, the input power, to the motor is higher than the power on the shaft, as you have a degree of efficiency to consider. The input power that you need for your analysis is not written on the motor but can be calculated from the rating information.

The active input power for a three-phase device is:

$$P_{in} = \sqrt{3} * U_H * I_L * \cos\phi$$

where

U_H = phase to phase voltage
I_L = line current
$\cos\phi$ = the displacement power factor, DPF

For the motor exemplified in Fig. 6.8, the supplied electric power can be calculated from the rated data:

$$P_{in} = \sqrt{3} * U_H * I_L * \cos\phi = \sqrt{3} * 400 * 27.5 * 0.9 = 17\ 150\ \text{W}$$

In reality, the motors are seldom running at the rated values, but on partial loads. If you don't know the real load and can't measure the load, you have to make some assumptions. The load factor differs substantially between different applications.

With the rated motor data, you can also calculate the rated efficiency of the motor:

$$\eta = P_{out}/\sqrt{3} * U_H * I_L * \cos\phi$$

With this information you can calculate the gain of replacing the motor with a new motor with higher efficiency. Though, measurements are also needed to note the actual load factor, which should also be considered—maybe the motor can be replaced by another motor size or a variable speed motor that better meets the need.

For many machines the active power is not written on the nameplate and not possible to calculate without additional information. For transformers, for example, the rated power is by definition the apparent power (IEC, 2019), and the rated current and voltage can only be used to calculate the apparent power, $S = U * I$ (VA). Both data from the manufacturer and user side measurements are needed to calculate the active input power.

6.9.3 Calculating the electricity use from instantaneous measurements

If you have measured the active power with a watt meter, you can simply multiply the measured value with the operating time. This method is useful when the power use is constant. Standby power is often constant, which means instantaneous measurements during nonworking hours may be used to calculate the standby energy use.

6.9.4 Calculating the electricity use from logged data

Logged data can be used for calculating

- Uptime,
- Average power for operating hours,
- Idling power, and
- Energy used to cover leakage in compressed air systems.

6.9.5 How to calculate the idling energy, uptime, and average power

In the following example the case is a hydraulic press, but the same procedure can be used for other equipment, for example, to calculate the energy used to cover leakage in compressed air systems. With some more measurements and analysis, it can also be used for a complete building or workshop.

The idling energy, uptime, and average power can be calculated in a few steps:

1. Measure the power factor at different loads,
2. Log the current,
3. Analyze the graph,
4. Convert the current to power, and
5. Analyze the data again.

How to do this will be shown with measurements from a hydraulic machine for sheet metal processing.

6.9.5.1 Measure the power factor at different loads

The power factor is measured while the machine is running. The machine has two well-defined modes of operation—one when the metal sheets are actually shaped and one when the hydraulic motors are running without any load, that is, idling. Table 6.3. displays the measured power and the power factor for a hydraulic press.

6.9.5.2 Log the current

The current is logged for about a week, to cover working days as well as the weekend. The original current logging data is presented in Fig. 6.24.

6.9.5.3 Analyze the graph

From Fig. 6.24, it is clear that the machine is not used Friday to Sunday and for some hours during the nights. The machine is turned off during nonproduction hours. Nonetheless it has some idling power during the weekend and nights. The conclusion is that there are in fact three different modes to consider: active, idling,

Table 6.3 Measured power and power factor for a hydraulic press.

Mode	Power (kW)	Power factor, λ
Active	<250	0.82
Idling	48	0.38

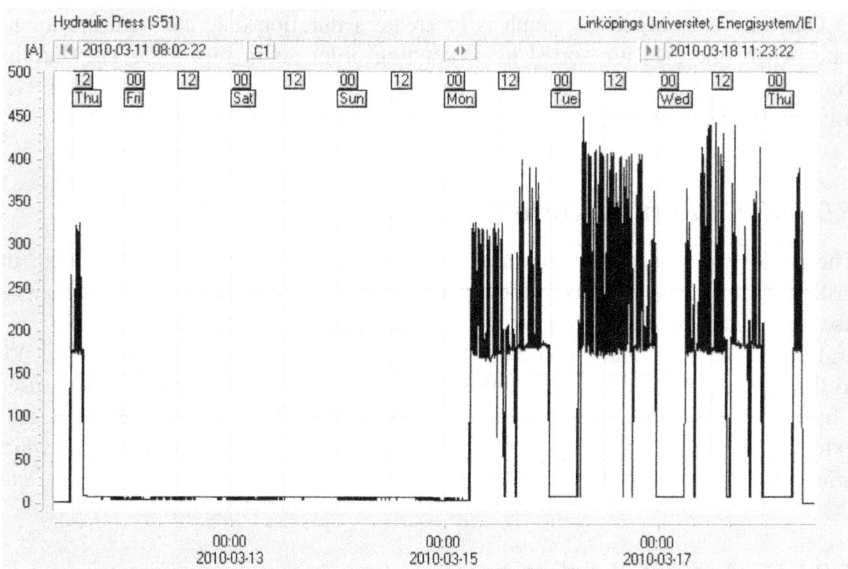

Figure 6.24 The current used by a hydraulic press for 1 week.

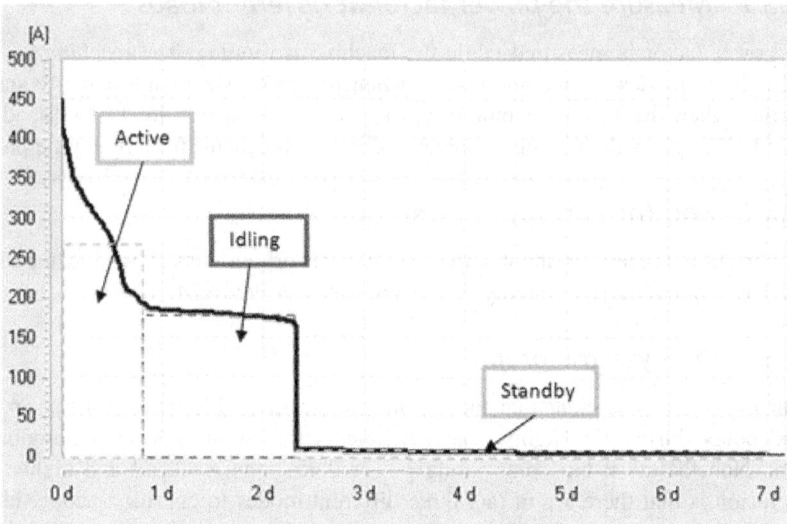

Figure 6.25 Duration diagram of the current used by a hydraulic press during almost 1 week (6.8 days). The active mode lasts for about 20 h, the idling mode for about 1.5 day, and the standby mode for about 4.5 days.

and standby. The periods with zero current in the beginning and the end of the graph is the time before the logger was connected to the machine and after it was disconnected.

One way to analyze the graph is to create a duration diagram with the logged data, where the data are sorted in descending order, as in Fig. 6.25. The duration diagram can be used to distinguish between the modes and to estimate the energy use for the different modes.

6.9.5.4 Convert the current to power

The power can roughly be estimated with the duration diagram together with the instantaneous measurements of the power factor. Table 6.4 shows power and energy use for a hydraulic press for 1 week.

In this case a calculation based on the raw data gave a similar result (5176 kWh) to the rough estimate, but that is mostly because the same assumptions were used. Changing the duration or average power just a little affects the result to a great extent. The power factor at partial load and at standby load is also unknown, which affects the accuracy of the calculations.

6.9.5.5 Analyze the data again

Together with information about the rate of production and the working hours, the annual energy use for this equipment can be calculated.

Table 6.4 Power and energy use for a hydraulic press for one week.

Mode	Power factor, λ	Current (A)	Voltage (V)	Power (kW)	Duration (h)	Energy (kWh/ week)
Active	0.82	270	230	150	20	3000
Idling	0.38	180	230	47	36	1700
Standby	Unknown	5	230	3.5	108	400
SUM						5100

Table 6.5 Power and energy use for a process for 1 week.

Mode	Power, average (kW)	Duration (h/year)	Energy (MWh/year)
Working hours (active and idling mode)	84	3456	290
Standby	3.5	5304	19
SUM		8760	309

If this is a representative "normal" production week, the result can just be multiplied with the number of working weeks per year. In this case it is not a typical week but information from the production manager tells us the working days are representative. The average power use during working hours can be calculated and multiplied with the number of working hours per year. In this case the rest of the year can be considered as standby time. In other cases, equipment can be completely turned off during longer holidays. Table 6.5 displays power, duration, and energy use for a process.

We know from previous calculations that about 35% of the energy use during working hours is really idling energy (see Table 6.4). That information is not needed for the allocation of the energy use to the right unit process, but it is important information for further analysis and to suggest energy efficiency measures. A closer look at the graph of the current used during a week also reveals that the machine sometimes is idling for more than 1 hour without producing (see Fig. 6.24). One example is Monday before midnight when it is idling for about 2 hours before it is at last turned off for the day. An automatic control system or changed routines for the operator can be used to switch to standby mode faster. The aim of the analysis was to be able to calculate the energy use, but on the way some possible improvement measures were also found.

6.9.6 Determine the operating time from temperature measurements

Temperature logging can be used to identify uptime for equipment. A temperature probe can be attached directly to the surface of a machine or in a flowing media

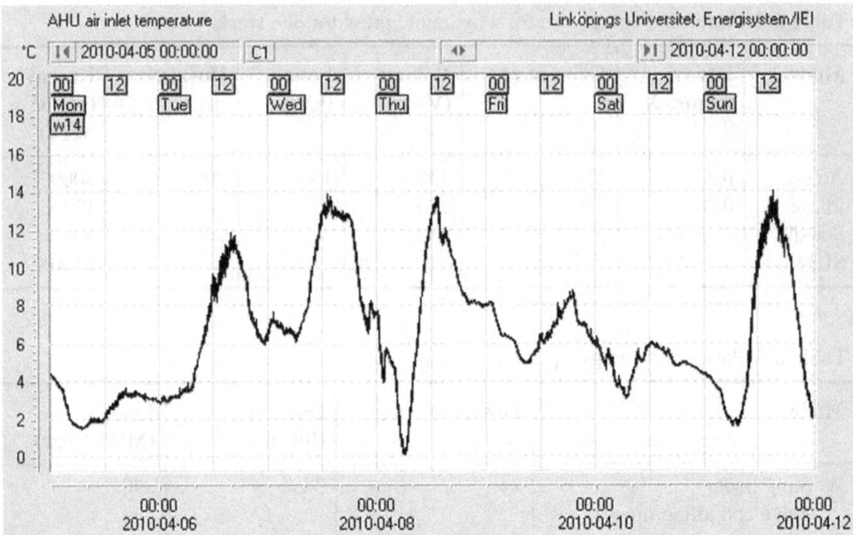

Figure 6.26 Outside air temperature supplied to an air handling unit during 1 week in April 2010, Mjölby, Sweden.

connected to the examined equipment. One example is an AHU where it was possible to make instantaneous power measurements, but no room for a current clamp logger and with no access to the control system. To determine the uptime for the AHU a temperature logger was placed in the duct taking in the outside air but inside the building. The graph clearly shows that the temperature is not at all affected by the indoor temperature, set to 20°C, which it would have been if the AHU was turned off during some period of the sampling (Fig. 6.26). When the graph is compared to a temperature graph from a nearby weather station and the graphs are found to look very similar, it is even more clear that outside air is taken into the AHU all the time (Fig. 6.27). The conclusion is that the uptime for the air handling unit is 24 hours a day (see Fig. 6.26).

Temperature loggers are useful to identify uptime for lighting. The lighting is often powered from more than one cabinet or case and from many groups in each case. As light sources also emit heat, temperature loggers attached directly on fluorescent tubes, luminaires, or the control gear give you information on when the light has been used (see Fig. 6.28).

6.9.7 How to calculate the energy used to cover leakage in a compressed air system

To identify how much air a compressed air system is leaking, you simply let the compressor or compressors run when there is no production going on, during the night, the weekend, or just an hour or two at the end of the day. The compressor will keep working to keep the pressure up. If there is an accumulator, it will turn on to fill the

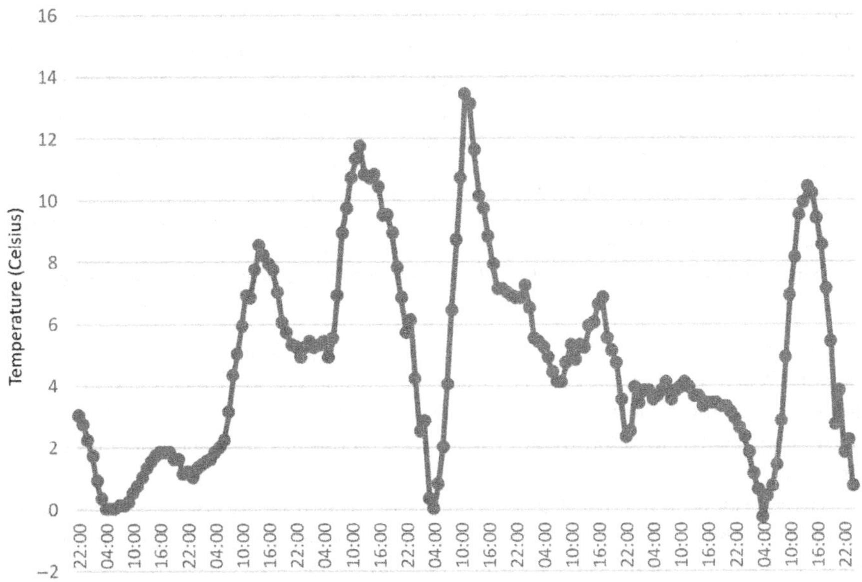

Figure 6.27 Outside air temperature during 1 week in April 2010, Linköping, Sweden (SMHI, 2019).

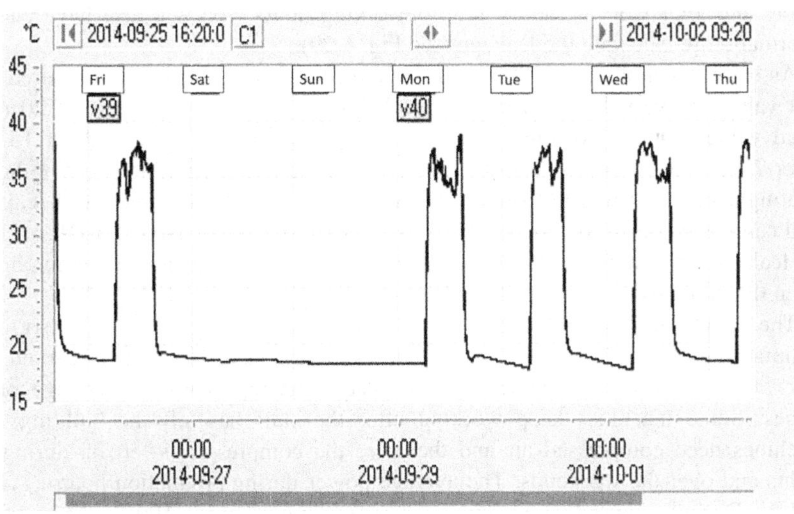

Figure 6.28 Temperature on luminaires with fluorescent tubes showing production-hours for a lunch restaurant kitchen in Ödeshög, Sweden.

accumulator tank and turn off when it is full. If it is a variable speed drive compressor and the leakages are substantial, it will run constantly. The current or power may be logged, and the energy use can be calculated the same way as in the example with the hydraulic press above (see Table 6.4). Rotary screw compressors which are commonly used in industries, with a few exceptions, have three clear modes:

- Loaded operation. The machine produces compressed air,
- No load operation. The machine is running, waiting to produce air. The screw is rotating but no air is produced, and
- Standby mode. The machine is turned on, waiting to produce air at demand, but the screw is not rotating.

The power used in no load mode is often about 25%–50% of the full load power. When the compressor is turned on during nonworking hours, it runs in loaded mode for a short while, shifts to no load mode when the pressure is up and then either switch to stand by after a while or goes back to loaded operation if the pressure drops too fast.

The energy used to cover leakages during working hours is not the same as during nonworking hours. The energy use is higher during nonworking hours because you have a greater part of unnecessary idling in no load mode. To get the proper value you must calculate how much air is produced (the time in loaded mode during nonworking hours) and calculate the energy cost to produce the same amount of air during working hours, when the compressor is working in loaded mode to a greater extent.

A current clamp and a logger provide you with good information on the variations in operation, but often the compressor has its own counter which records the number of hours worked in loaded and no load operation, respectively. By taking instantaneous power measurements in different loads and by reading the meters before and after working hours and nonworking hours, you will also have enough information to calculate the leakage (see Fig. 6.29).

As can be seen in Fig. 6.29, the line in the background shows the measured current values converted to power. The black line represents the average of 20 measured values. During working hours the compressor is often using about 15 kW. After 7 p.m. the compressor is turned off, as it usually is during nonworking hours. The night before it was left running, to give us an idea of the amount of leakage. You can see the result before 7 p.m. as the compressor needs about 1.5 kW to cover the leakage. The same amount of leakage can be assumed during production hours, when the compressor is normally running.

The graph in Fig. 6.30 displays the power use for an air compressor at a manufacturing industrial plant. The line in the background shows the measured current values converted to power and the black line represents the average of 20 values. Some machines keep working after the staff has left the building. The machines need compressed air and therefore the compressor is left on during the nights and over the weekends. The average power during production hours is about 13 kW. During nonproduction hours the power use is about 10 kW. Most of the energy supplied to this compressor is thus used to cover extensive leakage, a quite common situation for many industrial plants.

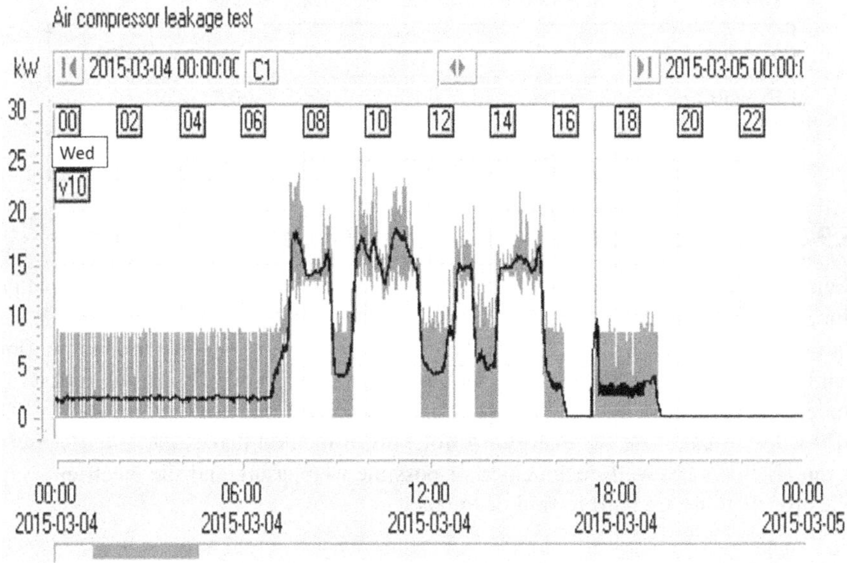

Figure 6.29 Power use for air compressor with variable speed drive at a manufacturing industrial plant during production hours and nonproduction hours. The black line represents the average of 20 measured values.

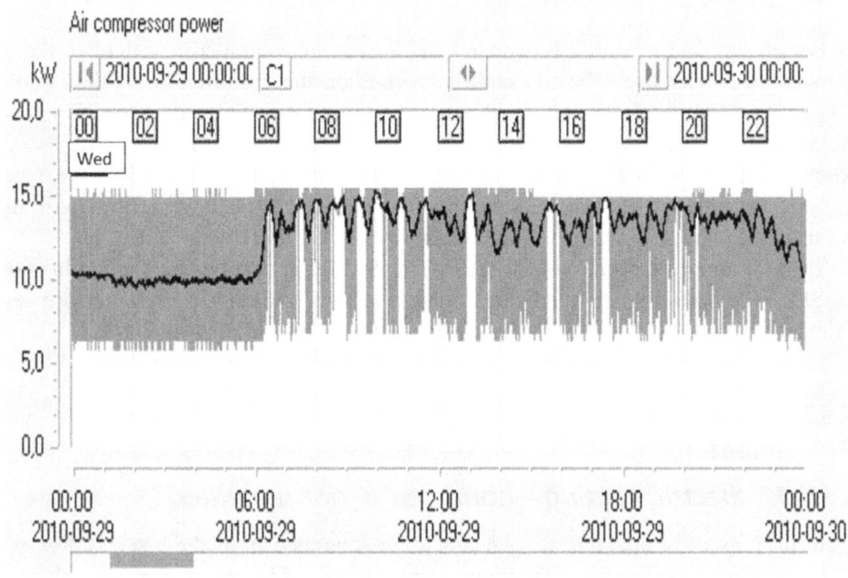

Figure 6.30 Power use for an air compressor at a manufacturing industrial plant. The black line represents the average of 20 measured values.

Table 6.6 Typical SFP-values for different ventilation systems.

Ventilation system	Typical SFP-value
Exhaust fan	0.5
Supply fan with heat coil	1.7
Air handling unit with heat-exchanger and heat coil	2.5

6.9.8 How to calculate air flow from power and vice versa

Sometimes the air flows in the air handling system can be found in the documentation, sometimes it can be measured in the ducts or in the air handling unit. If you know the power used by the fan motor from electrical measurements, the air flow can be calculated with the specific fan power value (SFP), which is used to describe the energy efficiency of a ventilation system. You can also use the air flow and the SFP-value to calculate the energy use for fans, a method that is particularly useful if the air flows are well documented or possible to measure and the electricity supplies for the fans are hard to find or to reach.

$$\text{SFP} = P_{\text{tot}}/q_{\text{max}}$$

where

SFP specific fan power (kW/(m^3/s))
P_{tot} total electric power for fan or fans in an air handling unit (kW)
q_{max} the highest air flow, either supply or exhaust (m^3/s)

For energy audits, the SFP in many cases must be estimated to calculate the air flow or electricity use. The SFP-value depends on the construction of the ventilation system. Short duct systems with large duct dimensions give a lower SFP-value. A hole in the wall with an exhaust fan can have an SFP value as low as 0.3. Narrow ducts with higher air speeds, typically for transporting particles or welding fumes, has a higher SFP-value, sometimes as high as 5.0. Clogged filters will also contribute to an increased SFP-value (Table 6.6).

There is more than one kind of SFP. The definition depends on the application, as SFP_v for performance verification test, SFP_e for building energy performance calculations, and SFP_{AHU} for air handling units (Schild and Mysen, 2009).

6.9.9 Analyzing energy supply graphs to separate comfort heating

6.9.9.1 Electric heating—hourly data, not separated

Electricity used for space heating needs to be separated from the rest to allow you to allocate the energy use to the proper unit process. The shape of the graph from the electricity supply can reveal the presence of electric heating, at least if electricity is the main source of comfort heat. The energy use will be much higher in the winter and form a clearly outdoor temperature dependent graph. If the studied object uses heat pumps or if the use of electric heating is limited to parts of the buildings, the slope of the graph will be less dramatic but still be recognized. The difference can be shown by comparing the variations in energy use during a summer week and a winter week (see Fig. 6.32).

Mostly the electricity for heating is not separately measured in the buildings today. One reason is that the electric supply for heating often is divided between many distribution boards in different parts of the building. For energy audit purposes the measurements are usually carried out for 1 week. Even if you are able to actually measure the electricity use for heating for 1 week, you can assume that this particular week cannot represent all weeks of the year.

If you have successfully identified the electricity use of all the other processes, you can simply regard the rest as space heating emanating from electricity. If the other processes are not depending on seasonal variations, the average electricity use during the warmest months will give you a good indication of the energy needed for nonheating purposes (see Fig. 6.31).

Basically, the same method that is used for calculating the hot water part of the district heating can be used (see Section 6.9.9.2). With more detailed data, more accurate values can be calculated. If you have day by day or hour by hour resolution data for the electricity use, you can use the electricity use during weekends and nonproduction hours together with outdoor temperature data to calculate the heat

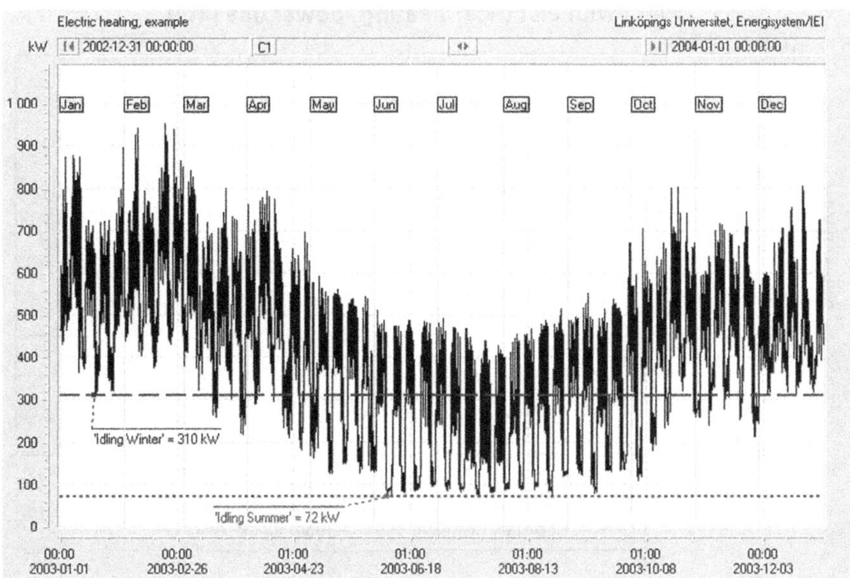

Figure 6.31 Electric power use, hour by hour data for 1 year, for a plant in southern Sweden with electric heating.

losses. In that case, information on electric equipment placed outdoors, like outdoor lighting, is useful. The uptime for air handling units can affect heat losses to a great extent, which makes information on operating hours for ventilation valuable as well. Fig. 6.31 displays electric power use, hour by hour for 1 year for an industrial plant.

As can be seen in Fig. 6.31, the base load clearly depends on the outdoor temperature—it is at its lowest during the summer months (June–August) and substantially higher during the winter. Fig. 6.31 also shows that the production dependent components of the electricity use show a pattern that recurs every week, with a low point every weekend when there is no production. Fig. 6.32 provides yet another example comparing electricity use between a week in February and one in June.

As can be seen in Fig. 6.32, the total power use is higher in the winter, even during nonproduction hours, which indicates some kind of electric heating. The production dependent component of the electricity use is similar for the 2 weeks.

The energy use is clearly higher during the nights than during the weekend. From this several questions arise such as:

- Are they running part of the production at night?
- Are machines and support processes left on standby during the nights, but turned off during the weekends?

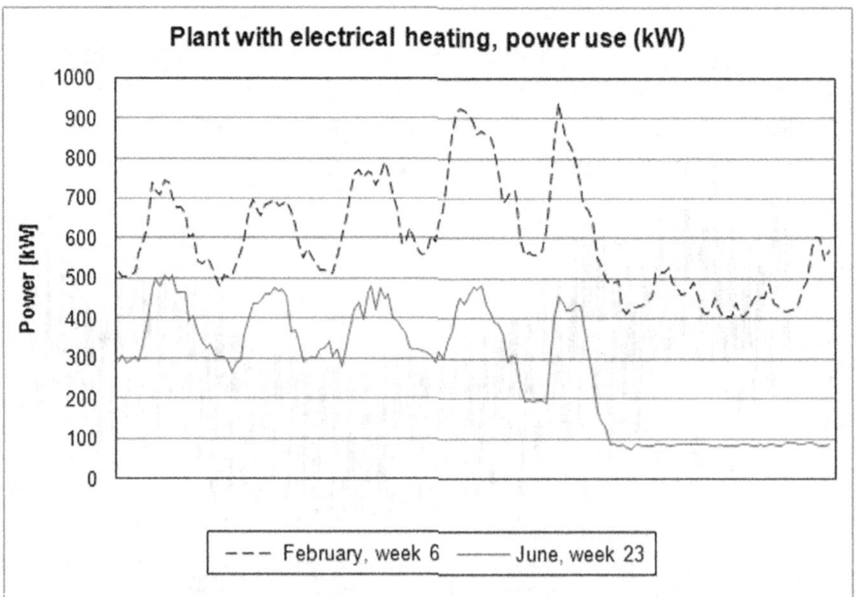

Figure 6.32 Comparison of 2 weeks electric power use, hour by hour data, for a plant in southern Sweden with electric heating.

6.9.9.2 District heating—monthly data, heating separately

When the district heating is used for space heating purposes only, the analyzing is easy. Often the district heating is also used for hot tap-water and sometimes also for production processes, which means some calculations are needed.

When the district heating is used for hot tap-water and heating, the usage during the warmest months can be considered representing the hot tap-water use.

Example 1: A printing plant in Linköping, Sweden, uses 610 MWh/year of district heating. The monthly energy use forms a typical graph of space heating energy use in Sweden (Fig. 6.33). In August the energy use was 5.5 MWh. The staff tell us the production has been running, causing internal heat gains, and also the comfort cooling has been running. The need for heating has been minimal. Presuming that the control systems have been working properly, the district heating use in August can be used as a template for the hot tap-water use for the whole year. Hot tap-water use thus is 5.5*12 = 66 MWh/year.

Example 2: The monthly energy supply for a manufacturing company in Mjölby, Sweden, 2010 is shown in Fig. 6.34 The curve for the district heating appears to be completely outdoor temperature dependent. This suggests that district heating is used for heating only—not for hot tap-water and not for process heating. The district heating use is very low during May–September. This suggests that the premises have either a lot of excess heat from the production processes or are well insulated or possibly both. The curve for electricity use seems to be dependent on production—the only low point is during July when the factory is closed for summer holidays. We can also state that monthly data are not enough to determine how

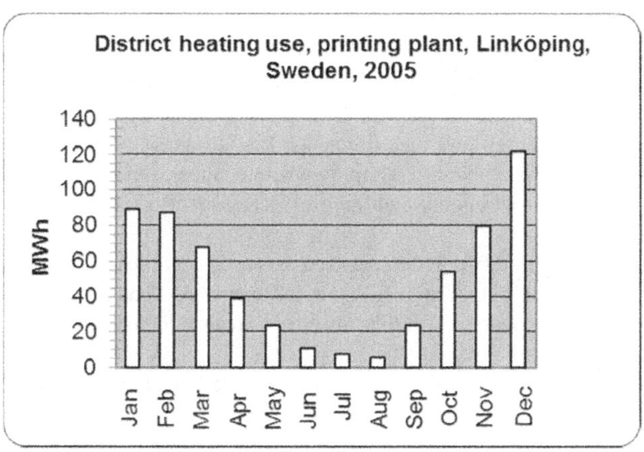

Figure 6.33 Monthly district heating supply for heating and hot tap-water at a printing plant in Linköping, Sweden 2005. The total use is about 610 MWh.

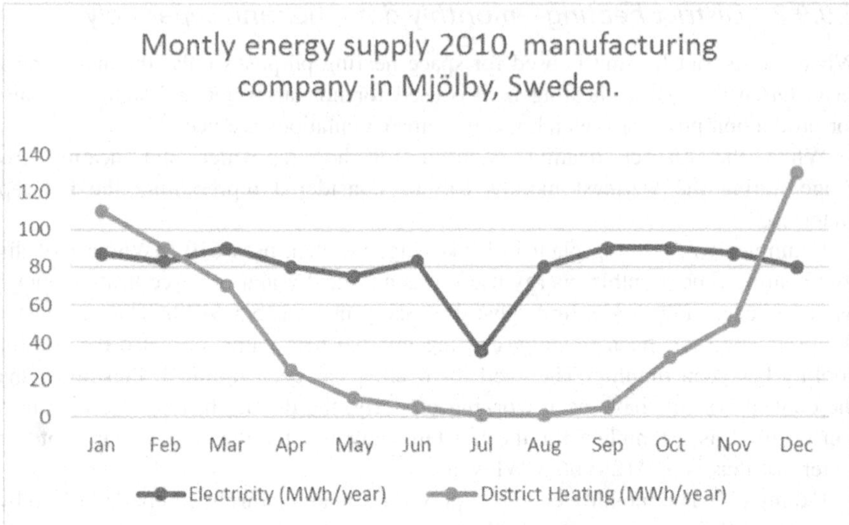

Figure 6.34 The monthly energy supply for a manufacturing company in Mjölby, Sweden, 2010. The U-shaped line is showing the outdoor temperature dependent use of district heating.

much of the electricity is idling energy. To analyze idling, average values for hours or at least days are needed.

References

IEC, 2019. Electropedia, the International Electrotechnical Vocabulary Online. http://www.electropedia.org (accessed 06.25.19.).

Jernkontoret, 2018. Jernkontorets energihandbok (Jernkontoret's Energy Handbook). Available at: http://energihandbok.se (accessed 02.28.18.) [in Swedish].

Rosenqvist, J., Thollander, P., Rohdin, P., 2012. Industrial energy auditing for increased sustainability: methodology and measurements. In: Sustainable Energy: Recent Research. InTech Publisher. Available at: <http://www.intechopen.com/books/sustainable-energy-recent-studies/industrial-energy-auditing-for-increased-sustainability-methodology-and-measurements>.

Schild, P.G., Mysen, M., 2009. Recommendations on Specific Fan Power and Fan System Efficiency, Technical Note TN65, IEA Air Infiltration & Ventilation Centre. Available at: <http://www.aivc.org/sites/default/files/members_area/medias/pdf/Technotes/TN65_Specific%20Fan%20Power.pdf>.

SMHI, 2019. Hourly temperature data. Malmslatt weather station. Available at: smhi.se (accessed 06.25.19.).

Investments, nonenergy benefits, and conservation

7

Chapter Outline

7.1 Investment assessments

7.1.1 Investment calculations

Although the environmental gains from improved energy efficiency are large, in practice it is often difficult or impossible to enforce proposals that are not economically profitable. Improvement proposals therefore often require an economic evaluation of the investments that are intended to lead to efficiency improvements of the energy system. It is of great importance that this assessment is done correctly, as new more energy-efficient technologies and processes will be used for a long period of time, sometimes for 10 years or longer.

This chapter describes four primary methods for economic calculations: pay-back, net present value (NPV), internal rate of return (IRR), and life cycle cost (LCC). Furthermore, the reason why the NPV method is the most suitable method for the purpose is justified, and how the method considers taxation.

7.1.2 Pay-back method

The pay-back method is easiest to use and provides answers to how long it takes before the investment's payment surplus covers the basic investment cost. The pay-back time is

Introduction to Industrial Energy Efficiency. DOI: https://doi.org/10.1016/B978-0-12-817247-6.00007-9

$$\text{Pay-back-time} = \frac{I}{R - C} \tag{7.1}$$

where I is the Investment cost (EUR); R is the Revenue (EUR/year); C is the Cost (EUR/year).

7.1.3 Net present value method

With the net present value method, one can evaluate whether an investment is profitable by comparing the investment with the present value of all future deposits and payments. If the net present value is greater than zero, the investment is profitable. Future deposits and disbursements are discounted to the present value of money using a selected interest rate. The present value, NPV, is:

$$NPV = \sum_{i=0}^{N} a_i(1+r)^{-i} \tag{7.2}$$

where a_i is the net profit (revenue R−cost C) during period i (EUR); N is the number of years; r is the interest rate; $(1 + r)^{-I}$ is the discount factor.

With the net present value method, payment consequences that occur at different times are valued. The method is based on a reference time and all future incoming and outgoing payments are recalculated to the monetary value of this point in time using an interest rate. This is done in order for it to be possible to compare the value of the payments. An investment is profitable if its present value is greater than zero. An example of an investment process is shown in Fig. 7.1.

If year zero is the reference time and the calculation interest rate (r) is set to 10%, the net present value in the example according to Eq. (7.2) becomes:

$$NPV = -300 + \frac{150}{1+0.10} + \frac{200}{(1+0.10)^2} + \frac{100}{(1+0.10)^3} = 77$$

The net present value of the above example is positive, which means that the investment is profitable. This means that it is more profitable to make the investment than to invest the money elsewhere at the interest rate of 10%. The interest rate should be chosen so that it corresponds to the return that the best alternative

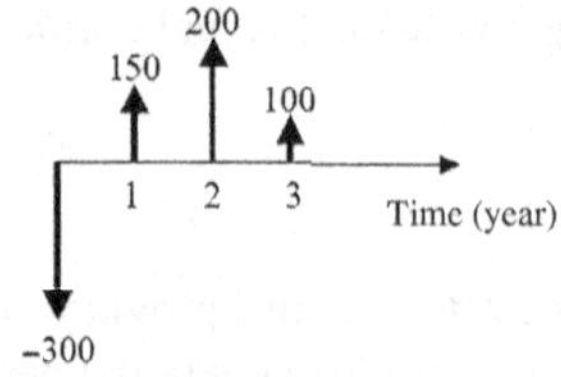

Figure 7.1 An example of an investment process.

that other investments can provide. The interest rate is difficult to determine as one does not always know the best alternative investment. There is then another possibility to choose the return on investment that the company wants to achieve instead. By weighing the interest and requirements of lenders and shareholders, the cost of capital can be estimated.

If a comparison between several different investment alternatives is to be made, a type of the net present value method, the so-called net present value quota method can be used. With this method, different investment options can be ranked, and the method is often used when a company is short of capital. The present value ratio is

$$\text{Net present value quota} = \frac{\sum_{i=0}^{N} a_i(1+r)^{-i}}{|a_0|} \tag{7.3}$$

where a_i is the net profit (revenue R−cost C) during period i (EUR); a_0 is the investment (EUR); $(1 + r)^{-I}$ is the discount factor; N is the number of years.

In the method, the net present value is related to the basic investment, and the larger the quota is, the higher the investment alternative is ranked.

7.1.3.1 With regard to tax

If a company gains profit, tax must be paid to the government in most cases. If tax is taken into account, the calculations according to the net present value method will be slightly different, since, for example, depreciation for investments and rest values are taken into account. The profitability of investments can be affected in two ways if tax considerations are taken. Firstly, the ranking between the investments can change if different depreciation rules are used and secondly, investments that were unprofitable in calculations without regard to tax can become profitable. However, it is basically impossible that the opposite occurs, that is, that a profitable investment calculated without tax considerations becomes unprofitable when tax considerations are taken. Fig. 7.2 shows an example where tax considerations are taken into account.

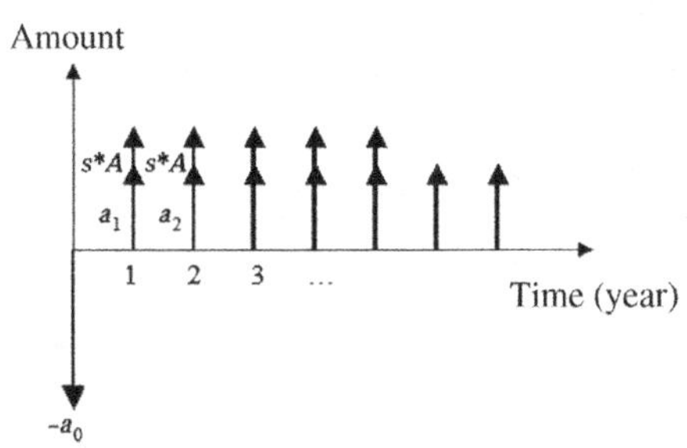

Figure 7.2 A payment processes when tax considerations are taken into account.

If the interest rate is r_b before tax, the interest rate after tax is, r_a

$$r_a = (1\text{-}s) * r_b \tag{7.4}$$

where s is the the calculated tax rate, which differs from country to country.

The depreciation, D., of the investment that is made means that the company's profit is reduced by the same amount. This means that the company pays less tax due to the reduced profit, provided that the company gains profit.

With respect to tax, the net present value can be expressed as

$$NPV_{\text{tax}} = -a_0 + s \cdot D\sum_{i=1}^{K}(1+r_a)^{-i} + \left(1-s\right)\sum_{i=1}^{N} a_i(1+r_a)^{-i} \tag{7.5}$$

7.1.4 Internal rate of return

The internal rate of return method is to identify the internal rate of return that gives an investment the net present value of zero. In the same way as for the net present value method, the investment is evaluated according to Eq. (7.2), with the difference that the net present value is set to zero and the internal rate of return(s) instead is calculated. If the calculated internal rate of return exceeds the company's cost for capital, for example, the interest rate, the investment is considered profitable.

7.1.4.1 Comparison between pay-back, net present value, and internal rate of return

In order to determine which method to use, the three methods, the pay-back, the net present value, and the internal rate of return methods, are compared with regard to:

1. All relevant payments are taken into account
2. Valuation is made of payments over time
3. The alternative cost of capital is taken into account
4. There is a possibility to rank the investments
5. The method is additive

Table 7.1 compares the three investment methods.

Table 7.1 Comparison of three investment methods (Grubbström and Lundquist, 1996).

Criteria for comparison					
	1	2	3	4	5
NPV	Yes	Yes	Yes	Yes	Yes
IRR	Yes	Yes	Yes	Yes	No
Pay-back	No	No	No	Yes	No

Based on the comparison in Table 7.1, the net present value method is the model that gives the most accurate result in an investment assessment.

7.1.5 Life cycle cost

LCC stands for life cycle cost and means that costs for the entire life cycle of the investment are included in the investment assessment. The method aims to compare different alternatives in order to determine which alternative has the lowest total cost over the entire lifetime.

In an LCC calculation, all major costs are normally included during the investment lifetime. For products that require energy, it is not uncommon for the energy costs and the investment cost to be the major expenditures.

The method is usually used in the real estate sector and is less commonly used in the industry.

The life cycle cost is calculated as

$$\text{LCC} = \text{Investment} + \text{LCCenergy} + \text{LCCmaintenance} + \text{LCCothers} - \text{rest value} \tag{7.6}$$

where

LCCenergy = annual energy cost * net present value
LCCmaintenance = annual maintenance cost * net present value
LCCothers = other costs cost * net present value
Rest value = rest value * net present value

7.1.5.1 Investment

The investment is a cost item that includes all costs that arise during the investment phase. This can be, for example, material purchase, installation costs, and so on.

7.1.5.2 LCCenergy

The energy cost that an action gives rise to is called LCC energy. The energy cost is determined by a net present value calculation where all purchased energy during the life cycle is included.

7.1.5.3 LCCmaintenance

Maintenance costs during the life cycle of the investment are called LCC maintenance. This cost is also calculated using the net present value method.

7.1.5.4 LCCothers

Other costs that can be derived from the investment are called LCC others. This cost is also calculated using the net present value method.

Rest value

In some cases, an investment has a rest value. If this is the case, this is added to the calculation using the present value method.

7.1.5.5 Net present value

This can either be derived from a present value table or calculated in accordance with Eq. (7.2).

7.2 Conservation supply curves

Conservation supply curves is a means of visualizing various type of energy efficiency measures and its respective magnitude in energy savings (Andersson et al., 2018). The cost of conserved energy (CCE) for a measure is calculated according to Eq. (7.7):

$$\text{CCE} = \frac{\text{Annualized capital cost} + \text{Annual change in operation and management costs}}{\text{Annual energy savings}} \tag{7.7}$$

Where

$$\text{Annualized capital cost} = \text{Investment cost of measure} * \left(\frac{d}{1-(1+d)^{-n}}\right) \tag{7.8}$$

d = discount rate
n = measure lifetime

In a study of energy efficiency measures from the Swedish national energy audit program, the discount rate was set to 7% for all measures, and the energy efficiency measures lifetime was set to 12 years for technology measures and 5 years for management measures (Andersson et al., 2018).

In Fig. 7.3 below, a conservation supply curve is exemplified. In the actual case, the discount rate was set to 33 % which is equivalent to a 3-year pay-back time.

7.3 Behavior and energy conservation

Within energy management, the company's staff are the key actors in initiating improvements and maintaining improvements and excelling further into new improvements. Despite so, energy management in relation to how people behave is a seldom covered research area. In Section 7.3.1, a non-exclusive presentation of some contributions into the area of energy-efficient behavior is presented.

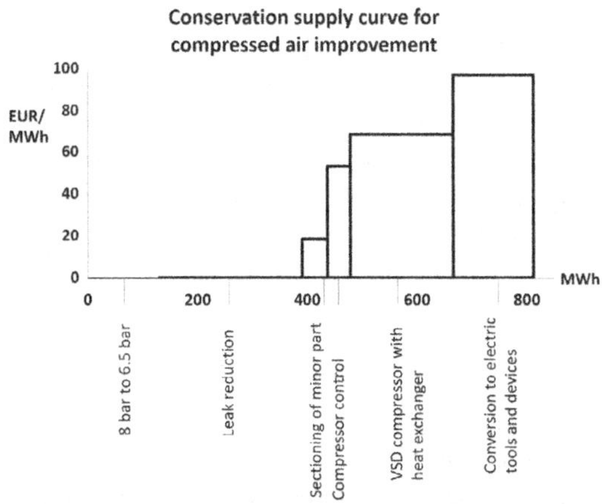

	Energy efficiency improvement (MWh/year)	CCE (EUR/MWh)
8 bar to 6.5 bar	130	0
Leak reduction, continously	261	0
Conversion to electric tools and devices	146	97
Sectioning of a minor part of the compressed air system	46	18
Compressor control	42	53
New VSD air compressor with water borne heat exchanger	187	68

Figure 7.3 Conservation supply curve for an actual carried out action plan for compressed air energy efficiency improvements in a vehicle factory.

7.3.1 Environment and health-based messaging as a driver for energy conservation

In a paper by Asensio and Delmas (2015), investigating nonprice information strategies to drive energy conservation behavior, they found that using environment and health-based messaging such as environmental and public health externalities of electricity production (e.g., own electricity use give rise to pollutants, childhood asthma, and cancer), was a much stronger driver than monetary savings. In fact, it was found that such information outperformed monetary savings as a driver. Their study involved 118 households. The authors revolve around two mechanisms, first increased attention to energy conservation motivated by decreasing harm to others and secondly, increased attention from direct own private benefits such as lower risk of disease. In total, the participating households receiving health-based information reduced energy use with 8%. For households with children, the savings reached 19%. (Asensio and Delmas, 2015).

Relating to the difference between stated preferences and actual behavior—the attitude–behavior gap—the authors found that when asking a random sample of participants of what major things that would make them conserve energy, monetary savings was high-ranked, while the study's result later found out that this was not the primary driver (Asensio and Delmas, 2015).

In a study by Palm and Thollander (2010), relating to studies of barriers to energy efficiency in industry, the authors argue that social constructivism might in fact appear in barrier studies. This is due to social regimes where one perception, for example, too high costs or pay-back periods for energy efficiency investments, may be perceived as a "truth" even though such investments might be cost-effective (Palm and Thollander, 2010). These findings might indicate that perceived drivers (and barriers) need to be critically assessed, and even questioned.

Relating this to the study by Asensio and Delmas (2015), it appears that even though their study concerned households, the very same phenomenon, that is, using nonprice information strategies to drive energy conservation behavior, is most likely also an effective driver for operative energy efficiency actions among industry workers.

7.3.2 Energy conservation effects of monitoring

In a paper by Schwartz et al. (2013), they study the Hawthorne effect, which refers to a change in behavior due to the awareness of being monitored. In their study, a total of five postcards were being sent out on a weekly basis, informing 6,350 households that they participated in a study of household's electricity use. Their study indicated a statistically significant mean electricity saving among the participating households of 2.7 %. In a follow-up survey, the researchers found that the households who had been achieving the savings stated that they were aware that they did more in terms of energy conservation than their neighbors. The authors conclude that the Hawthorne effect might diminish, taken that no continuous attempts to inform persons that they are being monitored occur (Schwartz et al., 2013).

Relating the findings by Schwartz et al. (2013) to an industrial organization, it appears that monitoring, even in the slightest form, may greatly enhance energy conservation among industry workers.

7.3.3 Social influence as a driver

Changing behavior as a consequence of how other people behave and their opinion on your choice of behavior was studied by Abrahamse and Steg (2013). In their review of 29 studies of social influence approaches it was found that social norms indeed affect the way people act and behave but that various magnitudes of change occurred, depending on which group that was studied. The authors outline six social influence approaches from the psychology literature, namely:

1. Social norms in information
2. Block leaders and social networks
3. Public commitment making
4. Modeling
5. Social comparison in feedback provision
6. Feedback provision about group performance

As regards social norms as a driver in relation to industrial energy efficiency, it appears that gaining commitment from a selected few within a group of individuals, for example, one person per shift, that advocates energy conservation, may enhance efficiency. Moreover, as such individuals also may have a mission to spread the word about the importance of energy conservation among colleagues in their network, effects may be even stronger. If such individuals also ask for a public statement, even greater effects may be seen.

As regards the fourth social influencer, modeling, people tend to do what other people do, more than what other people say. One example is given by Aronson and O'Leary (1983), where an attempt was made to shut off the hot water when putting soap on the body while showering in a university gym. When posting a sign, only 6 % made this change in behavior. When making an intrusive sign, figures rose to 19%. However, when instead another student was acting as role-model in the shower, figures went up to 49 %. Finally, when sending in two role-models, a figure as high as 67% was reached. (Aronson and O'Leary, 1983).

In relation to social comparison, people tend to judge their own performance by studying both other people perceived as better than themselves, as well as people perceived as performing worse than themselves. Thus sharing information feed-back of their own performance might increase energy conservation. Closely related to this is feed-back provision of a group's performance, which also may affect energy conservation positively.

Alcott (2011) found that in US programs providing letters to residential customers comparing electricity use with others, saved electricity use by 2%.

7.4 Nonenergy benefits and improved energy efficiency[1]

7.4.1 Nonenergy benefits in companies

In addition to direct energy savings and energy cost savings, energy-efficiency investments or measures can provide a number of additional benefits or added values, so-called nonenergy benefits (NEBs). Studies such as the one by Nehler et al. (2014) show that there are a number of NEBs linked to improved energy efficiency, such as improved health and well-being effects. For an industry, NEBs can include, for example, improved productivity and enhanced competitiveness. At national and international levels, NEBs such as new jobs, improved energy security, and reduced negative environmental impacts (e.g., reduced greenhouse gas emissions) are found. The most common NEBs are reported in Table 7.2.

7.4.1.1 Quantification of NEBs

Quantifying NEBs in monetary terms is often difficult. However, in order to see the full potential of energy efficiency investments and measures, NEBs must be assigned to monetary values wherever possible.

In a case study involving 77 cases of documented NEBs from 2001, it was possible in 52 cases to quantify and allocate monetary value to the NEBs, resulting in cutting the pay-back period with a factor two (from 4.2 years based solely on energy cost savings to 1.9 years when NEBs were included). In addition, the NEBs

[1] This text is revised and is based on a previous published article by Nehler et al., 2014.

Table 7.2 Categories of NEBs (nonenergy benefits) related to energy-efficient investments or measures in the industry.

Category	Examples
Production	Increased production Improved product quality Increased production reliability Increased equipment life Shorter process cycle time Reduced raw materials use
Operation and maintenance	Reduced maintenance Lower cooling requirements Reduced labor requirements Reduced need for engineering controls
Working environment	Increased worker safety Reduced noise Improved air quality Improved temperature control Improved lighting
Waste	Reduced waste water Reduced hazardous waste Use of waste fuel, heat, and gas Materials reduction
Emissions	Reduced emissions (CO, CO_2, $N0_X$, SO_X) Reduced dust emissions
Other	Improved public/corporate image Improved worker morale Increased sales levels

Sources: Based on Lilly and Pearson, 1999; Pye and McKane, 2000; Finman and Laitner, 2001; Laitner, Ruth and Worrell, 2001; Hall and Roth, 2003; Worrell et al., 2003; Lung et al., 2005.

in over half of the 52 cases was equal to or greater than energy cost savings. (Finman and Laitner, 2001).

In another study from the beginning of the 2000s, several NEBs were possible to categorize and quantify, including related improvements in production, operation and maintenance, waste reduction, and reduced emissions. The result showed that NEBs related to operation and maintenance and reduced emissions could be quantified and assigned monetary values, through measurements before and after the implementation of the measure. However, for other NEBs, it was not possible to assign a monetary value, wherefore these were omitted from the calculations. One of the studies found that the NEBs associated with the work environment was difficult to quantify (Laitner et al., 2001).

In yet another study, 81 energy efficiency measures were implemented in US manufacturing industry. In 54 out of 81 cases, NEBs could be quantified. Most of the quantified NEBs were related to operation, maintenance, and production. Improved working conditions (improved occupational safety, reduced noise, and improved air quality) and reduced emissions were not quantified. When quantified NEBs were included in the calculations, it was found that the pay-back period was shortened from 1.43 years to 0.99 years. (Lung et al, 2005).

It should be mentioned that there may also be negative consequences related to energy efficiency investments and measures, such as that excessive insulation of a building envelope can lead to moisture problems. These should of course also be taken into account in the same way as NEBs.

A study of 74 cases showed that some of the participating companies could quantify a large part of the ten suggested NEBs. The following NEBs were taken into account in the study; increased sales levels, productivity, operating costs for energy, duration of operation, maintenance costs, waste, staffing needs, injuries or illnesses, quality deficiencies, work ethic, and improved work environment. Of the 10 possible NEBs, one third could be quantified and assigned monetary values. In four cases, a change in the number of injuries and/or diseases could also be reported, but in none of these four cases, this could be given a monetary value. Overall, the study showed that the NEBs in monetary terms was about 2.5 times greater than energy cost savings. (Hall and Roth, 2003).

Given the above-stated studies of NEBs, it is evident that NEBs should be included in assessing energy efficiency measures. If not included, this faces the of underestimating the energy efficiency potential. On the contrary, an inclusion of other added values (which are possible to assign monetary values) in addition to solely energy cost savings means that a measure becomes significantly more profitable to implement.

One example of NEBs can be found in an iron foundry that introduced a routine to charge the melting furnace immediately after emptying it. This energy efficiency routine resulted in reduced electricity consumption for iron melting by 10% per melted tons of iron. An NEB for this measure was that the number of melts before changing lining (i.e., the bricks inside the furnace) could be increased from just over 100 to 400. The cost of changing lining is essential because the furnace has to cool for a number of days before bricks could be replaced.

References

Abrahamse, W., Steg, L., 2013. Social influence approaches to encourage resource conservation: a meta-analysis. Global Environ. Change 23 (2013), 1773–1785.

Alcott, H., 2011. Social norms and energy conservation. J. Public Econ. 95 (2011), 1082–1095.

Andersson, E., Karlsson, M., Thollander, P., Paramonova, S., 2018. Energy end-use and efficiency potentials among Swedish industrial small and medium-sized enterprises – a dataset analysis from the national energy audit program. Renew. Sustain. Energy Rev. 93, 165–177.

Aronson, E., O'Leary, M., 1983. The relative effectiveness of models and prompts on energy conservation: field experiment in a shower room. J. Environ. Syst. 12, 219–224.

Asensio, O., Delmas, M., 2015. Nonprice incentives and energy conservation. PNAS 112 (6), 510–515.

Finman, H., Laitner, J.A., 2001. Industry, energy efficiency and productivity improvements. In: Proceeding of the 2001 Summer Study on Energy Efficiency in Industry, pp. 561–570.

Grubbström, R.W., Lundquist, J., 1996. Investering och finansiering – metodik och tillämpningar (Investments and Finance – Method and Applications). Academia Adacta, Lund [in Swedish].

Hall, N.P., Roth, J.A., 2003. Non-energy benefits from commercial and industrial energy efficiency programs: Energy efficiency may not be the best story. In: Proceeding of the 2003 International Energy Program Evaluation Conference, pp. 689–702.

Laitner, J.A., Ruth, M.B., Worrell, E., 2001. Incorporating the productivity benefits into the assessment of costeffective energy savings potential using conservation supply curves. In: Proceeding of the 2001 Summer Study on Energy Efficiency in Industry, pp. 597–608.

Lilly, P., Pearson D., 1999. Determining the full value of industrial efficiency programs. In: Proceedings ACEEE Summer Study on Energy Efficiency in Industry, pp. 349–362.

Lung, R.B., McKane, A., Leach R., Marsh D., 2005. Ancillary savings and production benefits in the evaluation of industrial energy efficiency measures. In: Proceedings ACEEE Summer Study on Energy Efficiency in Industry.

Nehler, T., Thollander, P., Ottosson, M., Dahlgren, M., 2014. Including non-energy benefits in investment calculations in industry – empirical findings from Sweden. In Proceedings ECEEE Industrial Summer Study – Retool for a Competitive and Sustainable Industry, 711–719.

Palm, J., Thollander, P., 2010. Interdisciplinary perspective on industrial energy efficiency. Appl. Energy 87 (10), 3255–3261.

Pye, M., McKane, A., 2000. Making a stronger case for industrial energy efficiency by quantifying non-energy benefits. Resour. Conserv. Recycl. 28 (3–4), 171–183.

Schwartz, D., Fischhoff, B., Krishnamurti, T., Sowell, F., 2013. The Hawthorne effect and energy awareness. PNAS 110 (38), 15242–15246.

Worrell, E., Laitner, J.A., Ruth, M., Finman, H., 2003. Productivity benefits of industrial energy efficiency measures. Energy 28 (11), 1081–1098.

Further reading

Allcott, H., Mullainathan, S., 2010. Behavior and energy policy. Science 327 (5970), 1204–1205.

Energy efficiency in steam and pumping systems

8

Chapter Outline

8.1 Steam systems

8.1.1 Introduction to steam systems

Steam can be defined as a process whose purpose is to produce steam at a pressure required for a process.

The heat content of steam consists of (see also Fig. 8.1):

- Liquid heat (sensible heat), that is, the heat content of the water up to the boiling point.
- (Latent) Heat of vaporization, that is, the energy used to make steam (the so-called saturated steam) of water. When the steam condenses, the heat of vaporization can be recovered.
- Superheating, that is, the energy used to raise the temperature of the steam above 100°C. This amount of energy is supplied to the saturated steam. When all the water in the steam has become steam (the so-called dry steam), the superheating begins. The superheating

Introduction to Industrial Energy Efficiency. DOI: https://doi.org/10.1016/B978-0-12-817247-6.00008-0

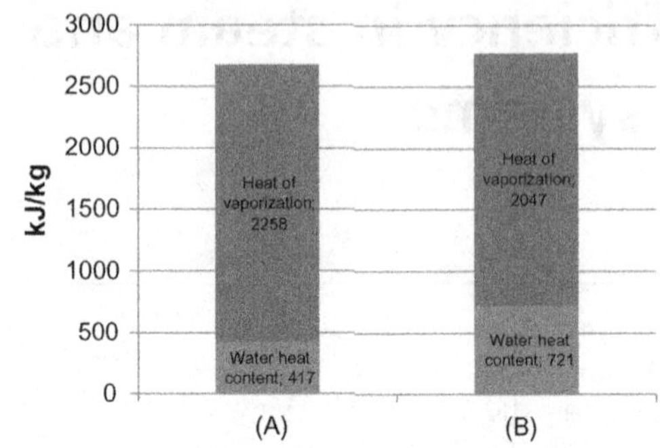

Figure 8.1 The heat content of steam at 1 bar (A) and 8 bar (B), respectively.

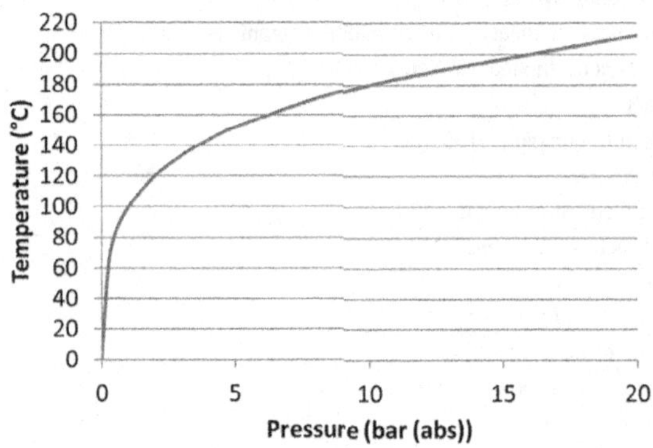

Figure 8.2 Steam saturation curve.

takes place completely separate from the water in the boiler (in a so-called superheater) since the energy otherwise will be used to evaporate more water. Essentially, superheated steam is used at pressures above 30 bar and usually for:

- electricity production in the so-called steam turbines;
- drying;
- transport of steam in long pipes, as superheating suppresses condensation in machines and pipes, which reduces the risk of corrosion and increases the thermal efficiency.

As shown in Fig. 8.1, great amounts of energy are absorbed during evaporation. This entails small transport volumes for large amounts of energy, which also means a relatively small cost for piping compared to if water were to be used. It also results in high efficiency in the heat transfer, which results in small units, lower costs, and less space requirements than if water were to be used. Temperature and pressure are linked together for steam, which means that for a given pressure the temperature is always the same. Pressure is relatively easy to control.

The so-called steam saturation curve is shown in Fig. 8.2. The figure shows that at the pressure of 1 bar (A), the boiling point is 100°C.

Table 8.1 shows the properties of steam and water.

Table 8.1 Properties for steam and water.

Enthalpy at the saturation curve					Enthalpy for superheated steam			
		Liquid heat	Total heat	Heat of vaporization	Total heat	Total heat	Total heat	Total heat
Pressure	Temperature	h'	h''	r	200°C	300°C	400°C	500°C
bar (abs)	°C	kJ/kg	kJ/kg	kJ/kg	kJ/kg	kJ/kg	kJ/kg	kJ/kg
0.5	81.3	340.6	2646.0	2305.4	2877.7	3075.7	3279.0	3488.7
1	99.6	417.5	2675.4	2257.9	2875.4	3074.5	3278.2	3488.1
2	120.2	504.7	2706.3	2201.6	2870.5	3072.1	3276.7	3487.0
3	133.5	561.5	2724.7	2163.2	2865.5	3069.7	3275.2	3486.0
4	143.6	604.6	2737.6	2133.0	2860.4	3067.2	3273.6	3484.9
5	151.8	640.1	2747.5	2107.4	2855.1	3064.8	3272.1	3483.8
10	179.9	762.6	2776.2	2013.6	2826.8	3052.1	3264.4	3478.3
20	212.4	908.6	2797.2	1888.6	–	3025.0	3248.7	3467.3
30	233.8	1008.4	2802.3	1793.9	–	2995.1	3232.5	3456.2
40	250.3	1087.4	2800.3	1712.9	–	2962.0	3215.7	3445.0
50	263.9	1154.5	2794.2	1639.7	–	2925.5	3198.3	3433.7
60	275.6	1213.7	2785.0	1571.3	–	2885.0	3180.1	3422.2
70	285.8	1267.5	2773.5	1506.0	–	2839.4	3161.2	3410.6

Source: Based on SFI (Swedish Forest Industry), 1990. Energikompendium för massa och pappersindustrin (Energy Compendium for Pulp and Paper). Swedish Forest Industry, Markaryd (in Swedish).

8.1.2 *Parts in a steam system*

The steam system basically consists of four parts (explained further in the coming subsections):

1. boilers,
2. steam distribution system,
3. steam users, and
4. condensate return system.

8.1.3 *Boilers*

In the boiler different fuels are burned. Solid fuels, for example, are incinerated on a grate of different types that may, for example, be inclined and moving, or in a fluidized bed. The goal of the combustion techniques is to ensure that the fuel that is combusted has burnt completely before it passes the combustion area. The various steps in the combustion area are drying, gasification, combustion of gases, and final combustion of carbon residues. For oil, burners of various kinds are used (e.g., rotary oil burners, pressure jet oil burners, low-pressure oil burners, high pressure oil burners, and blue flame oil burners) whose purpose is to ensure that the oil is atomized so that it is mixed in the combustion air. If thick oil is used, it must first be preheated.

The boilers, in turn, consist mainly of four parts, see Fig. 8.3: (1) a feedwater pump that pumps up the feedwater to the boiler pressure, (2) an economizer that preheats incoming feedwater with energy from the flue gases, (3) surfaces in the boiler that collect the heat from the fuel being combusted (tubes with water, where steam is produced, are placed inside the walls around the fireplace), and (4) a steam drum that acts partly as a liquid vessel from which feedwater is taken into the boiler as boiler water, and partly as a vessel where the steam generated is separated from the boiler water. The steam from the steam drum is passed on to processes with steam demand, possibly the steam is passing through a superheater and possibly also via a steam turbine.

The design of the heat-absorbing surfaces in a boiler is shown in Fig. 8.4, where the following are shown: radiation surfaces in the fireplace (1), water-filled walls (in tubes) (2), radiation superheater (6), convection superheater (7), convection surfaces where water is evaporated to lower the temperature of the flue gases (8, 14), the latter (14) being a slag screen (containing numerous tubes) where the temperature is adjusted to absorb as much slag and ash as possible and which is placed at the outlet of the fireplace, economizer preheating feedwater (9), and air preheaters (11) that are mainly used for fuels that are moist and which have too low heating value for the fuel to dry and to obtain a combustion temperature that is sufficiently high for combustion to be satisfactory (if back-pressure turbine is installed, it is possible to use low-pressure steam). To further cool the flue gases, it is possible to install a scrubber (not shown in the picture) for hot water production. If oil is burned, the acid dew point limits the cooling of the flue gases, approximately 150°C.

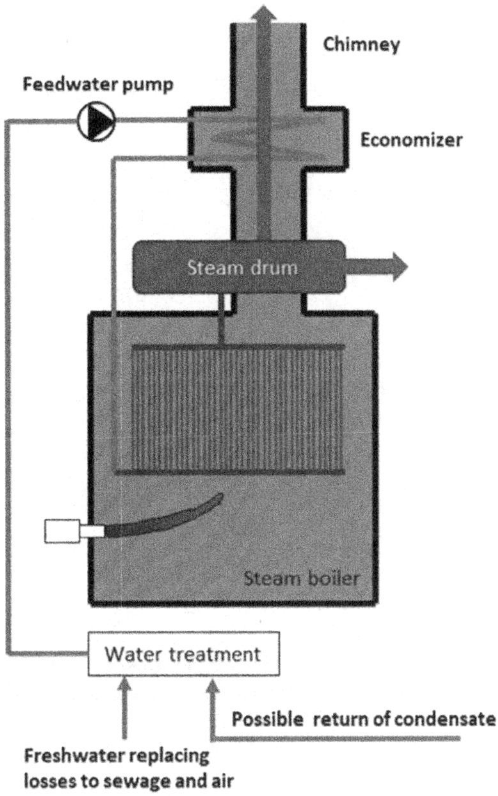

Figure 8.3 Schematic view of the construction of a boiler.
Source: Based on Jernkontoret, 2018. Available from: <http://www.jernkontoret.se/sv/publicerat/handbocker-och-utbildningskompendier/utbildningspaketet/> (28.05.18) (in Swedish).

If sulfur-free fuels are used, it is possible to reduce the temperature further and then the flue gas condensation temperature for steam limits the temperature.

The efficiency of boilers is generally between 70% and 90% and there are six different types of losses (Morvay and Gvozdenac, 2008). The largest losses are usually (1) flue gas losses that can amount to between 15% and 20%. This loss can be reduced by heat exchanging and utilizing the flue gases in, for example, the feedwater system via a so-called economizer. The temperature of flue gases is often between 180°C and 300°C. Losses due to (2) chemical incomplete combustion can range between 0.5% and 2.5%. Chemical incomplete combustion causes CO, H_2, and hydrocarbons in the flue gases, the latter at very poor combustion. The losses are very small at well-managed plants. (3) Mechanically incomplete combustion is due to unburned material in ash and fly ash and usually amounts to between 0.5% and 6%. For coal, these losses can amount to between 5% and 20%, but biomass combustion may also have large losses. (4) Radiation losses depend on the size of

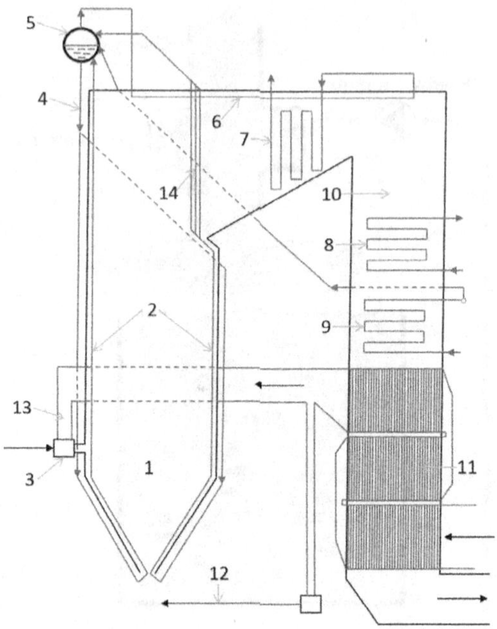

Figure 8.4 Heat-absorbing surfaces in a boiler are indicated by a number in the figure and are described in the text. The parts in the figure that are indicated by a number that do not relate to heat-absorbing surfaces are burners (3), downcomers (4), steam drum (5), flue gas channel (10), primary air (12), and secondary air (13). Based on Johansson (1995).

the boiler and usually amount to between 0.25% and 1.5%. (5) Slag formation depends on the slagging temperature, which in turn depends on the substance being burned. (6) Blowdown of the boiler is done continuously by removing feedwater from the system. Salts can form coatings on, for example, heat transfer surfaces, but these decreases when blowdown is done regularly. Heat losses due to blowdown can in some extreme cases amount to 5%, but usually they are between 0.2% and 3%. By heat exchanging the hot water before it is discharged, the losses can be reduced (Morvay and Gvozdenac, 2008).

The losses mentioned above apply to the combustion of solid fuels (Morvay and Gvozdenac, 2008). During the combustion of liquids, all losses occur as for solid fuels except for slag formation. The losses are generally smaller, except for the flue gas losses. For gaseous fuels only flue gas losses occur, chemically incomplete combustion (minor) and radiation losses (Morvay and Gvozdenac, 2008).

In the above paragraphs, possible efficiency improvements have been mentioned. Another way to improve efficiency is to preheat the combustion air and have variable speed of the fans for the combustion air (Morvay and Gvozdenac, 2008). Moreover, the design of the boiler (e.g., heat transfer areas, flue gas routes, and economizer) is crucial for the efficiency, but it is both difficult and expensive to change when the boiler is built. Generally, as high a load as possible on the boiler gives high efficiency. Furthermore, it is important to clean surfaces because coatings on heat exchanger surfaces, both on the flue gas side and water side, influence the efficiency. Despite blowdown and minimizing slag formation, continuous cleaning of these surfaces is necessary. Well-functioning operating controls influence the efficiency by, among other things, adjusting the mass and energy flows in an optimal manner and making sure that fans and feedwater pump start at the right time. The water quality is maintained with, among other things, blowdown, but the quality of the water must also be carefully analyzed in order to maintain sound energy efficiency (Morvay and Gvozdenac, 2008).

8.1.4 Steam distribution system

The steam distribution system transports the steam to the consumers. The pressure in the steam distribution system is usually between 2 and 12 bar, where steam with a pressure of between 2 and 5 bar is called low-pressure steam, and medium-pressure steam has a pressure of between 9 and 12 bar (SFI, 1990). In the steam distribution system condensate is formed, due to cooling of the steam. Steam traps are used to remove the condensate, and these are also used to ventilate the system during start-up and during operation. The steam traps must cope with varying capacity and pressure and close for steam (Armatec, 2004). There are basically two different types of steam traps: thermal and mechanical. The function of the thermal steam traps is based on the temperature difference between saturated steam and condensate that is somewhat cooled. There are several variants, of which capsule steam trap and bimetallic steam trap are two examples. The mechanical steam traps work from the basis that steam and condensate have different densities. A float steam trap is a common mechanical steam trap. To select the right steam trap, account must be taken to the type of installation, the amount of condensate, the pressure, and the temperature. Furthermore, consideration should be given to whether there should be individual drainage or group drainage, and the acceptable level of condensate to be impounded (Armatec, 2004).

When it comes to steam distribution systems, it is important to check for possible faults in steam traps, valves, and pipes and if there are leakages in the system. Fig. 8.5 shows the extent of heat losses from leakages of different hole sizes.

When very hot steam is transported in the steam distribution system, the insulation of the system is of the utmost importance. Table 8.2 shows examples of steam pipe losses for different status of insulation for different pipe dimensions.

It is also important to consider any pressure drop in the system as it requires energy to maintain the correct pressure, and some steam users require a certain pressure. The pressure drop increases with increasing flow.

The steam distribution system can also be supplemented with a steam accumulator, which in fact is a hot water accumulator because it is hot water stored. With the help of a steam accumulator, the boilers can operate at a more even load, which is advantageous from an efficiency perspective. It is also possible that the steam accumulator can reduce boiler capacity; since it can handle short-term peaks, capacity in the steam production is not required for these peaks and, hence, possibly increase of boiler capacity is avoided. The disadvantage is that the steam accumulator reduces any electricity generation because the steam supplied to the steam

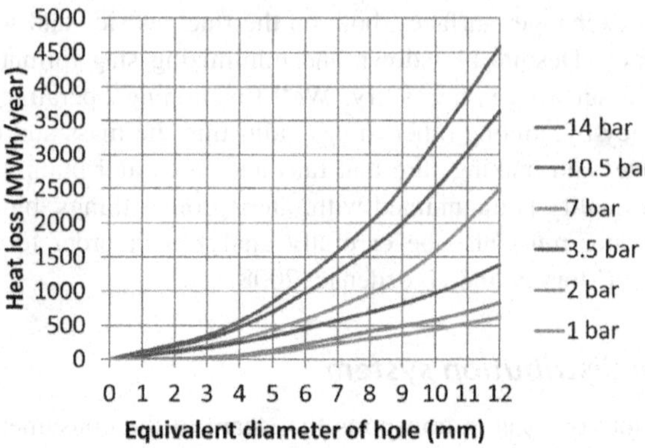

Figure 8.5 Heat losses from leakages of different hole sizes.
Source: Based on Morvay, Z.K., Gvozdenac, D.D., 2008. Applied Industrial Energy and Environmental Management. Wiley.

Table 8.2 Steam pipe losses for different status of insulation for different pipe dimensions.

Piping item	Pipe status	Unit	Heat loss	
			Steam	Condensate
All pipes	Well-insulated	W/m	35–130	15–35
Flat surface	Well-insulated	W/m^2	180	70
32 mm	Bare	W/m	340	110
50 mm	Bare	W/m	480	150
65 mm	Bare	W/m	580	185
80 mm	Bare	W/m	660	210
100 mm	Bare	W/m	820	260
125 mm	Bare	W/m	960	310
150 mm	Bare	W/m	1150	370
200 mm	Bare	W/m	1500	–
Flat surface	Bare	W/m^2	2300	700

Source: Based on Morvay, Z.K., Gvozdenac, D.D., 2008. Applied Industrial Energy and Environmental Management. Wiley.

accumulator must have a higher pressure than the pressure in the accumulator, and the steam delivered by the steam accumulator has a lower pressure.

8.1.5 Steam users

Steam is used, among other things, to heat buildings and processes, as well as to drive pumps and other mechanical artefacts. Steam users need steam with different pressure levels and there is a risk that suppliers of equipment are dimensioning for higher pressure than necessary. This is because a higher pressure yields smaller dimensions and consequently a lower manufacturing cost for the equipment.

For a steam user to function as good as possible, there should be no air or other noncondensable gases (e.g., free oxygen and carbon dioxide) or condensate in the steam users. The main problems are (1) reduced heat transfer, (2) reduced amount of distributed steam, (3) cold zones can be formed on the heating surfaces, and (4) possible corrosion (Armatec, 2004).

8.1.6 Condensate return system

To increase the efficiency of a steam system, a condensate recycling system is an alternative. The heat in the condensate contains a large part of the heat that originally existed in the form of saturated steam and may contain between 5% and 20% of the energy content of the steam, sometimes even more. Condensate can be obtained from steam users and from steam traps in the steam distribution system. If the condensate has high temperature this can be "flashed" and low-pressure steam can be produced. By using the condensate, the amount of fresh water added decreases, which ultimately also reduces the cost of chemicals. If contaminated steam occurs in any process, it is possible to use heat exchangers to be able to collect the heat of the condensate. It is not uncommon that a large proportion, or almost all, of the condensate goes to sewage (Jernkontoret, 2018). The reasons vary, for example, the condensate can be contaminated, or the energy prices may be too low to make it profitable to take care of the condensate.

8.1.7 Examples of possible measures when auditing steam systems

Below some examples of possible measures when auditing steam systems are shown. However, more measures are mentioned in the text above.

- Boiler, measures that are investigated and, if possible, carried out:
  - Preheat the combustion air.
  - Install an economizer.
  - Vary the speed of fans for combustion air.
  - Install a heat exchanger for blowdown.
- The steam distribution system, measures investigated and, if possible, carried out:
  - Ensure that steam traps do not leak and that there are no faults in the valves and pipes.
  - Check the insulation of the system.
  - Examine the possibility of installing a steam accumulator which, among other things, can provide better efficiency on boilers, due to more even load.
- Steam users, measures investigated and, if possible, carried out:
  - Examine the actual pressure requirement for steam users to see if it is possible to reduce the pressure.
  - Ensure that the condensate leaves the steam user so that the heat transfer does not deteriorate.

- Steam condensate recycling system, measures investigated and, if possible, carried out:
  - Examine to what extent the condensate is returned to the system from steam traps and from different steam consumers, increase if possible, the degree of recirculation.
- Cascading steam demand, measures investigated and, if possible, carried out:
  - Supply steam to the processes at the intended pressure by, if possible, cascading.
  - Examine the possibility of producing electricity.

8.1.8 Nonenergy benefits in steam systems

When assessing the available nonenergy benefits, the model developed by Nehler (2018) can be applied. Major nonenergy benefits for steam systems includes, but are not limited to:

- Production-related
  - Increased production rate
  - Increased production reliability
  - Increased equipment lifetime
  - Shorter cycle time of process
- Operation and maintenance
  - Reduced maintenance
  - Lower cooling demand
  - Reduced labor demand
  - Reduced need for engineering control
- Working environment
  - Increased worker safety, for example, lower risk of burns from surface of hot pipes
  - Reduced noise, reduction of noise can increase worker productivity but even more importantly, reduce days off work
  - Improved thermal comfort
- Waste
  - Reduction of wastewater
  - Reduction of waste heat
- Emissions
  - Reduced emissions. By improved energy efficiency, emission reductions are achieved, for example, CO, CO_2, NO_x, and SO_x
- Other
  - Improved public/corporate image, for example, by showing customers energy efficiency improvements
  - Improved worker morale, for example, by committing to improvements in working conditions
  - Increased sales levels, for example, by cutting energy costs, revenue increases on a one-to-one basis, 1 cent in energy cost means 1 cent more in revenue for the company

8.2 Pumping

8.2.1 Introduction to pumping

Pumping is a process whose purpose is to transport liquid. Pumps have many uses in industrial companies, for example, sewage pumping, water supply in industrial processes, cooling of equipment in the steel industry, pumping of sludge in the mining industry, and pumping of pulp. For example, the water requirement in a dairy amount to between 4 and 6 m^3 per m^3 of milk, and in the production of paper the water requirement amounts to between 25 and 100 m^3 per ton of paper (Alvarez, 1990). The Swedish industry is estimated to use about 10 TWh of electricity per year for pumping. This corresponds to about 18% of the total electricity consumption in Swedish industry (SEA, 2006). For some companies the energy demand for pumping may be up to 50% of the total demand (European commission, 2009).

In some industries a significate part of the total energy requirement is electricity for pumping, which means that a reduced energy requirement for pumping can result in great savings. An example that shows possible savings is taken from Sandvik Materials Technology (SMT) (Jernkontoret, 2018). At SMT, two pumps were previously used to pump water in a loop through two pickling plants. The water was also pumped when there was no need for water in the pickling plants. SMT rectified this and installed two separate pumps and pipelines, one for each pickling plant, and the pumps run only when needed. The electricity requirement before the measure amounted to 6.4 MWh per week (just over 300 MWh per year), while after rectification it amounted to 0.69 MWh per week (just over 30 MWh per year). Electricity consumption for the pump operation thus decreased by 89%, which in turn reduced electricity costs by about SEK 100,000 per year.

In existing plants, it is usually not possible to reduce the losses in the pipe systems, but it is usually where the large losses occur.[1] The pump losses are usually slightly less. The operation of a pump occurs almost exclusively with an electric motor. Electric motors often have efficiencies of over 90%. A rough estimate is that it is possible to reduce the electricity requirement in existing pump systems by 30% (pipe system not included) (Jernkontoret, 2018). This is accomplished by:

- turning off unnecessary pumps,
- adjusting the dimensioning of pumps,
- installing an energy-efficient control of the flow,
- installing an energy-efficient electric motor, and
- eliminating the wear in the pumps.

8.2.2 Pump types

For pumps it is possible to negate the fluid's compressibility without losing in accuracy. This distinguishes the process of pumping from, for example, compressed air, since air is a compressible medium. There are different types of pumps, for example, radial pumps (centrifugal pump), axial pumps (propeller pump), and diagonal pumps (intermediate between centrifugal and propeller pump). These pumps have slightly different uses. Propeller pumps are best suited for high volume flows (m^3/h) and low heads, while displacement pumps are suitable for low volume flows

[1] Losses in pipe systems can be calculated using Bernoulli's equation.

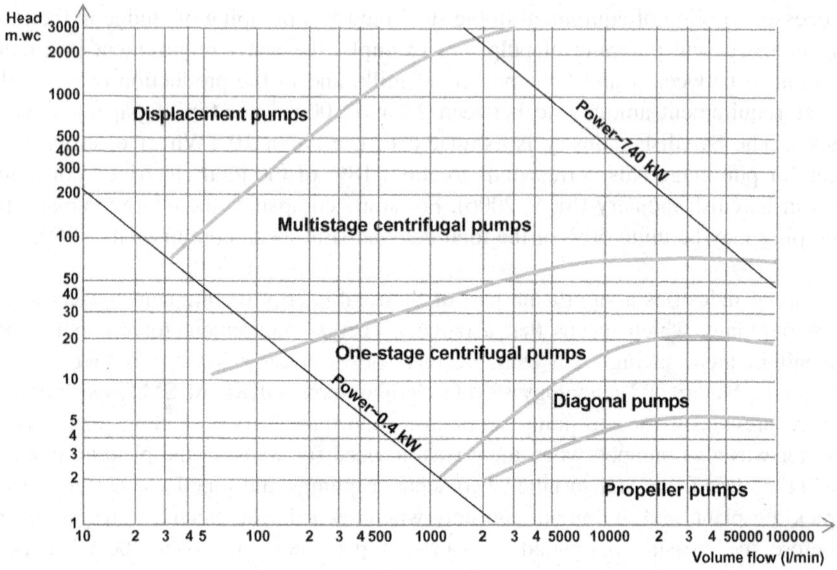

Figure 8.6 Field of applications for pumps. The power included in the diagram excludes the efficiency of the pumps.
Source: Based on Held, H., 2010. Svenskt Vatten Utbildning – Energieffektivisering (Swedish Water Education – Energy Efficiency). ITT Water & Wastewater. Linköping, 9–10 November (in Swedish).

and high heads. In between, there are centrifugal pumps where higher heads require multistage centrifugal pumps (the volume flow decreases). The diagonal pumps are an intermediate with the possibility of slightly higher head, but lower volume flow, compared to the propeller pump (see Fig. 8.6).

8.2.3 Power requirement of pumps

Pumps intend to increase the pressure energy of a liquid, potential energy or kinetic energy, or a combination of these. Pumps can be used to meet certain process requirements, such as controlling processes, keeping even pressure at varied flow, maintaining a uniform level in a tank or maintaining a constant flow.

A pump's power requirement amounts to

$$P_{\text{theoretical}} = Q * \Delta p \tag{8.1}$$

where $P_{\text{theoretical}}$ is the power (W); Q is the flow (m^3/s); and Δp is the pressure increase (Pa).

The real power requirement is

$$P = \frac{P_{\text{theoretical}}}{\eta} \tag{8.2}$$

where P is the actual power (W) and η is the efficiency of the pump.

Furthermore, it is

$$\Delta p = H * \rho * g \quad (8.3)$$

where H is the head (mvp); ρ is the density of the liquid (kg/m^3); and g is the gravity acceleration (about 9.81 m/s^2).

This gives

$$P = \frac{Q * H * \rho * g}{\eta} \quad (8.4)$$

$$\eta = \frac{P_{\text{Useful}}}{P_{\text{Supplied}}} \quad (8.5)$$

8.2.4 Important factors for pumps

Important factors for pumps are (1) head, (2) suction head, and (3) cavitation.

Head:

- The amount of energy to be supplied to the liquid to increase pressure, potential or kinetic energy from one level to another.
- The pump head = the system head (in a pipe system).
- If the system is unknown, it is possible to calculate the pump head.

Suction head (a pump tries to create a vacuum):

- The difference in level between the pump inlet and the liquid surface in the basin from which the liquid is withdrawn.
- The suction head can be a maximum of about 7 m for pumping cold water at atmospheric pressure.

Cavitation:

- The liquid boils if the pressure at the inlet of the pump is the same as, or less than, the liquid vapor pressure (at the current temperature). When the liquid boils, steam bubbles are formed which subsequently implode when the pressure is raised in the pump. This causes great damage to the pump.
- If the liquid has a high temperature, it must "flow down" to the pump instead of being "sucked up." This always applies if the liquid has a temperature above 100°C. It is also possible to raise the pressure in the suction basin to counteract this problem.

8.2.5 Pump curves

Figs. 8.7–8.9 show pump curves for a type of centrifugal pumps and cold water with a density of 1000 kg per m^3 at 2900 rpm (revolutions per minute). The figures show examples of the relationship between volume flow and head as well as volume flow

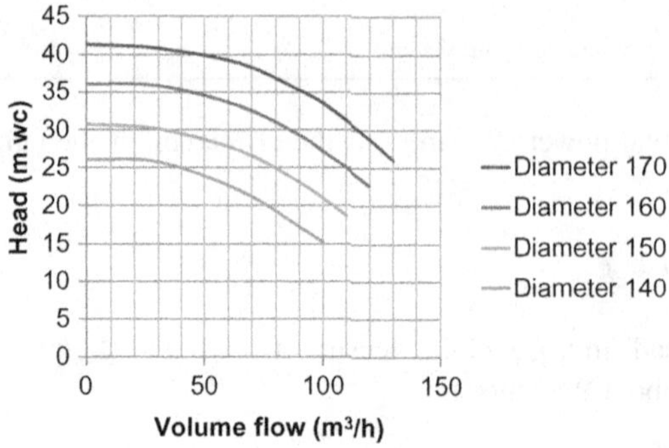

Figure 8.7 An example on the relationship between volume flow and head for different diameters (mm) of the impeller.
Source: Based on SFI (Swedish Forest Industry), 1990. Energikompendium för massa och pappersindustrin (Energy Compendium for Pulp and Paper). Swedish Forest Industry, Markaryd (in Swedish).

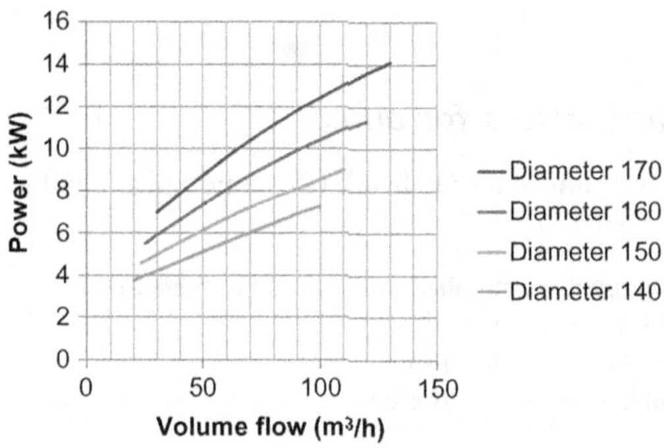

Figure 8.8 The relationship between volume flow and power for different diameters (mm) of the impeller.
Source: Based on SFI (Swedish Forest Industry), 1990. Energikompendium för massa och pappersindustrin (Energy Compendium for Pulp and Paper). Swedish Forest Industry, Markaryd (in Swedish).

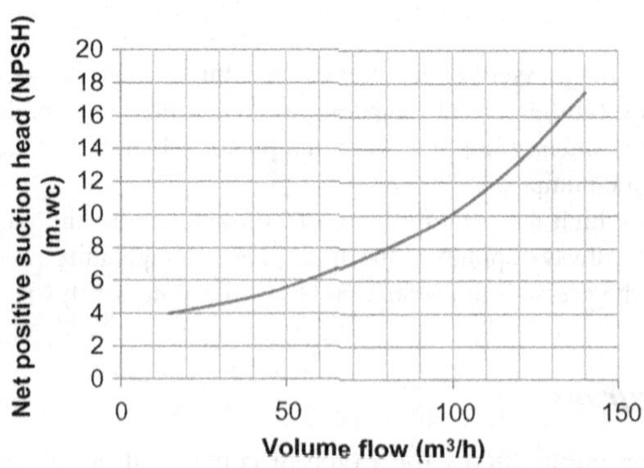

Figure 8.9 Example on curve to avoid cavitation for different volume flows.
Source: Based on SFI (Swedish Forest Industry), 1990. Energikompendium för massa och pappersindustrin (Energy Compendium for Pulp and Paper). Swedish Forest Industry, Markaryd (in Swedish).

and power, for different diameters of the impeller, as well as a curve to avoid cavitation. Fig. 8.7 shows the relationship between volume flow and head for different diameters of the impeller. Fig. 8.8 shows the relationship between volume flow and power for different diameters of the impeller. The motor selected must be large enough for operation at maximum volume flow. Fig. 8.9 shows a curve to avoid cavitation for different volume flows. It shows the overpressure required and is indicated as net positive suction head. This refers to the smallest vertical distance between the liquid surface on the suction side of the pump and the pump inlet for boiling liquids. To compensate for pressure losses in the suction pipe, a margin of 0.3–1 m should be added.

8.2.6 System curves

A system curve represents the necessary head at different volume flows. The system head of the system is the sum of (1) static pressure head and (2) the flow losses of the pipeline system. Static pressure head (H_s) (see also Figs. 8.10 and 8.11) is

$$H_s = H_2 - H_1 + \frac{p_2 - p_1}{\rho * g} \tag{8.6}$$

where $(p_2 - p_1)$ is the difference in pressure between point 1 and point 2; $(H_2 - H_1)$ is the height difference between point 1 and point 2; and H_s is the static pressure head, which can vary with level and pressure in the basins on either side of the pump, but is independent of the volume flow.

The flow losses (H_f) of the pipeline system are

$$H_f = \frac{\Delta p_f}{\rho * g} \tag{8.7}$$

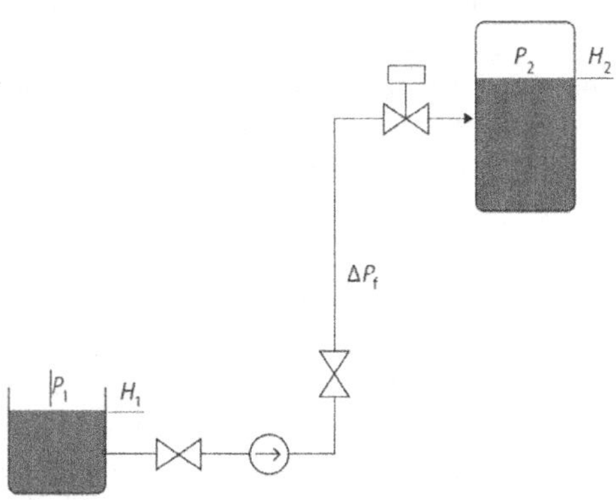

Figure 8.10 Schematic of a pump system.
Source: Based on SFI (Swedish Forest Industry), 1990. Energikompendium för massa och pappersindustrin (Energy Compendium for Pulp and Paper). Swedish Forest Industry, Markaryd (in Swedish).

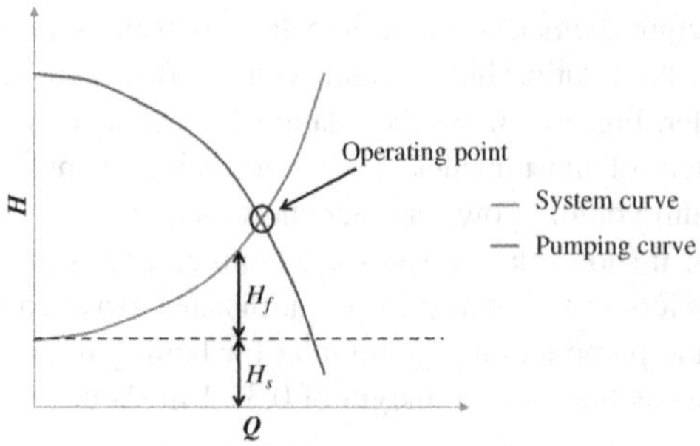

Figure 8.11 Pump curve, system curve, and operating point for a pump system.

where Δp_f is the pressure drop between point 1 and point 2 and H_f is the flow losses at unchanged position on any throttles, such as valves, approximately proportional to the volume flow in square. It is possible to influence H_f by changing the position of any throttles.

By placing the pump curve and system curve in the same diagram, the current operating point can be read as the point where the pump and system curves intersect. Fig. 8.11 shows the system curve, which consists of the two parts H_s and H_f, and the pump curve, which was previously reported in Fig. 8.7.

8.2.7 Pumping other than water

In many companies, other than water is pumped. In the pulp and paper industry, for example, a great deal of pulp suspensions is pumped where pulp is mixed with water. This influences the volume and demand curve (QH curve). Fig. 8.12 shows how both the pump curve and the system curve are modified by changed viscosity, due to pulp suspensions, that is, the liquid also contains solid particles (pulp fibers). More fibers in the pulp suspension increase the viscosity. The more fibers, the more difficult to get the pulp suspension moving. Pulp suspensions also contain air. The higher the pulp concentration, the more air in the pulp. This further deteriorates the QH curve. Therefore special pumps, other than conventional centrifugal pumps, are required to pump pulp suspensions with higher concentrations (SFI, 1990).

8.2.8 Parallel and series coupling of pumps

If the flow is to be increased, the appearance of the system curve has a great influence on whether parallel coupling (see Fig. 8.13) or series coupling (see Fig. 8.14) is preferred. In the case of parallel coupling, a large flow increase is obtained if the system curve is flat (i.e., if the system curve has a large proportion of static pressure head, H_s). A series coupling is suitable for increasing the flow if the system curve mainly consists of flow resistance (i.e., if the flow losses, H_f, are large). During series coupling of pumps, the pressure at the shaft seal on the last pump increases.

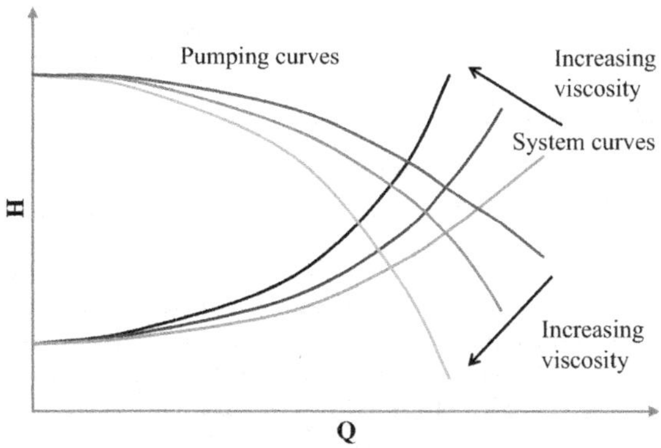

Figure 8.12 Modification of the pump curve and the system curve due to changed viscosity as a result of pulp suspensions.
Source: Based on SFI (Swedish Forest Industry), 1990. Energikompendium för massa och pappersindustrin (Energy Compendium for Pulp and Paper). Swedish Forest Industry, Markaryd (in Swedish).

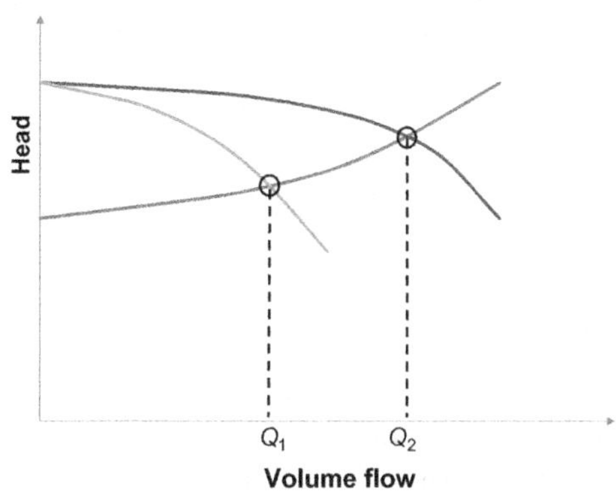

Figure 8.13 Parallel coupling of pumps.

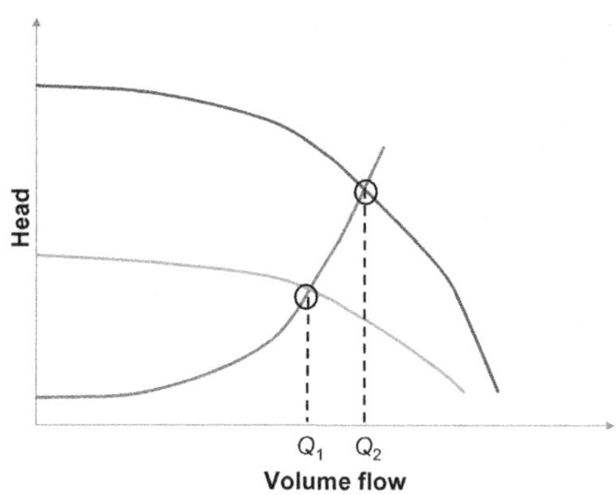

Figure 8.14 Series coupling of pumps.

8.2.9 Control methods

When a pump is selected for a system the most difficult operating case, with respect to flow and pressure, must be the design set (SFI, 1990). To control the flow, one of the following methods can be used:

- Throttle control. An extra resistance is introduced into the pipeline. In order to reduce the cavitation risk, the restriction must always take place on the pressure side of the pump. The throttling gives a smaller amount of heat to the pumped medium, which may be advantageous if heat is needed but worse if it is a cooling circuit. It is also important to be aware that throttling requires more energy. Therefore the economic outcome depends on the cost of the extra energy required for the pump (usually electricity) relative to the extra cost to heat the circuit with another energy carrier.
- Variable frequency drive (adjustable speed drive). Converting the frequency of the power to the electric motor operating the pump result in a change of the speed of the pump from n_1 to n_2, which according to the affinity laws gives the following influence (see also Fig. 8.15):

$$\text{Flow: } \frac{Q_2}{Q_1} = \frac{n_2}{n_1} \tag{8.8}$$

$$\text{Head: } \frac{H_2}{H_1} = \left(\frac{n_2}{n_1}\right)^2 \tag{8.9}$$

$$\text{Power: } \frac{P_2}{P_1} = \left(\frac{n_2}{n_1}\right)^3 \tag{8.10}$$

  The new speed required can usually not be calculated in advance; it is only possible if the system curve goes through the origin.
- By-pass control. The flow is controlled back via a three-way valve, after which the liquid is pumped again.
- Intermittent pump operation. The pump runs only occasionally, which works if continuous flow does not need to be maintained.
- Rotatable vanes. Can be applied, for example, to certain propeller pumps.
- Adjustable vanes. Can be placed before or after the impeller.

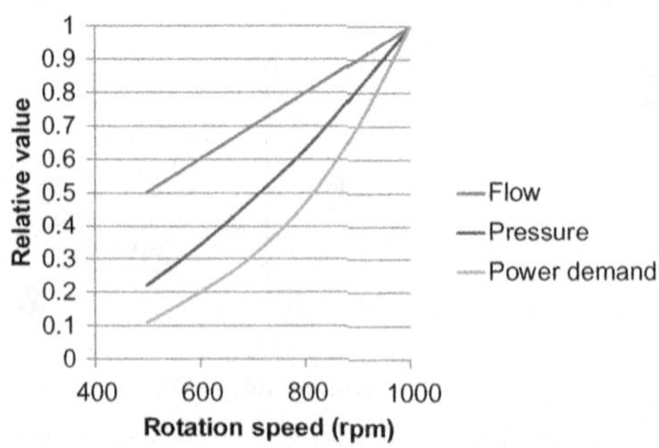

Figure 8.15 Influence on flow, pressure, and power demand when changing the rotation speed.

- Impeller trimming. Changing the impeller diameter is done by lathing the existing impeller or by changing to a new impeller. This is a nonreversible change.

The difference in power demand for different control methods for a change in the flow from 10 to 7 volume units is shown in Figs. 8.16–8.19. The figures also show the useful power and the wasted power. The figures show that at the original flow of 10 volume units, the head is 10 "head units" which gives a power of 100 "power units" (10 * 10). With a reduction of the flow from 10 to 7 volume units, the power demand for throttle control will decrease to 81.9 power units, for by-pass control the power demand will be 93.8 power units, for intermittent pump control the power requirement will be 70 power units, and for variable frequency drive

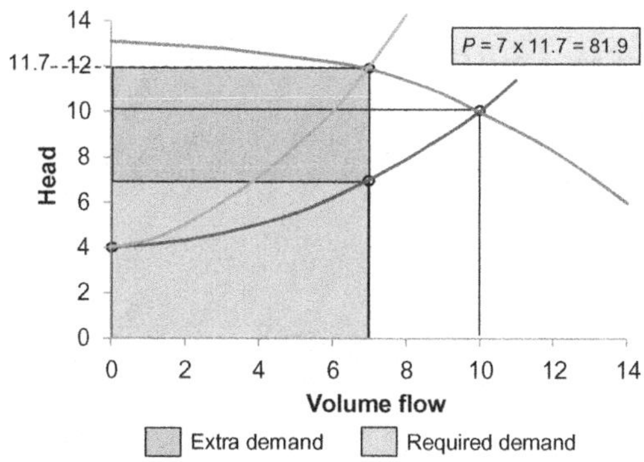

Figure 8.16 Power demand for throttle control when changing the flow from 10 to 7 volume units.
Source: Based on Held, H., 2010. Svenskt Vatten Utbildning – Energieffektivisering (Swedish Water Education – Energy Efficiency). ITT Water & Wastewater. Linköping, 9–10 November (in Swedish).

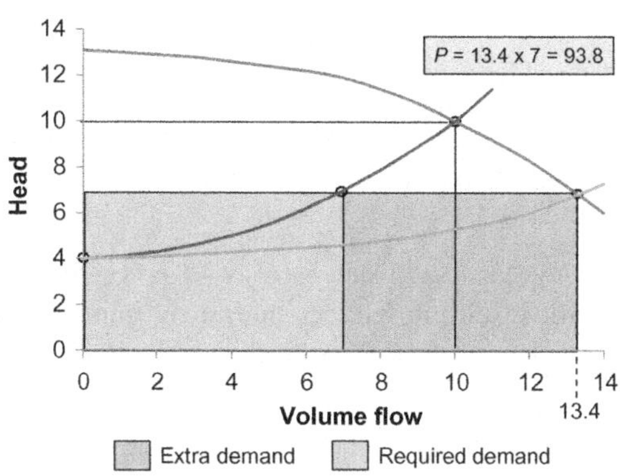

Figure 8.17 Power demand for by-pass control when changing the flow from 10 to 7 volume units.
Source: Based on Held, H., 2010. Svenskt Vatten Utbildning – Energieffektivisering (Swedish Water Education – Energy Efficiency). ITT Water & Wastewater. Linköping, 9–10 November (in Swedish).

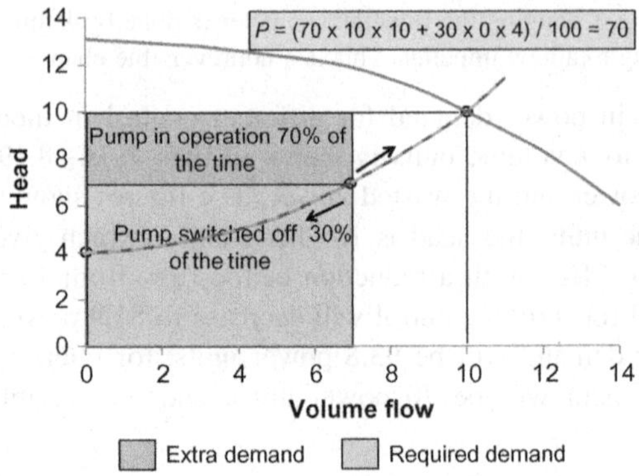

Figure 8.18 Power demand for intermittent pump operation when changing the flow from 10 to 7 volume units.
Source: Based on Held, H., 2010. Svenskt Vatten Utbildning – Energieffektivisering (Swedish Water Education – Energy Efficiency). ITT Water & Wastewater. Linköping, 9–10 November (in Swedish).

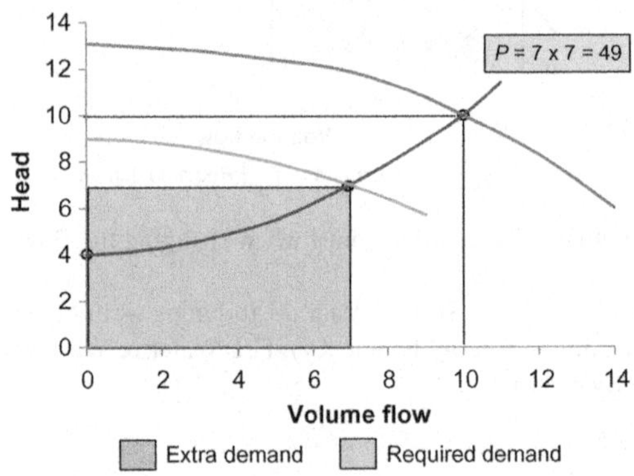

Figure 8.19 Power demand for variable frequency drive (adjustable speed drive) when changing the flow from 10 to 7 volume units.
Source: Based on Held, H., 2010. Svenskt Vatten Utbildning – Energieffektivisering (Swedish Water Education – Energy Efficiency). ITT Water & Wastewater. Linköping, 9–10 November (in Swedish).

control the power requirement will decrease to 49 power units. Thus the variable frequency drive control results in halving the power demand, while other control methods do not reduce the power demand to the same extent.

8.2.10 Costs for pumping

A pump is often operated continuously, which means that the operating cost is crucial for the total cost of the pump during its economic life. Fig. 8.20 shows the cost distribution for pumps over a 10-year period.

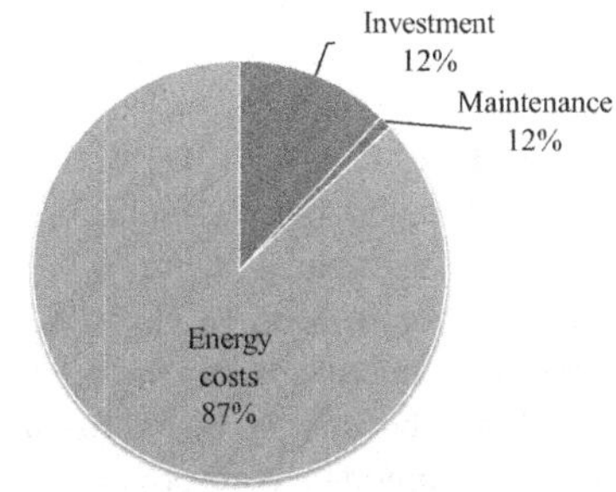

Figure 8.20 Cost distribution for pumps over a 10-year period.
Source: Based on SEA (Swedish Energy Agency), 2006. Krav på pumpar (Demand for Pumps). ET2006:11. Eskilstuna (in Swedish).

8.2.11 Investments in pumps and other examples of possible measures in pumping systems

Example of life cycle cost for pumping with different control methods:[2]

A 130 kW pump running in continuous operation for almost 1 year (8600 hours) requires electrical energy amounting to 1118 MWh (130 * 8600). An electricity price of SEK 0.30 per kWh gives a cost of SEK 335,400 per year (1,118,000 * 0.3). If the pump is operated for 10 years, it gives a cost of approximately SEK 3.4 million with the above stated electricity price. Electricity prices to consumers vary. The processing industry often has a lower electricity price than other industries, which in turn have lower electricity prices than private consumers.

Assume that the pump above half of the time run with a flow of about 1600 m^3 per hour and with the pressure head 22 m of water column, where the power amounts to 130 kW. During the remaining half-time, only a flow of 1200 m^3 per hour is required. For throttle control, 115 kW is then required at the lower flow rate. Electricity consumption during throttle control amounts to 1,053,500 kWh per year (4300 h * 130 kW + 4300 h * 115 kW). For variable frequency drive control, 63 kW is required at the lower flow rate. The electricity consumption for variable frequency drive control then amounts to 829,900 kWh per year (4300 h * 130 kW + 4300 h * 63 kW).

Electricity usage is thus higher in throttle control. On the assumption that throttle control is already installed, a life cycle cost estimate (LCC calculation) can be made to investigate whether it is economically justifiable to invest in variable frequency drive control. Assume the same electricity price as above and that the maintenance cost is SEK 5000 per year without variable frequency drive and SEK 8000 per year with variable frequency drive. The economic life of the variable frequency drive is estimated at 10 years, while the cost of capital is

[2] Examples from the Swedish Energy Agency (SEA) (2006).

assumed to be 10%. From the tables, the net sum factor is obtained for 10% and 10 years, which in this case is 6.14.

For throttle regulation, the LCC is SEK 1,971,000 (321,000 * 6.14), where energy costs amount to SEK 316,000 per year (1,053,500 * 0.3) and maintenance costs amount to SEK 5000 per year.

For variable frequency drive control, the LCC is SEK 1,698,000 (120,000 + 257,000 * 6.14), where the investment cost for an installed variable frequency drive amounts to SEK 120,000, the energy costs amount to SEK 249,000 per year (829,900 * 0.3 SEK) and maintenance costs amount to SEK 8000 per year.

The conclusion is that the total life cycle cost over a 10-year period decreases by SEK 273,000 in the case of an installation of a variable frequency drive. The annual operating costs are reduced by SEK 64,000 with an additional investment of SEK 120,000 for a variable frequency drive. The economic gain of variable frequency drive will of course be even greater if the electricity price is higher than SEK 0.3 per kWh.

Before deciding on a possible investment in a new pump, the existing pump should be examined. The engine condition is tested by measuring flow and pressure to find out possibilities to upgrade the old pump to the same performance as newly purchased. Perhaps the pump can be refurbished by changing seals and bearings or maybe composite coated. Possibly, the impeller diameter can be changed so that the pump fits better to the current operating point. Most often, the renovation costs are much lower than the investment costs of a new pump. During a renovation, neither the pump, the foundation, the piping system, nor the components must be changed. Before deciding on renovation or a new investment, the condition of the electric motor that drives the pump must also be examined.

If the existing pump is to be replaced with a new one, it is important to examine whether the conditions are the same as before. The flow through the existing pump should be measured for a long time and if this is not possible, the system should be examined to see if new installations have been made. It is also important to consider any changes in future needs.

When examining pumps, the following should be considered:

- Avoid unnecessary pump operation.
- Divide the system into separate sections.
- Install an energy-efficient electric motor.
- Install energy-efficient control of the flow (variable frequency drive where possible).

8.2.12 Nonenergy benefits in pumping

When assessing the available nonenergy benefits, the model developed by Nehler (2018) can be applied. Major nonenergy benefits for pumping includes, but are not limited to:

- Production-related
  - Increased production rate
  - Increased production reliability

- Increased equipment lifetime
- Shorter cycle time of process
- Operation and maintenance
  - Reduced maintenance
  - Lower cooling demand
  - Reduced labor demand
  - Reduced need for engineering control
- Working environment
  - Reduced noise, reduction of noise can increase worker productivity but even more importantly, reduce days off work
  - Improved thermal comfort
- Waste
  - Reduction of wastewater
  - Reduction of waste heat
- Emissions
  - Reduced emissions. By improved energy efficiency, emission reductions are achieved, for example, CO, CO_2, NO_x, and SO_x
- Other
  - Improved public/corporate image, for example, by showing customers energy efficiency improvements
  - Improved worker morale, for example, by committing to improvements in working conditions
  - Increased sales levels, for example, by cutting energy costs, revenue increases on a one-to-one basis, 1 cent in energy cost means 1 cent more in revenue for the company

References

Alvarez, H., 1990. Energiteknik (Energy Technology). Studentlitteratur, Lund.

Armatec, 2004. Handbok. Ånga och kondensat (Handbook Steam and Condensat) (in Swedish).

European Commission, 2009. Reference Document on Best Available Techniques for Energy Efficiency.

Held, H., 2010. Svenskt Vatten Utbildning – Energieffektivisering (Swedish Water Education – Energy Efficiency). ITT Water & Wastewater. Linköping, 9–10 November (in Swedish).

Jernkontoret, 2018. Available from: <http://www.jernkontoret.se/sv/publicerat/handbocker-och-utbildningskompendier/utbildningspaketet/> (28.05.18) (in Swedish).

Johansson, H. (1995) Ångpannor – en samling figurer att samtala till. Kurskompendium. [Steam boilers – a number of figures to speak about. Course compendium]. (in Swedish).

Morvay, Z.K., Gvozdenac, D.D., 2008. Applied Industrial Energy and Environmental Management. Wiley.

Nehler, T., 2018. A systematic literature review of methods for improved utilisation of the non-energy benefits of industrial energy efficiency. Energies 11 (12), 1–27.

SEA (Swedish Energy Agency), 2006. Krav på pumpar (Demand for Pumps). ET2006:11. Eskilstuna (in Swedish).

SFI (Swedish Forest Industry), 1990. Energikompendium för massa och pappersindustrin (Energy Compendium for Pulp and Paper). Swedish Forest Industry, Markaryd (in Swedish).

Energy efficiency in compressed air, ventilation, and lighting

9

Chapter Outline

Introduction to Industrial Energy Efficiency. DOI: https://doi.org/10.1016/B978-0-12-817247-6.00009-2

9.1 Compressed air

Compressed air is a process which purpose is to produce pressurized air (for industry, 7 bar, is a commonly used pressure). Compressed air is a necessity in almost all factories.

Compressed air is normally categorized as a support process, and its energy end-use usually accounts for a minor part of the total energy use for energy-intensive companies with a high proportion of energy use in the production processes. However, for nonenergy intensive companies with a high proportion of energy use in the support processes, the energy end-use of compressed air can be a significant part of the total energy use, but rarely more than 10%. For example, the Swedish Energy Agency states that within Sweden's engineering industry, about 8% of total energy use, emanates from compressed air (SEA, 1999).

The modern industrial compressed air technology was born around 1880. Pneumatic lifting blocks driven by a blower was introduced in the foundry industry. The literature does not reveal how or when the internal compressed air grids inside companies were developed. However, there is information available on how AB Atlas developed new tools. The development of tools exploded in the early 1900s and led to a plethora of new compressed air applications, many of which created significantly better work environments for industrial workers. In the book Compressed air in industry: an important rationalization tool, Möre (1953) describes the tools that existed in the early 1950s. Interestingly, no major changes in compressed air technology have occurred up until today, but the use of compressed air has become increasingly widespread in the manufacturing industry worldwide.

9.1.1 Industrial compressed air systems

An industrial compressed air system consists principally of one or more compressors, air drying, an air receiver, filters, a compressed air grid, and a number of compressed air users/applications. Compressors are normally powered by electricity but can also be operated with, for example, an internal combustion engine. The main network or grid consists of a pipe system that provides users with air compressed to between 6 and 8 bars of pressure.

Compressed air is commonly denoted a support process with a wide range of applications. For example, in a manufacturing industry, compressed air can be used to open and close valves and couplings, often integrated into a machine. In a manufacturing process, compressed air can act as a support process for various mechanical activities, such as pneumatic hand tools. Fig. 9.1 shows an overview of a compressed air system.

Compressing air gives rise to major heat losses because air is a compressible medium. In Fig. 9.2, these losses are illustrated in a so-called Sankey diagram. Out of the total demand, only as low as 5%–10% of electric power supply can be used as effective end-use power.

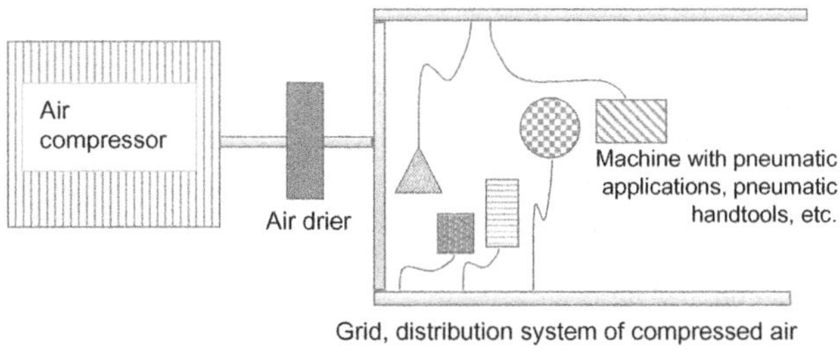

Figure 9.1 Schematic illustration of an industrial compressed air system.

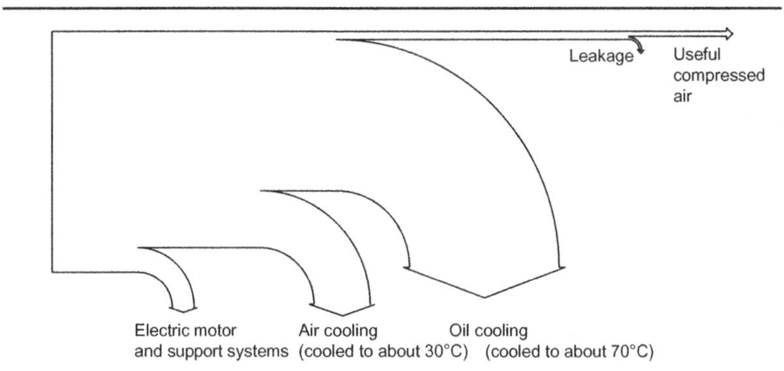

Figure 9.2 Sankey diagram showing the ration of supplied electricity and useful compressed air (Åberg et al., 2003).

In Fig. 9.2 the leakage losses appear to be relatively small, but in fact the compressor's energy use can in some cases be significantly reduced by sealing of leakage.

9.1.2 Air compressors

There are a variety of air compressors available on the market with principally different types of construction. In principle, two main means of compressing air are used, displacement compressors or dynamic compressors. While displacement compressor functions as shrinking the air volume in a compression chamber to a fixed pressure level, dynamic compression is enabled by impellers that increase the speed of the air and later is turned into increased pressure.

Displacement compressors are either of piston- or rotary types.

Furthermore, yet another important distinction between compressors are if they use oil lubricant or are oil lubricant free. Oil lubricant free compressors are preferred for systems that have a high air quality demand where small oil molecules can have negative effect on, for example, the products, or the room. Oil lubricant free compressors are often used in, for example, the food industry. The downside is that oil lubricant free compressors have a higher energy demand than oil lubricant injected compressors.

The report *Improving Compressed Air System Performance: A Sourcebook for Industry* provides a good overview of the various compressor types. The four most common types of air compressors in industry are:

- Piston (reciprocating) compressor
- Rotary vane compressor
- Rotary screw compressor
- Centrifugal compressor

9.1.2.1 Piston compressor

The piston compressor, also denoted reciprocating, is a displacement compressor type which consists of a moving piston which compresses the air. It has high efficiency both at full and partial loads, but less positive aspects are that it is noisy and moreover, demands more space than other types of compressors. Also, due to many moving parts in this type of compressor that may wear out, the cost of maintenance is higher than for other compressor types. Piston compressors can come both as oil lubricant free as well as oil lubricant injected compressors.

9.1.2.2 Rotary vane compressor

The rotary vane air compressor is a displacement compressor type which rotates around a shaft. It has a medium high efficiency at full load, with a lower efficiency at partial loads. The rotary vane compressor is the quietest type of air compressors, and has a very compact design, making it suitable to use in areas with lack of space for placing the air compressors. Also, due to many moving parts in this type of compressor that may wear out, the cost of maintenance is higher than other compressor types, however not as high as for piston compressors. Rotary vane compressors can come both as oil lubricant free as well as oil lubricant injected compressors.

9.1.2.3 Rotary screw compressor

The rotary screw compressor is a displacement compressor type where rotating screws are used to compress the air. Compared with a piston compressor, the screw compressor compresses air continuously without pulsation. It has high efficiency at full load and low efficiency when partially loaded. It is less noisy than a piston compressor and has a very compact design, making it suitable to use in areas with lack of space for placing the air compressors. The rotary screw compressor has very few parts that may wear out, wherefore the cost for maintenance is lower than for piston and rotary vane compressors. Rotary screw compressors can come both as oil lubricant free as well as oil lubricant injected compressors.

9.1.2.4 Centrifugal compressor

The centrifugal compressor is a dynamic compressor type where a rotating impeller is used to compress the air. It has high efficiency at full load and low efficiency when partially loaded. It is less noisy than a piston compressor and has a very compact design, making it suitable to use in areas with lack of space for placing the air compressors. The centrifugal compressor has very few parts that may wear out, wherefore the cost of maintenance is low. Furthermore, the centrifugal compressor is oil lubricant free.

9.1.2.5 On-/off-load compressors

Traditionally, air compressors have an on- and off-load function. This means that when the compressor is not working, it enters an off-load mode where no air is produced. In off-load mode, about 40%–50% of the power of the compressor is used compared with the power demand for on-load conditions. Since the introduction of variable speed drives (VSD), it is now possible to purchase VSD compressors. A VSD compressor is significantly more energy efficient than conventional on-/off-load compressors, provided that the compressed air demand varies, alternatively acts as a so-called top-load compressor. If this is not the case, but instead the compressor is made up of base load in a compressed air system, or the compressed air requirement does not vary to a great extent, a conventional on-/off-load compressor works as well.

An important efficiency routine for air compressors, in addition to preventive and operational maintenance, is that compressors are shut down at the end of the working day or when no demand exists.

9.1.2.6 Variable speed drive compressors

In many applications where electric motors are used, variable speed drives (VSD) have enabled large energy efficiency improvements. Air compressors are no different. However, in a compressor station which consists of both base load and top load compressors, it is important to assess if a VSD is needed or not. For a base load compressor that operates on full load most of the time, a VSD may not be as effective as if installed as the top load compressor.

9.1.2.7 Vacuum pumps and booster

Air compressors can also often be used as vacuum pumps, which usually operate at 0.1–1 bar pressure. In a compressed air system, even so-called boosters occur, that is, a booster placed locally in the compressed air grid which compresses the air to a higher pressure than in the other parts of the compressed air system.

9.1.3 Air dryers

To avoid corrosion and other problems with water in the distribution and use of compressed air, the moisture in the compressed air needs to be removed. This can be done in four fundamentally different ways, namely through:

- Cooling
- Over compression
- Absorption
- Adsorption

Cooling first takes place in the so-called after-cooler where the compressed air is cooled. In the after-cooler's water separator, between 80% and 90% of the precipitated condensate water is collected. The cooler is usually built into the compressor today.

In case of over compression, the air is compressed to a higher pressure than the operating pressure in the system. Then the air is cooled, and water condensed and separated. Due to high energy demand, this technique is only suitable for smaller airflows.

In absorption drying, water vapor binds to an absorption material. Common absorption materials are saline or sulfuric acid.

Adsorption dryers are divided into two types: cold and heat generating, where the former is best suited for smaller airflows, while the heat generating is better suited for larger airflows. For heat-generating adsorbent driers, the desiccant is regenerated either with compressor heat or with electricity, making heat-generating adsorbent more energy-efficient than cold-generating.

In Table 9.1 standard values for annual electricity consumption for different types of dryers are shown.

From an energy efficient point of view, a minimum of drying should take place, and if only one part of the compressed air system demands high level of dry compressed air, the most suitable solution may be to place the dryer at that specific location.

9.1.4 Air receivers

An air receiver is a buffer of air that enables the compressor to operate more smoothly and is used if the air compression grid is volume-vise, not large

Table 9.1 Scheduled values for annual electricity consumption for dryers in relation to the compressor's electricity consumption.[1]

Type of dryer	% of air compression of annual energy demand
Cooling dryers	3.7
Heat generating adsorption dryer	6
Cold generating adsorption dryer	11

[1]The above figures are based on a compressor of 90 kW nominal power and average values, but as a general template useful for both larger and smaller compressors (Larsson, 2009).

enough. It can be placed either directly after the air compression or close to the end-users.

9.1.5 Air filters

Air filters can be used both prior to the inlet air to the air compression, or after the compression. The use of filters is for removal of particulates and also to some extent oil lubricant and water. From an energy efficient point of view, if filters are used, it is of key importance to change filters on a regular basis, in order to avoid throttling that causes system inefficiencies.

9.1.6 Compressed air grid or distribution system

A compressed air grid or distribution system is often made of metal pipes and should have a large enough diameter in order not to cause high internal friction and to high velocities. According to Morvay and Gvozdenac (2008), the compressed air grid should not have a higher pressure drop than 0.5 bar with maximum air velocities of 6–10 m/s.

9.1.7 Heat recovery from compressor cooling

The major heat losses take place in the air compression where more than 80% of the energy turns into heat directly (DOE, 2018). Today, there are both air-cooled compressors and compressors with water-borne heat exchangers. There are thus two principal means of utilizing the heat from a compressor, either by producing:

- Hot air
- Hot water

If there is a need, the heated air can be blown into a room where there is a heating demand. This is of course only possible if the compressor is in close proximity to a room with heating demands. However, in for example the summer time, the heated air must be blown out from the room in which the compressor is located. In some cases, heat can be used even in the summer, provided there is a heat requirement somewhere in the production.

Another way is to install a water-borne heat exchanger that cools the compressor. This produces hot water. Water-borne heat exchangers on air compressors are one of the most commonly used energy efficiency measures for compressed air systems in industry. Despite so, such a measure should be undertaken with some degree of caution and is not entirely unproblematic. Because a large part of the energy demand in a compressed air system leads to extensive heat losses, and today there are often more power-efficient solutions than, for example, pneumatic tools, a heat exchanger installation may become too large if/when other energy efficiency measures are taken. For example, if a compressor with waterborne heat exchanger installation produces air with a 50% air leakage in distribution and use, it means that if the leak is reviewed and sealed, the heat exchanger will be extremely oversized, which, in addition to unnecessary capital costs of the investment, also means that the heat exchanger efficiency decreases.

9.1.8 Energy efficient compressed air systems

The following list of measures presents a methodology for improving energy efficiency in an industrial compressed air system from a systems perspective. The measures are organized in order of priority.

1. Examine alternatives to compressed air. Replace as much as possible of the compressed air demanding tools and processes.
2. Search, measure, seal, analyze, and set goals for maximum air leakage (continuous improvement).
3. Lower the working pressure in the system. If a process requires higher pressure, consider a locally positioned booster.
4. Section the industrial compressed air system (different pressure levels, different working hours).
5. Purchase, if needed, a new compressor of the correct size.
6. Control the compressors based on demand.
7. Recycle heat from the compressor.

9.1.8.1 Examine alternatives to compressed air

Today, there are several options to compressed air. In many cases, it is possible to switch pneumatic applications to electric power-driven applications, for example, DC motors are significantly more energy efficient than pneumatic driven applications. Certain processes, such as blast cleaning of work benches, can instead be performed with high-speed fans.

If a large part of the compressed air system applications can be converted, consider in the new down-sized compressed air grid, a locally placed compressor to pressurize the new compressed air system.

In a project funded by the Swedish Energy Agency, Volvo's factory in Olofström, Sweden, was analyzed in order to investigate whether compressed air could be eliminated fully or not. The project resulted in the report *The Pneumatic-Free Factory*, and the conclusion was that for most of the compressed air-driven processes, these could in fact be replaced. For economic reasons, some processes, however, were not beneficial to convert, but the report concluded that the assessment would most likely be different if the cost of compressed air grid, compressor station, and so on were included in the investment assessment (Åberg et al., 2003).

9.1.8.2 Search, measure, seal, analyze, and set goals for maximum air leakage (continuous improvement)

Leakage analysis of a compressed air system is a continuous process which must be performed at a regular basis. In rare cases leakage accounts for the overwhelming

Table 9.2 Cost of leakage for different sizes of holes. The calculation is based on a working pressure of 7 bar and an annual average energy cost of 80 EUR/kWh (SEA, 2006).

Diameter of hole (mm)	Leakage-flow (m^3/min)	Power demand, compressor (kW)	Energy cost per year (EUR)
1	0.06	0.4	280
5	1.5	10	7000
10	6	40	28,000
20	25	150	105,000

share of a compressor's electricity consumption. A reasonable level of leakage is about 10% of the compressor's power load.

A good first step to investigate the compressed air system is to install a flow meter, which also gives an idea of how the system works. Even idle loads are easily visualized using a flow meter.

Table 9.2 illustrates how costly a leakage may be depending on the size of the hole.

9.1.8.3 Lower the working pressure in the system

As a rule of thumb, below are average percentage of savings when lowering the pressure in the system from a single-stage compressor (Morvay and Gvozdenac, 2008):

- 8–7 bars, 8% energy savings
- 7.5–6.5, 9% energy savings
- 7–6 bars, 9% energy savings
- 6.5–5.5 bars, 11% energy savings
- 6–5 bars, 12% energy savings

In case of a pressure reduction, the working pressure is normally lowered in the range of 0.5–1 bar. After that it is suitable to wait for some time. If the production works as usual without interference, the procedure could be repeated. With this method, companies can manage to lower the pressure to below 4 bar in rare cases. If a process requires higher pressure, a locally positioned booster may be considered.

9.1.8.4 Section the industrial compressed air system

If different processes in the company require different pressures at different working hours, sectioning of the compressed air system can be an energy-efficient alternative.

9.1.8.5 Purchase, if needed, a new compressor of the correct size

After the compressed air system has been made more energy efficient, existing compressor or compressors may be oversized. Consider investing in a smaller compressor to avoid unnecessary energy use due to an oversized compressor.

9.1.8.6 Control the compressors based on demand

When purchasing a new compressor and with the aim of improving energy efficiency of the existing compressor stock, it is important that the compressor or compressors are controlled in an optimal manner. Fig. 9.3 below illustrates how to control the compressors.

9.1.8.7 Recycle heat from the compressor

The last option that should be investigated is heat recovery. If the previously outlined measures are not considered, the heat exchanger installation will be over-sized, given that the energy demand in the compressed air system has been significantly reduced.

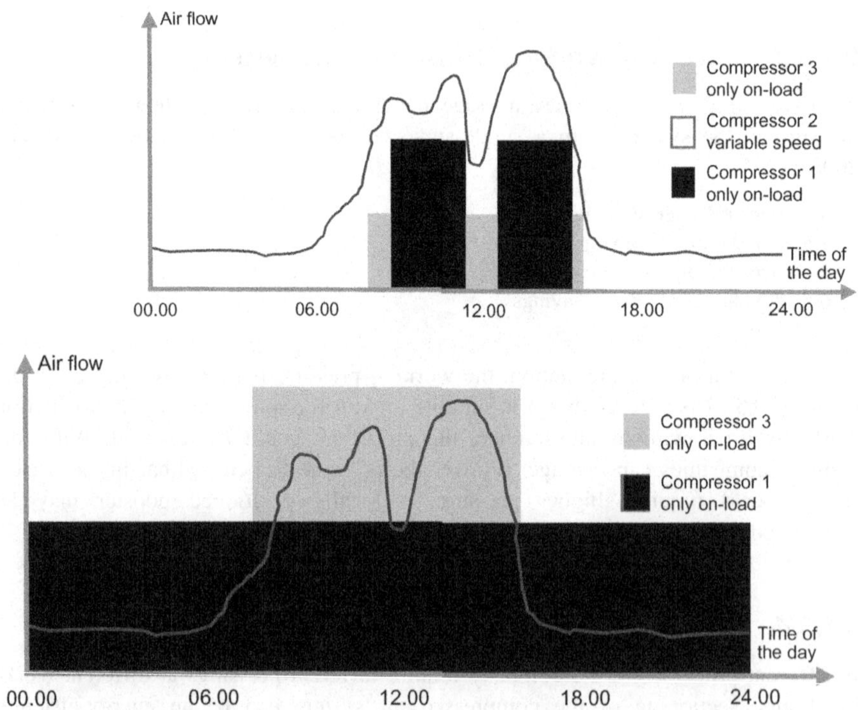

Figure 9.3 Different ways to control a compressor (SEA, 2006).

9.1.8.8 Other energy efficiency improvements

Other improvements for the compressed air system is to optimize that air inlet temperature for the compressor room and compressor cooling, regular filter changes, and overall compressed air system management measures (Nehler et al., 2018).

9.1.9 Practical implications, KPIs, and examples for improved compressed air energy efficiency

For compression of air to about 700 kPa or seven bars in a modern industrial air compressor, a normal specific power requirement is about 6.5 kW/m^3/min. This is equivalent to about 0.11 kWh/m^3 energy use for compressing air to seven bars. However, due to inefficiencies in various parts of the compressed air system, this level is normally not easily reached.

From a LEAN Energy Management point of view, setting the system's normal state performance level is the initial step in initiating an improvement process. In Figs. 9.4 and 9.5 as well as Table 9.3 an example from a vehicle factory's practical roadmap to improving energy efficiency in their compressed air system is presented. As for the sake of simplicity, the annual energy use was normalized to 1000 MWh/year before improvements were made.

As can be seen from Table 9.3 the total energy use of the compressed air system went from 1000 to 187 MWh/year over a period of a seven years, through both technology as well as management-related energy efficiency measures. The challenge that now remains is to maintain the new normal-state and find new ways for improvements.

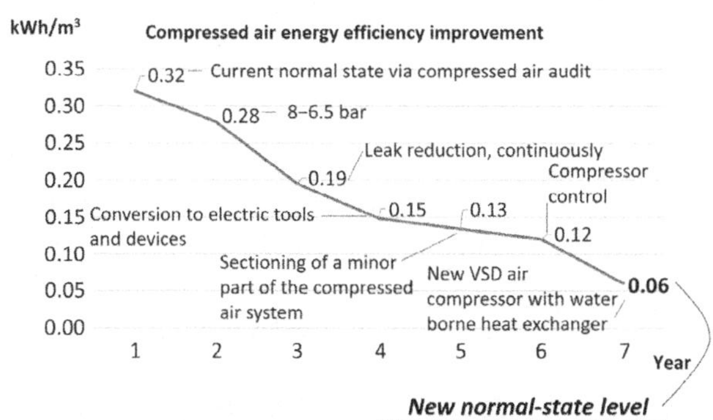

Figure 9.4 A vehicle factory's 7-year journey of efficiency improvements. Please note that the current factory has a heating demand, thus it can utilize heat exchange of compressor heat.

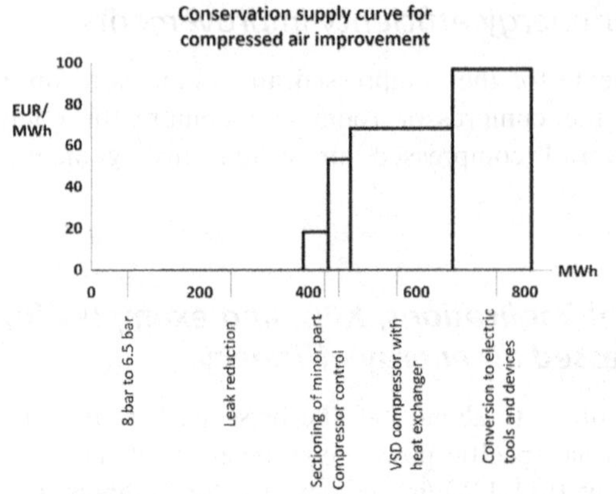

	Energy efficiency improvement (MWh/year)	CCE (EUR/MWh)
8 bar to 6.5 bar	130	0
Leak reduction, continously	261	0
Conversion to electric tools and devices	146	97
Sectioning of a minor part of the compressed air system	46	18
Compressor control	42	53
New VSD air compressor with water borne heat exchanger	187	68

Figure 9.5 Conservation supply curve for an actual carried out action plan for compressed air energy efficiency improvements in a vehicle factory.

Table 9.3 Measures and its effects on total energy end-use in the factory.

	MWh/year	Investment cost (EUR)	Comment
8–6.5 bar	870	0	By in-house staff
Leak reduction, continuously	609	0	By in-house staff
Conversion to electric tools and devices	463	11,500	Procure
Sectioning of a minor part of the compressed air system	417	2500	Procure
Compressor control	375	6500	Procure
New VSD air compressor with water-borne heat exchanger	187	37,500	Procure

9.1.10 Conservation supply curves for compressed air improvements

Conservation supply curves is a means of visualizing various type of measures and its respective magnitude in energy savings (Andersson et al., 2018). In Fig. 9.5 below, a conservation supply curve is exemplified. In the actual case, the discount rate was set to 33% which is equivalent to a 3-year pay-back time.

9.1.11 Nonenergy benefits in compressed air system

When assessing the available nonenergy benefits, the model developed by Nehler (2018) can be applied. Major nonenergy benefits for compressed air improvements include, but are not limited to:

Production-related
- Improved product quality, for example, reduction of oil via an oil lubricant free compressor
- Increased production reliability, for example, via efficiency improvements reducing maximum compressor load and thus system pressure drops
- Increased equipment life, for example, via less usage time for efficiency management of the compressor system

Operation and maintenance
- Reduced maintenance, for example, via a compressor with low maintenance
- Lower cooling requirements, for example, via reduced operation times of compressors, a lower cooling demand is achieved

Working environment
- Increased worker safety, by reducing pressurized parts in the company
- Reduced noise, reduction of noise can increase worker productivity but even more importantly, reduce days off work
- Improved air quality, with oil lubricant free compressor and good filters, air quality can be improved
- Improved temperature control, by heat exchanger on the compressor, the compressor can operate more efficiently

Emissions
- Reduced emissions. By improved energy efficiency, emission reductions are achieved, for example, CO, CO_2, NO_X, and SO_X
- Reduced dust emissions, for example, by reducing for example air leaks and reduction of pneumatic tools and devices

Other
- Improved public/corporate image, for example, by showing customers energy efficiency improvements
- Improved worker morale, for example, by committing to improvements in working conditions
- Increased sales levels, for example, by cutting energy costs, revenue increases on a one-to-one basis, 1 cent in energy cost means 1 cent more in revenue for the company

9.2 Ventilation

The importance of clean air in industrial work environments is well known. The consequences of ventilation losses also have a significant impact on the energy use in many industries (Rohdin, 2008). In Goodfellow and Tähti (2001), industrial air technologies are defined as "Air flow technologies that control workplace indoor environment and emissions." In this definition air flow technologies encompass efforts to achieve a safe, healthy, productive, and comfortable indoor environment for the persons working as well as a functioning product environment. The term air flow technologies can be divided into industrial ventilation, treating air conditioning systems, general ventilation, local ventilation and process ventilation, and process air technology, treating cleaning systems, pneumatic conveying systems, drying systems, and safety air technology systems (Goodfellow and Tähti, 2001). The focus in this chapter is primarily on industrial ventilation.

Industrial ventilation is used to ventilate industrial premises so that the air quality and to some degree the indoor climate meet the requirements. Industrial ventilation includes both what is called process ventilation and general ventilation (Goodfellow Tähti, 2001). Process ventilation and local ventilation are used to ventilate industrial processes more directly. For effective operation, most industrial plants also then require compensation air to replace the air exhausted by the local and process ventilation in creating a functioning indoor environment (Rohdin, 2008; AHRAE, 2003). This compensation air is often categorized as general ventilation. The aim of the process ventilation is in most cases to capture pollution and heat as close to the source as possible. Process ventilation is often a necessity but is generally an expensive and inefficient solution for solving problems with air pollution (Antonsson et al., 2005). Where there are alternatives, process ventilation should not be the primary method of solving the problem. First, it should be investigated whether hazardous substances can be replaced, working methods can be changed, or if the source can be completely removed. This analysis should investigate whether it is possible to reduce the occurrence of pollutants by, for example, (1) switching to another chemical product that emits a smaller amount of pollutants to the environment, (2) changing the working method or process so that as little pollution as possible is formed alternatively, and (3) choose other equipment that emits no or less amount of air pollution. Process ventilation is also used in some cases to reduce the amount of heat that is spread from process to local. In a similar way, it should then be investigated whether there is a possibility to reduce the need to ventilate amounts of heat by (1) switching to a working method or process that does not emit as much heat or (2) encapsulating or isolating hot processes more efficiently. General ventilation refers to the room ventilation not directly connected to the production processes.

The main idea is to take precautionary measures near the source of pollution so that air pollution does not spread to the occupied zone. This is done by installing process ventilation so that contaminated air is captured as close to the source as possible or to perform work or processes that can spread air pollution in a closed system or closed space such as in a special room or at a separate location. This may mean that the work is carried out, for example, from a control room or operation room with a special air supply. The air supplied to the separated room should then be as clean as possible and the room should have an overpressure against the zones around it. In this case, the air direction is important. In the case where the pollution from production is very high, it is recommended that the work is carried out at a time when only the staff performing the work is present. In addition, this often requires the use of personal protective equipment. When this time is over, it must be ensured that the occupied workplaces have been well ventilated and that the conditions are satisfactory to start normal production again.

9.2.1 Laws and regulations related to industrial ventilation - an example from Sweden

In Sweden the Work Environment Act is the law that primarily regulates indoor environment issues in industry. The Working Environment Act states that "The occupational hygiene conditions for air ... should be satisfactory." Many of the factors that influence ventilation in industrial premises are described in the publication "The workplace design" (AFS, 2005: 2). Examples of this are that (1) the ventilation should be arranged so that dispersion to other workplaces is avoided, (2) if the process ventilation is important in order not to create dangerous levels of air pollution, there must be some kind of control system that shows the ventilation's function. Where there is a risk of danger to life or serious injury if the process ventilation does not work, this must be connected to an alarm.

Limit values, or target values, for many in industry commonly used pollutants are gathered in the publication "Hygienic limit values" (AFS, 2015:7) issued by the Swedish Work Environment Authority. Here the regulation that governs pollution in the form of, for example, dust, smoke, fog, gas, or steam is specified. The regulation also states that the levels in the breathing zone should be as low as possible. This means that even if the value is below the limit value, measures must be taken if these are economically reasonable. Here, limit values are published for several hundred different substances and contaminants. The hygienic limit value describes the maximum acceptable average (time-weighted) of air pollutants in the inhaled air. A limit value can be either a (1) level limit value, that is, the average exposure for one working day, (2) a ceiling limit value, that is, the average exposure during a short time (5 minutes or 15 minutes), or (3) a short-term value, which is the same as ceiling limit but more a guideline than a limit value. If exposure to several or combinations of substances occurs, these should be weighed together. In international standards these limit values, also called Threshold Limit Values (TLVs), are also presented, see for example Goodfellow and Tähti (2001) to describe requirements on indoor and outdoor air quality.

9.2.2 Design of efficient process ventilation

In order to be able to design efficient process ventilation some basic understanding of how polluted air is moving and commonly used ventilation technology as well as a general understanding of common mistakes may be needed. In short, we can observe that the level of air pollution decreases with the distance from the source, while the air volume containing contaminated air increases significantly (Goodfellow and Tähti, 2001). In general, one can also say that polluted air, in most cases, moves like clean air. Which just means that highly contaminated air

Table 9.4 In order to capture contaminants, a certain local velocity is required. The capturing velocity for a number of processes (Antonsson et al., 2005).

Type of contaminant production	Lowest capturing velocity (m/s)	Examples of types of processes
Without noticeable motion	0.25–0.5	Exhaust for cleaning and degreasing
Low velocity	0.5–1.0	Slow internal conveyors and spray booths
Active generation	1.0–2.5	Shake outs in foundries, welding, and fast internal conveyor systems
High velocity	2.5–20	Polishing and milling
Processes	**Capturing velocity (m/s)**	
Steam, oil mist	0.5–1.5	
Welding fumes	0.5–5.0	
Shavings in wood industry	About 20	
Grinding	15–18	

does not sink to the floor. Large particles (0.02 mm) will fall to the floor, but with a low velocity of only about 1 cm/s. Air containing a high content (400 mg/m3) of air pollution is less than 0.03% heavier than the ambient air (Antonsson et al., 2005). This has only marginal significance for how the air moves. Hot air rises as we know upwards, which makes the placement of hoods and point extraction above the source effective, not next to the source. Cold air has a negative buoyancy and thus moves downward, which results in for example that a poorly insulated roof can influence the contaminant level in the occupied zone. Air jets also move air, which causes the air flow to increase while reducing concentration and speed of the air jet. Contaminated air streams or high-speed air jets can be thrown out of machines and tools, such as grinders. Then the location of the process ventilation is critical. Table 9.4 provides a summary of approximate capturing velocities and associated examples of processes. The capture velocity is defined as the velocity required to capture the contaminant with, for example, a local exhaust device.

This capturing velocity is influenced by the velocity and the weight of the particle. The large spread in the capturing velocities at high speed shows how important it is to have a well thought out design of the exhaust air's location.

9.2.3 Design of local and process ventilation

In ASHRAE (2003) four different strategies of controlling the air quality in a building are outlined. The strategies are eliminating the source, local hooding or screening with exhaust or recirculated air cleaning, dilution by increasing the general ventilation, and when none of the above is feasible general ventilation air cleaning with or without increased flows. The fourth option is generally difficult as the contaminant is fully dispersed in the facility (ASHRAE, 2003). The most important aspects when designing process ventilation may be considered trivial but are still very important (Antonsson et al., 2005; Goodfellow and Smith, 1982; Rohdin, 2008):

1. Proximity to the contaminant's source. This is because the concentration of the contaminant decreases with increasing distance from the source and thereby a larger air volume must be captured to capture the contaminant. It is perhaps most important to minimize the capture distance (Goodfellow and Smith, 1982).
2. Using gravitational force as contaminated air moves in the same manner as clean air. It is a common belief that contaminated air falls to the floor. This is however only true for large air containing particles which deposit on the floor (ACGIH, 2004; Goodfellow and Smith, 1982).
3. Air jets move large quantities of air. The airflow in the jet increases with increasing distance from the supply point and at the same time the volume increases and the concentration of contaminants decreases.
4. Air movement is created by ventilation, moving objects, and buoyancy effects from heat sources such as machines and persons. Persons moving in close proximity to a contaminant source may affect the airflow pattern and make contaminants move in proximity to the head region if the process ventilation is poorly designed (Antonsson et al., 2005).
5. Contaminants are ejected from machines such as blasting or grinding equipment. This implies a need for specially designed local exhaust units to capture these contaminants (ACGIH, 2004). In Goodfellow and Smith (1982) it is also recommended that a static pressure gauge and testing system be installed and that the fan be the last component to be chosen. In Cascetta and Rosano (2001) an assessment of the velocity field in the vicinity of the rectangular exhaust hood is made and several empirical formulas are presented. In Rueegg et al. (2004) the local capture of contaminants is outlined for different exhaust systems and a reinforced exhaust system is shown to give higher capture efficiency than conventional exhaust systems.

Goodfellow and Tähti (2001) outline three factors that govern the process of designing industrial ventilation in general and local exhausts in particular. (1) The toxic target levels and the amount in the working environment, which states how effective the ventilation needs to be. (2) The occupied zone and how the personnel move in the facility, this to know if they are constantly by the contaminant source or only temporarily. (3) Whether the source is widely spread or dispersed from a small area (Goodfellow and Tähti, 2001; Antonsson et al., 2005). The design principles stated above may seem easy to implement but several problems still exist in solving the problem. The efficiency of different processes are possible to predict or measure but how the overall systems function together requires either full-scale measurements or in the form of validated computational fluid dynamic simulations (CFD) (Rohdin, 2008). The use of CFD for complex industrial ventilation projects is recommended in ASHRAE (2003) as well as Goodfellow and Tähti (2001) and Rohdin (2008).

The most common types of nonproduction integrated process ventilation are:

- Local exhausts
- Welding exhausts and hoods

- Tool integrated process ventilation
- Ventilated tables and fume hoods
- Encapsulation
- Glove-box systems

Local exhausts and welding exhaust and hoods are placed close to hot processes above the source of pollution to best capture the plume that is created. With tool-integrated process ventilation, it is often possible to get closer to the source, which reduces the flow needed to remove the contaminant. Ventilated tables are mainly used for cool processes, and the height of safety is determined by the air flow. At 280 l/s.m^2, a working height of 15–20 cm is typically obtained. Encapsulation means that most of a process is enclosed. An under pressure in the enclosure is used to guarantee the air direction. A glove box creates a completely separate handling of the product. A more detailed description for the design procedure of the above-introduced types of process ventilation is found in, for example, ACGIH (2004) and Goodfellow and Tähti (2001).

9.2.4 General ventilation

General ventilation is a collective term for the part of the ventilation that is not directly related to the processes. This may be, for example, ventilation in office parts but also referrers to the supply and exhaust air not directly connected to the processes. In rooms with large process ventilation flows, general ventilation usually mainly accounts for the compensation air, that is, the supply air to compensate for the large exhaust air flows in a room, so that a reasonable pressure difference is achieved against surrounding premises (Goodfellow and Tähti, 2001). In addition to this, if it is difficult to obtain the desired indoor environment conditions using process ventilation and local ventilation, an increase in the requirements on the general ventilation may be a feasible way to solve the problem. The downside of using the concept of increased requirements on the general ventilation is the risk of large increases of supply airflows resulting in an increase in energy use. In applications where the process ventilation demands large amounts of exhaust air, the general ventilation may be limited to supplying air just to obtain balanced ventilation in the facility. The industrial ventilation is highly dependent on the actual process and should be treated as such (Goodfellow and Tähti, 2001; Hedlund, 2005). This makes general airflow recommendations difficult if not impossible to give (Rohdin 2008).

Effective diffusion of clean air in industrial facilities as well as in other rooms and the proper amount of conditioned air is essential for creating acceptable working conditions, for removing contaminants, and for minimizing the initial and operation cost of a ventilation system. Generally speaking the air supply system must supply air of proper temperature, velocity, turbulence intensity, humidity, and with contaminant concentration within the desired limits (ASHRAE, 2003). However, in ASHRAE Application (2003) the object in most cases is referred to as "to provide tolerable (acceptable) working conditions rather than comfort (optimal) conditions."

Ventilation can be categorized according to how the system is structured. The six most common categories are:

- Natural ventilation
- Hybrid ventilation
- Exhaust air ventilation
- Exhaust air system with heat pump
- Supply and exhaust ventilation
- Supply and exhaust ventilation with heat exchange

Of these, supply and exhaust ventilation with heat exchanger and supply and exhaust ventilation are the most common systems in industrial environments. This is related to the fact that it is important to be able to heat the compensation air to the premises as the process ventilation generates large flows. If this is not done the thermal comfort will be poor during the cold season.

At room level, ASHRAE Applications (2003) divide air supply methods into localized ventilation, mixing air distribution, and displacement ventilation. The localized ventilation may be designed as, for example, a piston flow which is sometimes cited to be a separate ventilation principle. A fourth principle, that is not something desirable and should be seen as a malfunction, is short circuit. In mixing ventilation systems, the air is generally supplied with velocities far greater than accepted in the occupied zone, but the air is blended with the room air by entrainment before it reaches the occupied zone. The system can handle both over-heated air as well as air cooler than room temperature depending on the heating or cooling load in the occupied zone. In displacement ventilation which is based on stratified flow, air with lower temperature than room temperature is supplied with a low velocity directly into the occupied zone from a supply device located at floor level. The cooled air rises as it meets heat sources in the rooms, thus creating temperature and density stratification, see Mundt (1994) and Mundt (1990). In localized ventilation air is supplied locally in the occupied region or at specific work stations. The conditioned air is supplied directly to the breathing zone, making these zones much less contaminated than the surrounding parts of the facility. The piston flow principle may be used.

Displacement and mixing ventilation have been compared in several publications such as Lin et al. (2005a), Lin et al. (2005b), and Breum and Ørhede (1994), and different aspects are discussed such as ventilation efficiency and thermal climate. All these supply systems have their advantages and disadvantages: mixing ventilation with its poor ventilation efficiency; displacement ventilation with the inability to effectively supply heated air during the cold season and to supply air to the occupied zone in facilities with high thermal loads and in a multiple buoyancy source environment; and the piston flow system with its susceptibility to disturbance with low airflows or high energy use for large flows. However, they also have their advantages. Mixing ventilation offers even temperature distribution, flexibility, and the ability to be used for both heated and cooled air. Displacement ventilation has the advantage of its stratification characteristics enabling low pollution concentrations in the occupied zone and its ability to use less active cooling than the

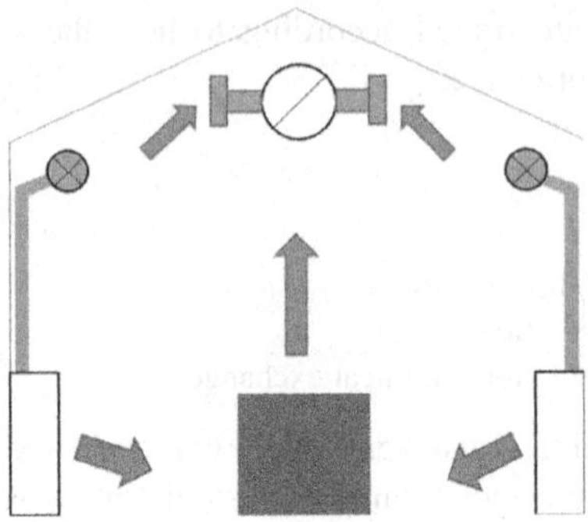

Figure 9.6 Displacement ventilation in an industrial room. Under-tempered air is supplied low and the exhaust air is removed high. This principle is suitable for premises with heat generation.

mixing system. It also has low air velocities in the occupied zone and low turbulence intensity, enabling high comfort. The piston flow principle has the advantages of high efficiency in terms of pollution transports and is mainly used in pharmaceutical and other critical applications.

Piston flow is a principle where the flow over a room or part of a room is laminar and goes from, for example, roof to floor. This requires large flows to maintain but provides very good control of how the pollution spread. Displacement ventilation, see Fig. 9.6, is based on the fact that the supply air is under-tempered, that is, with a lower temperature than the room temperature (approx. 2°C–5 °C). The exhaust air is removed high. And the air flow is based on the fact that pollution is transported upward by the heat generation in the room. This principle is effective in rooms with heat generation.

Mixing ventilation is based on the principle that supply air with high speed (impulse) is supplied to the premises in such a way that an even concentration of pollutants is created. The same applies to the temperature that gets even in the room. The advantage of this principle is that even over-tempered supply air can be supplied to the premises. Fig. 9.7 presents a number of common principles for mixing ventilation in industrial premises.

Short circuit is, as previously mentioned, not a principle that is sought but should be seen as a malfunction that occurs when the ventilation air does not reach the occupied zone, or when the supply air goes straight out with the exhaust air, hence the term short circuit. Short circuit has no ventilation efficiency. When designing ventilation in industrial environments where larger amounts of pollutants are generated, the air movements are very important. The air must then have the right direction locally so as not to spread out in the room, and the balance between both supply air and extract air as well as its location is important. In the design phase of complicated production facilities, advanced computational fluid dynamic (CFD) models are now often used to describe these air movements, in combination with measurements and local design when the premises are to be commissioned. In

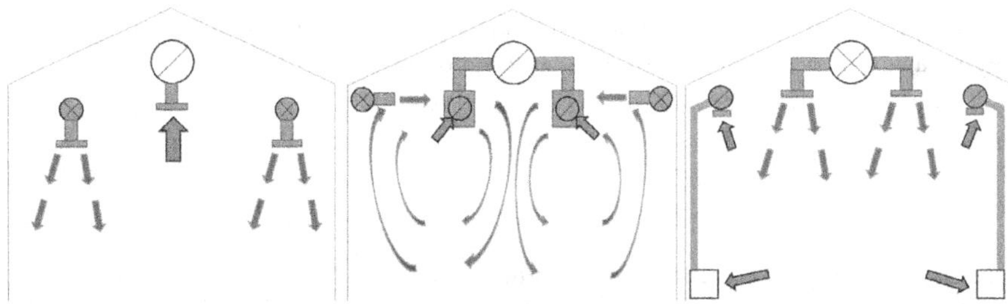

Figure 9.7 Mixing ventilation, with high exhaust air, the "tornado-method" where exhaust air is placed on one short side of the room or mixing ventilation with low and high placed exhaust air (Antonsson et al., 2005).

Rohdin and Moshfegh (2007) and Rohdin and Moshfegh (2011) examples of integrated use of CFD in industrial ventilation design are found.

9.2.5 Power requirement for fans and air flow control

The power requirement for a fan can be calculated equally for a pump described as in Eq. (9.1):

$$P = \frac{q \cdot \Delta p}{\eta_{\text{tot}}} \tag{9.1}$$

where P is power (W), q the volume flow (m^3/s), Δp the pressure increase (Pa), and η_{tot} the total efficiency. Control of the flow, like pumps, can be described using the affinity laws presented in the Chapter 8, Eqs. 8.8 to 8.10. Control is done with dampers, adjustable guide rails or using variable fan speed. Adjustable guide vanes can be mounted in the fan inlet.

9.2.6 Performance metrics

The measures of performance for industrial ventilation are numerous, but often used indices are ventilation efficiency, air exchange efficiency, mean age of air, and ventilation effectiveness for heat distribution or heat removal. The relative ventilation efficiency may be defined as in Sandberg (1981) as

$$\varepsilon_c = \frac{C_o - C_i}{C_m - C_i} \tag{9.2}$$

where C represents the concentration of the contaminant and subscripts o, i and m denote outlet, inlet, and mean value for the occupied zone. For ideal mixing ventilation $\varepsilon_c = 1$, if the facility has a short-circuit between supply and exhaust $\varepsilon_c < 1$ and for a contaminant source in the proximity of, for example, a local exhaust $\varepsilon_c > 1$. In Sandberg (1981) the index is defined locally in the facility where C_m indicates a point or a zone. The capture efficiency of a local exhaust system, that is, the percentage of released contaminants extracted by a local exhaust system, is also a key measure of performance for local ventilation systems.

The air exchange efficiency is defined as [see Etheridge and Sandberg (1996)]:

$$\eta_a = \frac{\tau_n}{2 \cdot \tau_i} \tag{9.3}$$

where $\tau_n = \frac{V}{q}$ represents the nominal time constant of the facility, V is the volume and q is the airflow, and τ_n represents the mean age of the air in the facility. The local mean age of the air or the mean age of air can be used to describe the freshness of the air. These indices can be found either by measurements such as using tracer gas or by calculation using, for example, CFD.

The ventilation's effectiveness for heat distribution or heat removal (ε_t) is a measure of how effective the ventilation system is in removing the heat produced internally in the occupied zone and is defined as

$$\varepsilon_t = \frac{t_o - t_i}{t_{oz} - t_i} \tag{9.4}$$

In Eq. (9.4), t_i and to are the supply and exhaust air temperatures respectively and t_{oz} represents a mean value in the occupied zone. The value of ε_t is dependent on the method of room air distribution, room characteristics, heat, and sources.

9.2.7 *Typical measures during an audit—ventilation*

Common questions and measures when auditing industrial ventilation are as follows:

- Has the ventilation followed the process development?
- Has more efficient process ventilation resulted in lower general ventilation requirements?
- Analysis of ventilation principle.
- Sectioning.
- Time control of fans.
- Variable speed drivers (VSD) for ventilation control.
- Is heat exchange possible?

9.2.8 *Nonenergy benefits in ventilation*

When assessing the available nonenergy benefits, the model developed by Nehler (2018) can be applied. Major nonenergy benefits for ventilation include but are not limited to:

Production-related

- Improved product quality, for example, reduction of air emissions that in turn negatively affect the products
- Increased equipment life, for example, via usage time for efficiency management of the ventilation system

- Increased production reliability, for example, via efficiency improvements reducing maximum ventilation loads and thus system pressure drops

Operation and maintenance

- Reduced maintenance, for example, via a ventilation unit with lower maintenance

Working environment

- Improved air quality, with more efficient ventilation systems
- Increased worker safety, by reducing air emissions in the company
- Reduced noise, reduction of noise can increase worker productivity but even more importantly, reduce days off work
- Improved thermal comfort, which in turn reduce risk of accidents and improves the general perceived comfort of the production environment

Emissions

- Reduced emissions. By improved energy efficiency, emission reductions are achieved, for example, CO, CO_2, NO_X, and SO_X
- Reduced dust emissions

Other

- Improved public/corporate image, for example, by showing customers energy efficiency improvements
- Improved worker morale, for example, by committing to improvements in working conditions
- Increased sales levels, for example, by cutting energy costs, revenue increases on a one-to-one basis, 1 cent in energy cost means 1 cent more in revenue for the company

9.3 Lighting

The definition of lighting is a process which purpose is to improve and adjust the brightness of a room in accordance with the demand. Lighting is a necessity for operation in different industries and is categorized as a support process. Its energy use accounts for only a few percent of the total energy use for energy-intensive companies with a high proportion of energy use in the production processes. However, for nonenergy intensive companies with a high proportion of energy use in support processes, the electricity use of lighting can account for 30% or more of total energy use.

Several different types of lighting occur in industry, such as

- Fluorescent lamps
- High pressure and low-pressure sodium lamps
- Mercury lamps
- Halogen lamps
- Incandescent lamps
- LED

Before these lighting types are described in detail, a review of several related lighting concepts will be presented in Sections 9.3.1–9.3.5.

9.3.1 A lighting fixture and its components

A lighting installation consists of four primary parts: the actual fixture, the light source which is the part of the lighting installation that generates the light itself. In a fixture for fluorescent lighting, the light source itself refers to the fluorescent lamp. The third part is the electrical appliance that enables the light source to work, the so-called ballast. The fourth part is the reflector that reflects the light emitted by the light source. For some fixtures, lenses are also included, for example, a plastic sheet that covers/protects the lamps.

9.3.2 R_a-index

An important factor within lighting is the so-called R_a-index which measures color rendering from the light source, that is, how well the emitted light represents daylight. The R_a index varies between 0 and 100, where 100 corresponds to the light obtained from an incandescent light bulb or from daylight. The R_a-index depends on how much of the full spectrum of wavelengths the light source displays. If the R_a-index is just over 90, there is no risk of color changes. An R_a-index of between 80 and 85 represents some risk of changes in sensitive colors, but most of the colors are reproduced well. An R_a-index below 70 means there is a high risk of insufficient color representation.

9.3.3 Color temperature

The color temperature is measured in Kelvin and is a measure of the color of the light. Hot light is defined by a value below 4000K, and cold light has a value above 4000K. Fig. 9.8 shows color temperatures for some light sources.

9.3.4 Light output and luminous efficiency

The light output measures how much light is emitted and is measured in lumen (lm). The luminous efficiency, in turn, measures the light output in relation to the supplied electricity (in Watt) and is measured in lumens per watt (lm/W).

9.3.5 Illuminance

Illuminance is a measure of how much lumen that is emitted over an area. Illuminance is measured in lux (lx) and is defined as lumen (lm) per m^2. Another common measure used is installed electric power per square meter (W/m^2). The latter measure should however be used with level of caution as it does not directly take into account illuminance, R_a-index or color temperature. In Table 9.5 an example of a guideline is shown.

The following sections describe the different types of lighting that can be found in industry:

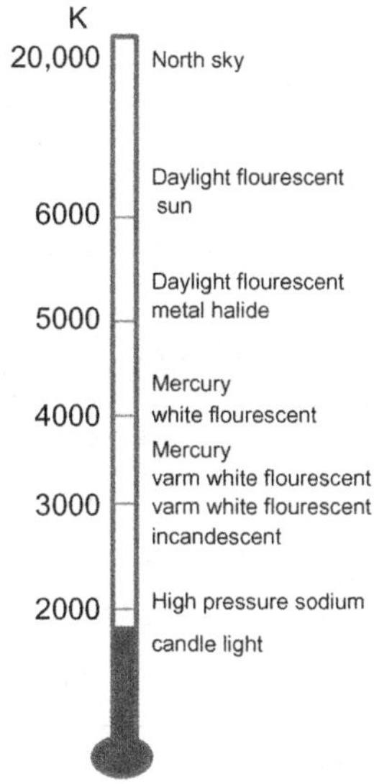

Figure 9.8 Color temperature for some common light sources (Starby, 2006).

- Incandescent lamps
- Halogen lamps
- Fluorescent lamps
- High pressure and low-pressure sodium lamps
- Mercury lamps
- Metal halide
- Ceramic metal halide
- Induction lamps
- LED

9.3.6 Incandescent lamps

9.3.6.1 Incandescent

The incandescent lamp is made of a socket, a piston, a filler gas (often argon or nitrogen), and a filament, usually made of tungsten. In September 2012, the conventional incandescent light bulb was banned after being used for more than 100 years, and it may no longer be sold within the European Union. The light output for an incandescent light bulb is in the range of 10–30 lm/W, with a color temperature of between 2500 and 3500K and a R_a-index of 100. The life span is between 1000 and 4000 hours.

9.3.6.2 Halogen lamp

The halogen lamp is like the incandescent lamp and made of a socket, a piston, a filler gas, and a filament. The difference to an incandescent lamp is that it has a

Table 9.5 A guideline for lighting (Ljuskultur, 2018).

Industry	Guidelines installed power (W/m^2)
Rough, 300 lx	
High *Low*	8 6
Average, 500 lx	
High *Low*	13 11
Excellent, 750 lx	
High *Low*	19 16
Limited area, 300 lx	
High *Low*	8 6
Painting	
High *Low*	19 15
Storage	
High *Low*	6 5
Storage aisle	
High *Low*	15 13

"Rough" means an area with low demand for good lighting. "Average" means an area which has higher demands but still not very high. "Excellent" means an area with high demand for good lighting.

regenerating process where evaporated tungsten from the filament is caught by the filler gas and then reabsorbed by the filament. This means that it will have a longer life-time than a standard incandescent lamp. The light output for a halogen lamp is in the range of 20–40 lm/W, with a color temperature of between 3000 and 3500K and a R_a-index of between 80 and 100. The life span is between 2000 and 5000 hours.

9.3.7 Fluorescent lamps

9.3.7.1 T8-flourescent lamp

The most common fluorescent lamp has historically been the so-called T8 lamps. A T8 fluorescent lamp consumes about one fifth as much electricity as a conventional incandescent lamp. The lamp is reliable and has historically a conventional ballast technology, that is, magnetic ballasts. There are also modern T8 fluorescent lamps that are equipped with electrical ballasts or HF (high frequency) ballasts. These more modern ballasts make these fluorescent lamps more energy efficient. An electrical ballast uses about 10% of the fluorescent lamp's electrical power, while a conventional magnetic ballast uses about 20% of the fluorescent lamp's electrical power. The light output for a T8 lamp is in the range 60–90 lm/W, with a color temperature of between 2500 and 5500K and a R_a-index of less than 80 for single color and over 80 for full color. The life span is between 9000 and 17,000 hours.

9.3.7.2 T5-fluorescent lamp

T5 fluorescent lamps represent a new generation of fluorescent lamps. Unlike the T8 fluorescent lamp that can use either conventional magnetic ballast technology and electrical ballasts, T5 fluorescent lamps can only use electrical ballasts. In addition to the electrical ballast's lower power consumption compared to the magnetic ballast, electrical ballasts provide a flicker-free light. The light output for a T5 fluorescent lamp is in the range 70–110 lm/W, with a color temperature of between 3000 and 7000K and an R_a-index of less than 80 for single color and over 80 for full color. The life span is approximately 17,000 hours or up to 40,000 hours for so-called long-life T5 fluorescent lamps.

9.3.7.3 Compact fluorescent lamp

The technology of compact fluorescent lamp is the same as for a conventional fluorescent lamp, but the big advantage is that compact fluorescent lamps can replace incandescent light bulbs due to their size. The light output for a compact fluorescent lamp is in the range of 50–80 lm/W, with a color temperature of between 2500 and 4000K and a R_a-index of more than 80. The life span is between 10,000 and 20,000 hours.

9.3.8 Sodium lamps

9.3.8.1 Low pressure sodium lamps

The low-pressure sodium lamp is by far the best light exchange with up to 200 lm/W, at the expense of a very limited color reproduction, that is, The R_a-index is very low, providing a yellowish light. The color temperature is approximately 2000K and the life span is about 12,000 hours.

9.3.8.2 High pressure sodium lamps

The high-pressure sodium lamp with its higher pressure has better color properties than the low pressure sodium lamp. The R_a-index is just over 80. The light output is between 70 and 130 lm/W, with a color temperature of between 2000 and 2500K and a life span of between 10,000 and 30,000 hours.

Sodium lamps is an energy-efficient alternative for industrial premises where the requirement for color rendering is not particularly high.

9.3.9 Mercury lamps

Mercury lamps are the perhaps one of the historically most used light sources in industrial premises. The lamp gives a white light of between 3000 and 3700K, with a R_a-index of between 50 and 60 and a light output of between 40 and 60 lm/W. The life span is between 6000 and 12,000 hours.

Mercury lamps were banned from 2015 and is no longer available for sale within the European Union.

9.3.10 Metal halide lamps

The metal halide lamp is a further development of the mercury lamp. The lamp has a color temperature of between 2800 and 4200K, a R_a-index of between 75 and 95, and a light output of between 75 and 100 lm/W. The life span is between 6000 and 12,000 hours.

9.3.11 Ceramic metal halide lamps

The difference between a ceramic metal halide lamp and a metal halide lamp is the burner, which in ceramic metal halide lamps is the same as in sodium lamps. This gives a higher light output and longer life. Ceramic metal halide lamps have color temperatures between 3000 and 4200K, a R_a-index of between 90 and 95 and a light output of between 90 and 100 lm/W. The life span is between 9000 and 15,000 hours.

9.3.12 Induction lamps

The induction lamp is a relatively new type of lamp that offers very good color rendering and long life. The color temperature is between 2700 and 4000K, with a R_a-index of over 80 and a light output of 80–130 lm/W. The life span is between 60,000 and 100,000 hours.

9.3.13 Light emitting diode

Light emitting diode (LED) consists of a semiconductor material that rectifies the current. The LED has attracted great attention over the last two decades since its potential has been shown to be very large. LEDs have color temperatures between 3000 and 5000K, a R_a-index of between 40 and 95, and a light output of between 80 and 150 lm/W. The life expectancy is at least 50,000 hours.

9.3.14 Energy efficient lighting

The standard EN 12464-1. Light and lighting—Lighting of work places—Part 1: Indoor work places is a very good guide when planning for lighting installations when delamping and relamping as well as for new installations. Other guidelines can also be found on adequate illuminance and energy efficiency levels for lighting, for example, the one shown in Table 9.5.

In addition to installing energy-efficient luminaires, electricity consumption resulting from the control of the lighting can also be affected by

- Dimming
- Load reduction
- Constant light level control
- Presence control
- Dynamic lighting control
- Sectioning

9.3.14.1 Dimming

Dimming means that the lux values are lowered, for example, when no one is occupying the room. However, all types of fittings are not dimmable.

9.3.14.2 Load reduction

Load reduction is a type of dimmer where the voltage is lowered by an energy-saving module, which reduces electricity consumption. This method can be applied to, for example, sodium lamps, mercury lamps, metal halides, and fluorescent lamps, but just as is the case for dimming, it does not work for all types of lamps. This type of load reduction as a method is doubtful for new lighting installations but can be a cost-efficient choice in areas with low occupancy, for example, in storage halls, where cost for brand new lighting installations cannot be justified.

9.3.14.3 Constant light level control

Constant light level control, also known as daylight control, means that the lamps are regulated with the help of sensors and controlled by the amount of daylight entering the room's windows.

9.3.14.4 Presence control

Presence control means that the lamps are controlled by means of sensors that respond to whether or not the room is occupied.

9.3.14.5 Dynamic lighting control

Dynamic lighting control combines dimmable lamps with presence control and constant light level control, which works well in premises with many short visits.

9.3.14.6 Sectioning

Sectioning, together with dimming and presence control, are suitable in, for example, larger storage halls or production areas where you work in shifts. Then it is possible to control the occupied parts of the hall separately, so that the lighting can, for example, be turned off in areas which are unoccupied. On larger industrial sites where a few machines operate in shift, the savings with this method can be very large.

Specific lux values for various types of industrial rooms can be obtained from standard EN 12464-1: 2011.

9.3.15 Nonenergy benefits in lighting

In addition to energy cost savings, energy-efficient lighting can also create other benefits, such as reduced maintenance costs and increased productivity. Good lighting is also an important factor that positively affects the company's work environment. When assessing the available nonenergy benefits, the model developed by Nehler (2018) can be applied.

Major nonenergy benefits for lighting include, but are not limited to:

Production-related
- Improved product quality, for example, by adequate lighting levels enabling misprints, and other product defects to be detected in the production
- Increased equipment life, for example, via long-life T5-lamps or LED, the time between relamping can be extended

Operation and maintenance
- Reduced maintenance, for example, via longer life-time of the light source and ballasts
- Lower cooling requirements, for example, via lower power consumption of the lamps and ballast, less cooling demand is achieved
- Reduced labor requirements, for example, via plastic lenses that cover the lamp, removal of oil and dust can be done quickly by using a long stick and a cloth, and maintenance without a time-consuming ladder can be carried out. Also, lamps and ballasts with longer life-time has a similar very positive effect and reduce maintenance staff labor hours

Working environment
- Increased worker safety, by adequate illuminance in the industrial plant
- Improved lighting

Emissions
- Reduced emissions. By improved energy efficiency, emission reductions are achieved, for example, CO, CO_2, NO_X, and SO_X

- When relamping using lamps with, for example, no mercury content, mercury waste can be eliminated

Other

- Improved public/corporate image, for example, by showing customers energy efficiency improvements
- Improved worker morale, for example, by committing to improvements in working conditions
- Increased sales levels, for example, by cutting energy costs, revenue increases on a one-to-one basis, 1 cent in energy cost means 1 cent more in revenue for the company

References

Åberg, B., Gralén, K., Björk, C., Räftegård, O., 2003. Den tryckluftslösa fabriken (The compressed air free factory). Linkoping University, Linköping [in Swedish].

ACGIH, 2004. Industrial ventilation – a manual of recommended practice. In: American Conference of Governmental Industrial Hygienists, 25th ed.

AFS, 2005. Hygieniska gränsvärden och åtgärder mot luftföroreningar (Target limits for airborne contaminants), 2005:17, The Swedish Work Environment Act.

AFS, 2015. Workspace Design. AFS 2015:7. The Swedish Work Environment Act.

Andersson, E., Karlsson, M., Thollander, P., Paramonova, S., 2018. Energy end-use and efficiency potentials among Swedish industrial small and medium-sized enterprises – a dataset analysis from the national energy audit program. Renew. Sustain. Energy Rev. 93, 165–177.

Antonsson, et al., 2005. Process-och Industriventilation (Process and Industrial Ventilation). JOMA grafiska Produktion AB, Prevent, Stockholm.

ASHRAE, 2003. Application Handbook. American society of Heating, Refrigerating and Air-Conditioning Engineers, Atlanta.

Breum, N.O., Ørhede, E., 1994. Dilution versus displacement ventilation: environmental conditions in a garment sewing plant. Am. Ind. Hygiene Assoc. J. 55 (2), 140–148.

Cascetta, F., Rosano, F.M., 2001. Assessment of velocity fields in the vicinity of rectangular exhaust hood openings. Build. Environ. 36 (10), 1137–1141.

DOE, 2018. Improving Compressed Air System Performance: A Sourcebook for Industry. Available from: https://www.compressedairchallenge.org/data/sites/1/media/library/sourcebook/Improving_Compressed_Air-Sourcebook.pdf (accessed 11.25.2018.).

Etheridge, D., Sandberg, M., 1996. Building Ventilation – Theory and Measurement. John Wiley & Sons, Chichester.

Goodfellow, H.D., Smith, J.W., 1982. Industrial ventilation – a review and update. Am. Indust. Hygiene Assoc. J. 43.

Goodfellow, H., Tähti, E. (Eds.), 2001. Industrial Ventilation Design Guidebook. Academic Press.

Larsson, U., 2009. Oral communication with compressed air expert.

Lin, Z., Chow, T.T., Fong, K.F., Wang, Q., Li, Y., 2005a. Comparison of performances of displacement and mixing ventilation. Part I: Thermal comfort. Int. J. Refrig. 28 (2), 276–287.

Lin, Z., Chow, T.T., Fong, K.F., Wang, Q., Li, Y., 2005b. Comparison of performances of displacement and mixing ventilation. Part II: Indoor air quality. Int. J. Refrig. 28 (2), 288–305.

Ljuskultur, 2018. Riktlinjer (Guidelines). Ljuskultur, Stockholm. Available from: <https://ljuskultur.se/wp-content/uploads/2016/04/ljus-och-rum_riktlinjer-fr-installerad-effekt.pdf>(accessed 12.16.2018). [in Swedish].

Möre, O., 1953. Tryckluft inom industrin: Ett viktigt rationaliseringshjälpmedel (Compressed air within industry: an imortant rationalization instrument). Stockholm. [in Swedish].

Morvay, Z.K., Gvozdenac, D.D., 2008. Applied Industrial Energy and Environmental Management. Wiley.

Mundt, E., 1990. Convection flows above common heat sources in rooms with displacement ventilation, Roomvent 1990 paper 38, Oslo.

Mundt, E., 1994. Displacement ventilation systems – convection flows and temperature gradients. Build. Environ. 30 (1), 129–133.

Nehler, T., 2018. A systematic literature review of methods for improved utilisation of the non-energy benefits of industrial energy efficiency. Energies 11 (12).

Nehler, T., Parra, R., Thollander, P., 2018. Implementation of energy efficiency measures in compressed air systems: barriers, drivers and non-energy benefits. Energy Effic. 11 (5), 1281–1302.

Rohdin, P., 2008. Energy efficiency and ventilation in Swedish industries barriers, simulation and control strategy, Linköping Studies in Science and Technology Dissertation No. 1223.

Rohdin, P., Moshfegh, B., 2007. Numerical predictions of indoor climate in large industrial premises: a comparison between different k-ε models supported by field measurements. Build. Environ. 42 (11), 3872–3882.

Rohdin, P., Moshfegh, B., 2011. Numerical modelling of industrial indoor environments: a comparison between different turbulence models and supply systems supported by field measurements. Build. Environ. 46 (11), 2365–2374.

Rueegg, T., Gubler, D., Sprecher, P., Moser, A., Weber, R., 2004. Local capture of contaminants. ASHRAE J. 46 (1), 26–29.

Sandberg, M., 1981. What is ventilation efficiency? Build. Environ. 16 (2), 123–135.

SEA (Swedish Energy Agency), 1999. Programkrav tryckluft (Program demand, compressed air). Eskilstuna, Sweden [in Swedish].

SEA (Swedish Energy Agency), 2006. Krav på tryckluftssystem(Demand on compressed air systems). Eskilstuna, Sweden [in Swedish].

Starby, L., 2006. En bok om belysning (A Book About Lighting). Ljuskultur, Stockholm [in Swedish].

Energy efficiency in space heating and cooling systems and hot tap water production

10

Chapter Outline

10.1 Space heating and cooling

The term space heating refers to the activities that make it possible to achieve the desired temperature in a room or space. In winter, most industrial premises must be heated. In many cases, industrial premises are heated by heating the incoming ventilation air.

During an energy audit, however, space heating and ventilation should be separated as the two processes by definition do not have the same purpose, although sometimes premises are heated only by the ventilation system, so-called indirect heating or air-heating. Radiator system is another way of heating a room. Local heat is a typical nonelectric process. If there are electric radiators or heating batteries in a ventilation unit that is powered by electricity, these should if possible, be replaced if a more efficient energy system is to be achieved.

Generally, however, many premises are self-sufficient in terms of space heating during most part of the year, this being dependent on machines and people generating heat. The problem is in many cases rather over temperature and the design of a production space is thus often more focused on how to remove heat during the warmer months of the year.

In practice, however, industrial premises are still not self-sufficient in terms of heat as the incoming ventilation air still needs to be heated for comfort reasons,

Introduction to Industrial Energy Efficiency. DOI: https://doi.org/10.1016/B978-0-12-817247-6.00010-9

mainly draught, and as the hot exhaust air is continuously transported out of the space. If the main design factor is to remove heat from a room, different forms of heat recovery can sometimes be counterproductive.

10.1.1 Thermal comfort

Humans want and seek thermal comfort wherever they are, whether at work or at home. The term thermal comfort may be defined as "that condition of mind that expresses satisfaction with the thermal environment" (ASHRAE, 2003). Clothing, activities, posture, location, and shelter are chosen, altered, and adjusted to minimize discomfort both consciously and unconsciously. Discomfort is related to human health, number of mistakes, productivity, and industrial accidents, as discomfort is a form of physiological strain. This strain can be in the form of sweating, muscle tension, stiffness, shivering, and loss of dexterity (Goodfellow and Tähti, 2001). Thus thermal climate is an important factor and it is desirable to minimize thermal discomfort. In general it is often cited that a facility that provides a satisfactory environment will be financially more successful as well as more desirable for the people working there, as it contributes to a safer and more comfortable environment with less loss of productivity and better comfort (Goodfellow and Tähti, 2001).

When making statistical correlations of comfort in the thermal environment one may distinguish some reproducible primary factors and some secondary factors of less importance. The less important factors are factors such as age, local climate, physical fitness, food, and illness. These secondary factors have a lesser effect on the perceived thermal climate and are not discussed further in this chapter. The primary factors directly affect the heat transfer of the person, where the objective of a person's thermoregulation is to maintain the body's core temperature vital for the internal organs to function properly (Goodfellow and Tähti, 2001).

One way to quantify comfort and discomfort is through empirical environmental indices. These studies are based on a model of the heat exchange between the body and its surroundings, which is made by subjecting a large population to different known activities and clothing to a range of environmental conditions and recording the response to these conditions. Probably the most used model is the predicted percentage dissatisfied (PPD) index created by Fanger and his co-workers in Denmark (Fanger, 1972). This work as well as other more zone-specific indices are included in the ISO 7730 standard of "Ergonomics of the thermal environment" (ISO, 2005).

The human experience of the thermal climate depends on a number of climate factors: (1) air temperature, (2) the average radiation temperature of the surrounding surfaces, (3) the speed of the air and turbulence, and (4) humidity. Clothing level and physical activity (metabolism) also affect this experience of the thermal climate.

As an example, in the Swedish regulation Workplace design, the Work Environment Authority, it is stated that if the temperature in light and sedentary work varies outside the range of 20°C and 24°C in winter and between 20°C and 26°C in summer, the thermal climate should be investigated more closely.

Since it is not only the air temperature that affects the thermal climate, it may be appropriate to determine PMV and PPD according to ISO 7730: 2005.

Furthermore, the heating power requirement should be calculated so that in occupied spaces one can normally maintain an air temperature in the residence zone of at least 20°C during sedentary, physically less strenuous work or between 14°C and 15°C during moving or physically more strenuous work. If the work is mainly carried out in part of a room, the heating can be limited to that part. Examples of this are control rooms for sawmills or other activities where heated cabins are suitable.

Experience shows that air velocities below speeds between 0.15 and 0.2 m/s are generally perceived as draught free for sedentary work.

The importance of air humidity to the experience of the thermal climate is limited for indoor temperatures of about 20°C or above. On the other hand, if the air temperature is raised by 2°C or more, discomfort and complaints concerning dry air often appear, especially for air heated rooms or buildings. For this reason, measures should primarily focus on the temperature conditions. In case of complaints on dry air, it may also be advisable to go through the cleaning routines and examine the flooring materials to avoid problems associated with dust and static electricity. High humidity increases the risk of mold growth, and also increases gas emissions from materials and equipment.

To quantify the consequences in terms of comfort three major indices are often used: the PD_{Draught} (predicted dissatisfied due to draught), PMV (predicted mean vote), and its related PPD index.

To relate the physical parameters such as temperature, velocity, and turbulence intensity to the predicted comfort of the personnel, the PD Draught index can be used. The PD Draught index is defined as:

$$PD_{\mathrm{Draught}} = (34 - t_{\mathrm{oz}})(U_{\mathrm{oz}} - 0.05)^{0.6223}(3.143 + 0.3696U_{\mathrm{oz}} \cdot TI_{\mathrm{oz}}) \tag{10.1}$$

where PD_{Draught} represents the percentage of dissatisfied people due to draught in the head region.

In Fanger (1970) human thermal comfort has been described by the predicted mean vote (PMV) index. The PMV index represents the mean votes of a large group of individuals describing their thermal sensation using a seven-point thermal scale that ranges from hot to cold, with neutral at zero (ISO, 1994; ISO, 2005). The equation for the PMV index is defined by

$$\begin{aligned}
\mathrm{PMV} = {} & \left(0.303e^{-0.036M} + 0.028\right) \\
& \times \left[(M - W) - 3.05 \cdot 10^{-3} \times \left[5733 - 6.99(M - W) - p_{\mathrm{a}}\right]\right. \\
& - 0.42 \times [(M - W) - 58.15] \\
& - 1.7 \cdot 10^{-5} M(5867 - p_{\mathrm{a}}) - 0.001 \cdot 4M(34 - t_{\mathrm{a}}) \\
& \left. - 3.96 \cdot 10^{-8} f_{\mathrm{cl}} \times \left[(t_{\mathrm{cl}} + 273)^4 - (\bar{t}_{\mathrm{r}} + 273)^4\right] - f_{\mathrm{cl}} h_{\mathrm{c}}(t_{\mathrm{cl}} - t_{\mathrm{a}})\right]
\end{aligned} \tag{10.2}$$

where

$$t_{cl}=35.7-0.028(M-W)-I_{cl}\{3.96\cdot10^{-8}f_{cl}\left[(t_{cl}+273)^4-(\bar{t}_r+273)^4\right]+f_{cl}h_c(t_{cl}-t_a)\};$$

$$h_c=\begin{cases}2.38(t_{cl}-t_a)^{0.25} \text{ for } 2.38(t_{cl}-t_a)^{0.25}>12.1\sqrt{v_{ar}}\\ 12.1\sqrt{v_{ar}} \text{ for } 2.38(t_{cl}-t_a)^{0.25}<12.1\sqrt{v_{ar}}\end{cases};$$

$$f_{cl}=\begin{cases}1.00+1.290I_{cl} \text{ for } I_{cl}\leq0.078\text{m}^2\cdot{}^{\circ}\text{C}/W\\ 1.05+0.645I_{cl} \text{ for } I_{cl}>0.078\text{m}^2\cdot{}^{\circ}\text{C}/W\end{cases};$$

and

$$\text{PPD}=100-95\cdot e^{-(0.03353\times\text{PMV}^4+0.2179\times\text{PMV}^2)}$$

It is also possible to transform the PMV index into a statistical expression for the predicted percentage of dissatisfied (PPD) people [Eq. (10.2)]. The expression for the PPD index is based on empirical correlations and the equation reflects the fact that in large groups of people there will be at least about 5% dissatisfied no matter how the overall thermal climate is arranged.

10.1.2 Requirements

In the standard for the PMV and PPD index (ISO, 1994; ISO, 2005) it is suggested that the requirements for the physical variables: temperature, mean radiant temperature, relative air velocity, and partial water vapor pressure should be solved for PMV = 0. This is because when most people are predicted to be neutral in terms of their climate sensation, neither too hot nor too cold. The use of this concept may, however, result in large installations and high energy use. However, in the newer standard (ISO, 2005), this concept is less in focus. The use of the controlled indices instead of more adaptive models for free running buildings is due to the occupants' inability to control their climate, which is often the case in industry. In industrial cases the occupants often have been entitled to wear certain types of clothing, special clean clothing, and hairnets in the dairy case and boiler suits in the foundry case. Furthermore, the occupants don't have any way to control the indoor temperature or airflows in the premises.

10.1.3 The heat balance of a building

A building's heating requirement for local comfort is primarily determined by four main components: (1) transmission, (2) infiltration, (3) ventilation, and (4) internal gains. The transmission losses are related to the thermal performance of the construction such as insulation level, windows, doors, and gates. The infiltration of a building is related to how airtight the construction is, and to some extent how it is oriented toward the wind and local wind conditions. The ventilation losses are related to the need for process ventilation and outdoor air. The internal loads describe how much heat is emitted from processes and people in the premises. A buildings heat demand, excluding hot tap water, for example, for a year is described in Eq. (29):

$$E = Q_{\text{tot}} \cdot G_t = \left(\sum (U \cdot A) + \sum (\psi \cdot l) + (1-\sigma)\dot{v}_v \rho c_p + \dot{v}_{\text{inf}} \rho c_p\right) \cdot G_t \tag{10.3}$$

where Q_{tot} describes the building's specific loss term (W/°C) and G_t (°Ch) is the number of degree-hours for the current year. The building's specific loss term (Q_{tot}) includes the sum of the building's insulation performance (U * A), thermal bridges, $\sum(\psi \cdot l)$, ventilation losses, $(1-\sigma)\dot{v}\rho c_p$, and infiltration losses, $v_{\text{inf}}\ \rho c_p$. To determine the number of degree hours, we need to know the so-called balance temperature, T_b, which describes the temperature that the heating system needs to heat the building up to before the internal gains cover the rest of the heat demand. This is a way to include the effects of internal gains. P_g is the delivered heat output to the premises as a result of internal gains (i.e., the excess heat of the processes), and T_i is the indoor temperature:

$$T_g = T_i - \frac{P_g}{Q_{\text{tot}}} \tag{10.4}$$

Gt, degree hours, can be found in tables or can be calculated from a duration chart, and depends on the balance temperature, T_g, and the annual normal mean local temperature where the building is located.

10.1.4 Heating systems and components

Technical systems require continuous maintenance to function as intended, and, at the same time, the production processes often changes in industrial premises over time. This puts new demands on the technical systems. Energy efficiency of heating systems often works best if the organization sees it as cyclical. A possible way of including this process is recommended for large heating systems (>20 kW).

At local level, industrial heating systems can be constructed in a large number of different ways. For the office parts or other public spaces with normal ceiling height that are tempered, the systems are similar to the systems used in general office buildings both for heating and cooling. These can then often be designed as two-pipe systems, with room appliances such as radiators and chilled beams.

One of the most common ways of heating industrial premises is with the aid of aerotempers. In principle, an aerotemper acts as a convector. Either the aerotemper is supplied with heat from the hot water circuit or from an electric heating coil. When the temperature measured in the thermostat goes below the set value, the aerotemper fan (convector) starts blowing out heat in the room. For large industrial premises with high ceiling height heating strips are often used or the room is largely heated with ventilation. Also spot heating or incorporation of temperate spaces occurs.

Heat strips is a ceiling heating system, in the form of a panel. This type of technology is adapted to cope with high ceiling heights, where the purpose is to largely heat the occupied area. The heat is distributed mainly by radiation. The technology can be combined with any ventilation system.

10.1.5 Change-point model describing the performance of a building

Change-point models used for following or predicting building performance are becoming more and more of interest as the availability of measured data with high time resolution is increasing. One of the earlier articles on this topic is Hammarsten (1987). Hammarsten (1987) describes how a change-point model can be used to model building performance, how the time resolution of the included measurements affects the result, and how a model can be supplemented with parameters for solar and wind impact. The conclusion already at that point was that a change-point model is a good tool for estimating energy use.

Another major change is that information on energy use in buildings has gone from being used mainly for invoices to being part of the analysis of building energy performance and the validation of energy conservation measures (Rohdin et al., 2018). In Claridge et al. (1992) a description of how energy use data can be presented to see whether the building's energy systems are functioning properly, evaluate energy efficiency measures and investigate whether energy use can be reduced by changing the system's regulation. In ASHRAE (2002), a list of guidelines for the process of change-point models, based on two to five parameter breakpoint models, is described and categorized depending on use. Cooling for comfort purposes is usually modeled with two parameters (slope and breakpoint on the *y*-axis), while heat demand is analyzed with three parameters (slope, breakpoint, and temperature independent base load), but may also have a fourth parameter for slope after the balance temperature if there is a heat recovery system. A building or system with both heating and cooling is modeled with five parameters (two slopes, two breakpoints, and a base load). The presented models are illustrated in Fig. 10.1.

In Acquaviva et al. (2015) a method is presented, the Energy Signature Analysis (ESA), to collect, store, and process large amounts of data (so-called Big Data) in order to monitor energy efficiency. The method (1) investigates the efficiency of the heating system by comparing heating power with previously observed power at the same state, which may, for example, be at the same outdoor temperature, and (2) ranks the building compared to similar and nearby buildings with respect to energy performance. In Hitchin (2017), the current ISO standard for energy performance in buildings is reviewed, and Hitchin (2017) also compares calculated energy use based on a change-point model with the ISO-standard calculation and shows that the accuracy for total energy use is good but that the change-point model

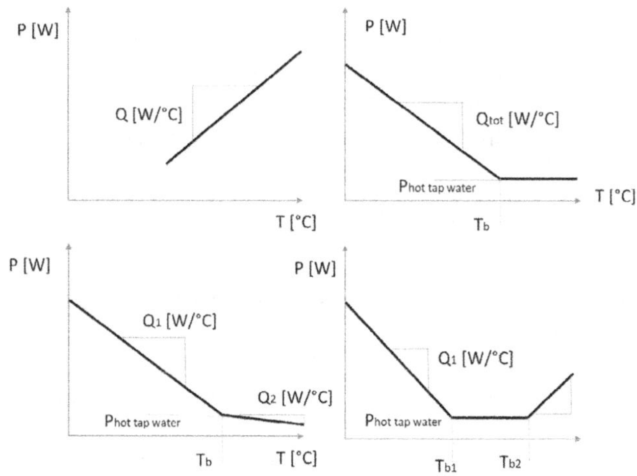

Figure 10.1 Illustration of change-point models with two to five parameters (ASHRAE, 2002; Rohdin et al., 2018).

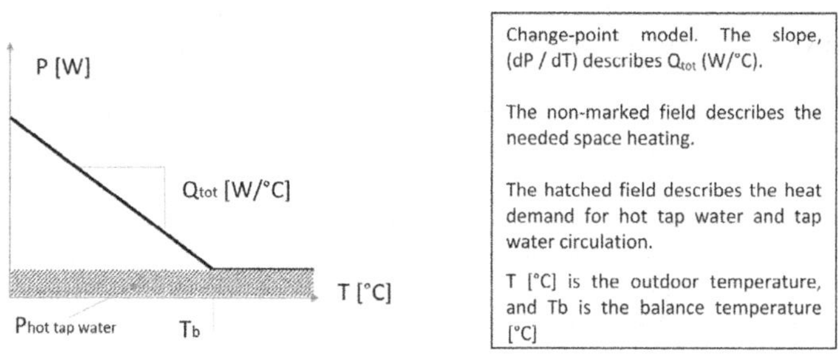

Figure 10.2 Description of the various factors in a change-point model (Rohdin et al., 2018).

provides more information, including the building's balance temperature, and is therefore preferred. The standard calculation generally under-predicts energy use and has larger variations. In Vesterberg et al. (2014), the first step in using a change-point model as a calibration tool to increase the accuracy of building performance simulation programs (BES) is investigated. A linear regression model is used to generate the specific heat loss and heat loss to the ground. The model is based on district heating data and electricity consumption for 2 years.

As an example, based on the top-right model in Fig. 10.1, with three breakpoints, two key factors can be quantified. The factors are (1) the hot tap water and hot water circulation, and (2) the total loss term for the building, Q_{tot} (W/°C). The quantified parameters are illustrated in Fig. 10.2.

The consequences of being able to calculate Q_{tot} are important, as Q_{tot} is an indicator of the building performance, as described in Eq. (10.3). Q_{tot} describes the sum of the (1) transmission losses through walls, windows and roofs, and cold bridges, (2) infiltration, and (3) ventilation. Thus, all the losses for the building. This means that this is a much better indicator of a building's energy performance than knowing the specific energy use in, for example, kWh/m^2 per year which is commonly used for benchmarking non-industrial buildings.

10.1.6 Common energy efficiency measures for space heating

Typical measures found for industrial energy audits are:

- Investigation of temperature requirements in the premises. Heat or cool only the parts where personnel or product require this.
- Heat or cool to a suitable temperature. (See the Swedish Work Environment Authority's statutory collection for suitable setpoints in different room types.)
- Temperature control. Heat or cooling, where appropriate, only those times during the day when needed.
- Type of room supply. Review if the systems followed with the change of the room's use and requirements. Is the right type of room system used for current operations?
- Segmentation. Divide the premises into parts where different requirements exist. Alternatively, if appropriate, move control and monitoring functions into the control room.
- Investigate measures related to the thermal performance of the building envelope, where appropriate. This can be, for example, window changes, roof insulation, or insulation of process pipes or other equipment.
- Reducing infiltration and improving air tightness, where appropriate. Reduce infiltration through industrial doors or climates. Modern industrial doors with steering make it often possible to operate the gates faster while maintaining safety. Review the use of air curtains.
- Review whether simultaneous heating and cooling occur. This is a common problem with often large potential.
- Strive for low temperature systems. Take a strategic decision within the organization, for example, max. 55°C for heating systems.
- Installation of reference sensors, to get feedback about the heating system and how it works.

10.1.7 Common energy efficiency measures for space cooling

Typical measures from an energy audit are:

- Investigation of temperature requirements in the premises. Heat or cool only those parts where personnel or product require this.
- If cooling units are installed, these should only be used when free cooling does not cover the need for cooling.
- Review whether simultaneous heating and cooling occur.
- Examine the possibilities of sectioning the cooling system.

- If there is a need for simultaneous heating and cooling demand, where the cooling requirement cannot be met with free cooling, one can investigate whether it is possible to activate the warm side on a heat pump and use the cold side to cool.

10.1.8 Nonenergy benefits in space heating and cooling

When assessing the available nonenergy benefits, the model developed by Nehler (2018) can be applied. Major nonenergy benefits for space heating and cooling include but are not limited to:

Production-related
- Improved quality of products
- Increased production reliability
- Increased equipment lifetime

Operation and maintenance
- Reduced maintenance

Working environment
- Improved worker safety
- Improved thermal comfort
- Reduced noise, reduction of noise can increase worker productivity but even more importantly, reduce days off work

Emissions
- Reduced emissions. By improved energy efficiency, emission reductions are achieved, for example, CO, CO_2, NO_X, and SO_X

Other
- Improved public/corporate image, for example, by showing customers energy efficiency improvements
- Improved worker morale, for example, by committing to improvements in working conditions
- Increased sales levels, for example, by cutting energy costs, revenue increases on a one-to-one basis, 1 cent in energy cost means 1 cent more in revenue for the company

10.2 Hot tap water

For buildings in northern Europe, there are mainly two ways of heating hot tap water. The most common way is with accumulator preparation. These are often used in single-family homes with their own boiler as well as in older district heating houses. The second most common type is using cross-flow heaters that are found in most district heating connected buildings. For industries, both types occur.

One of the reasons for regulating the temperature in our domestic hot water systems is the Legionnaires' disease. Legionnaires' disease is a type of pneumonia caused by a bacterium that is spread through small airborne and inhalable water droplets. The most common spreading cause is considered to be water mist which occurs in showers and air intakes on faucets. For this reason, a minimum temperature for hot water pipes or hot water circulation lines must be at least 50°C. Due to the scalding risk, the temperature should not exceed 65°C. The recommended temperature in hot water heaters is 60°C, and for cross-flow heaters 55°C.

10.2.1 Energy use for hot tap water production

In "Building guidance 8 – Energy conservation and heat insulation" it is stated that the heat consumption for heating hot water can be approximated to 55 kWh/m^3. An alternative way of calculating this is to start from the local water temperature. The expression can then be written as

$$E = \frac{\rho \cdot c_p \cdot (T_{vv} - T_{kv}) \cdot V}{3600} \tag{10.5}$$

where ρ is the water density (about 1000 kg/m^3); the C_p is the water's specific heat capacity (approximately 4.18 kJ/kg°C), the two hot water temperature (approximately 55°C), raise the cold water temperature (approximately 10°C or annual average temperature at the location), and the V is the volume of water during the analyzed time, for example, annually (m^3). This equation does not include hot water circulation or other losses.

In several of the projects carried out, related to energy use for hot water production in offices, it has been seen that the hot water circulation losses constitute a large part of the total tap water energy use for premises where there is a low tap water requirement. This also applies to industrial premises. Hot water circulation losses can then be at least as great as the heat requirement for hot water preparation in total. An alternative to conventional hot water preparation with hot water circulation can then be to install electric water heaters at different locations close to where the use is located, rather than having a long hot water circulation loop from a centrally located connection point.

10.2.2 Suggested points to check for hot tap water

- Examine the demand for hot tap water
- Measure the hot water circulation losses and determine whether local heating can be relevant
- Adjust the temperature of the tap water (55°C)

10.2.3 Nonenergy benefits in tap hot water

When assessing the available nonenergy benefits, the model developed by Nehler (2018) can be applied. Major nonenergy benefits for tap hot water include but are not limited to:

Operation and maintenance
- Reduced maintenance

Working environment
- Improved worker safety

Waste
- Reduction of wastewater

Emissions
- Reduced emissions. By improved energy efficiency, emission reductions are achieved, for example, CO, CO_2, NO_X, and SO_X

Other
- Improved public/corporate image, for example, by showing customers energy efficiency improvements
- Increased sales levels, for example, by cutting energy costs, revenue increases on a one-to-one basis, 1 cent in energy cost means 1 cent more in revenue for the company

References

Acquaviva, A., et al., 2015. Energy signature analysis: knowledge at your fingertips. IEEE International Congress on Big Data.

ASHRAE, 2002. ASHRAE Guideline: Measurement of Energy and Demand Savings. 140–143.

ASHRAE, 2003. Application Handbook. American Society of Heating, Refrigerating and Air-Conditioning Engineers, Atlanta, GA.

Claridge, D., Haberl, J., Sparks, R., Lopez, R., Kissock, K., 1992. Monitored Commercial Building Energy Data: Reporting the Results. ASHRAE Transactions: Symposia, pp. 881–889.

Fanger, P.O., 1970. Thermal Comfort: Aalysis and Application in Environmental Engineering. Danish Technical Press, Copenhagen.

Fanger, P.O., 1972. Thermal Comfort. McGraw-Hill, New York.

Goodfellow, H.D., Tähti, E., 2001. Industrial Ventilation Design Guidebook. Academic Press.

Hammarsten, S., 1987. A critical appraisal of energy-signature models. Appl. Energy 26, 97–110.

Hitchin, R., 2017. Monthly utilization factors for building energy calculations. Build. Serv. Eng. Res. Technol. 38, 318–326.

ISO, 1994. Moderate Thermal Environments—Determination of the PMV and PPD Indices and Specification of the Conditions for Thermal Comfort, ISO 7730: 1994. International Standard Organization, Geneva.

ISO, 2005. Ergonomics of the Thermal Environment—Analytical Determination and Interpretation of Thermal Comfort Using Calculations of the PMV and the PPD Indices and Local Comfort Criteria, ISO 7730: 2005. International Standard Organization, Geneva.

Nehler, T., 2018. A systematic literature review of methods for improved utilisation of the non-energy benefits of industrial energy efficiency. Energies 11 (12).

Rohdin, P., Milic, V., Wahlqvist, M., Moshfegh, B., 2018. On the use of change-point models to describe the energy performance of historic buildings, The 3rd International conference on Energy Efficiency in Historic Buildings (EEHB), Visby, Sweden.

Vesterberg, J., Andersson, S., Olofsson, T., 2014. Robustness of a regression approach, aimed for calibration of whole building energy simulation tools. Energy Build. 81, 430–434.

Further reading

Abel, E., Ekberg, L., 2002. Energieffektivitet, Rapport Effektiv 2002:01, Borås, Sweden.

Abel E., Elmroth, A., 2007. Buildings and Energy—A Systematic Approach. Formas.

Commtech, 2003. Achieving the Desired Indoor Climate: Energy Efficiency Aspects of System Design. Studentlitteratur, Lund.

Fanger, P.O., 2006. What is IAQ? Indoor Air 16 (5), 328–334.

Energy efficiency in internal transports and administration

11

Chapter Outline

11.1 Internal transport

Internal transport is a process which purpose is to transport raw materials, products under manufacture, and finished products inside an industrial plant. Internal transport is a necessity for a well-functioning production flow in an industrial company.

Internal transport is categorized as a support process, and its energy use constitutes only in a few exceptions a significant part of the energy use of a company, normally the internal transport energy use is low. For companies where, for example, a large part of manufacturing is automated, internal transport can, however, be a significant part of the company's energy use. Examples of different types of internal transport are rolling belts, traverses, cranes, and other lifting devices and various types of trucks. Several energy carriers can be used for internal transport where electricity and diesel are the most common.

Examples of energy efficiency measures for internal transport are as follows:

- Replacement of electric motor
- Replacement of electric drive
- Operations control
- Conversion from, for example, diesel to electric-driven trucks
- Correct pressure in the truck tires
- Improved production planning, which can lead to a reduced need for internal transport
- Optimized placement of storage rooms to minimize internal transport
- Measures to reduce waste disposal or reduce the level of post-processing, which means reduced energy use for internal transport
- automated transportation systems, e.g. in storage rooms.

Important to note is how the route affects the transports energy use. For example, if a route is established that involves a truck bringing material back from the unloading point to the loading site, a large part of the energy use for transport can be eliminated.

Introduction to Industrial Energy Efficiency. DOI: https://doi.org/10.1016/B978-0-12-817247-6.00011-0

Another important aspect is the truck's degree of filling. With today's often narrowly stocked storages, trucks often drive with low filling rates. This means that the amount of tied-up capital in storage can be reduced, but at the same time, it increases energy use for internal transport in the company.

11.1.1 Nonenergy benefits for internal transport

When assessing the available nonenergy benefits, the model developed by Nehler (2018) can be applied. Major nonenergy benefits for internal transport include but are not limited to:

Production-related
- Increased production rate
- Increased production reliability
- Increased equipment lifetime
- Shorter cycle time of process
- Raw materials use reduction

Operation and maintenance
- Reduced maintenance
- Reduced labor demand

Working environment
- Improved worker safety
- Reduction of noise
- Improved quality of air

Emissions
- Reduced emissions. By improved energy efficiency, emission reductions are achieved, for example, CO, CO_2, NO_X, and SO_X

Other
- Improved public/corporate image, for example, by showing customers energy efficiency improvements
- Increased sales levels, for example, by cutting energy costs, revenue increases on a one-to-one basis, 1 cent in energy cost means 1 cent more in revenue for the company

11.2 Administration

Administration is a process which purpose is to manage administrative purposes in an industrial plant. It is a necessary process for all industry activities. Administration is categorized as a support process and its energy use accounts for only a minor part of the energy use of a company. Examples of different types of administration that require energy are computers and servers, printers, and copy machines, but also coffee machines and microwaves. In Table 11.1 below, the most common operations for the support process administration are found.

Table 11.1 The most common administrative operations in a company with typical electric power demands (Nordenaudit, 2018).

Administration	Power (kW)
Computer (stationary)	0.065
Computer (lap-top)	0.020
Computer (server)	0.165
Printer	0.020
Copy-machine	0.020
Coffee brewer	2.000
Kaffebryggare	0.140
Soft drink dispenser	0.570
Vending machine	0.166
Water cooler	0.080
Freezer and fridge	0.033
Microwave oven	0.800

11.2.1 Nonenergy benefits for administration

11.2.1.1 Nonenergy benefits in administration

When assessing the available nonenergy benefits, the model developed by Nehler (2018) can be applied. Major nonenergy benefits for administration include but are not limited to:

Other

- Improved public/corporate image, for example, by showing customers energy efficiency improvements
- Improved morale of workers
- Increased sales levels, for example, by cutting energy costs, revenue increases on a one-to-one basis, 1 cent in energy cost means 1 cent more in revenue for the company

References

Nehler, T., 2018. A systematic literature review of methods for improved utilisation of the non-energy benefits of industrial energy efficiency. Energies 11 (12).

Nordenaudit, 2018. Energy audit software for industry. Nordic Energy Audit.

Energy efficiency through time control and energy deliveries

12

Chapter Outline

12.1 Time control – Load management and energy storage

12.1.1 Load management

Load management (alternatively power management) suggests that the company has strategies for controlling the load profile. There may be several reasons to control the load profile, such as limited energy transmission (the power grid can be a bottleneck) or limited power at the production plant (a boiler may be at its maximum power). Load control does not always result in reduced energy use at the company, but can, on the other hand, lead to reduced energy costs.

Load management may be divided into different techniques, such as peak clipping, valley filling, and load shifting (Pirkey and Eto, 1993). In the case of peak clipping, the peaks of the load profile are removed, for example, by turning off the engine preheaters for cars located on the staff parking lot. In the case of valley filling, the valleys of the load profile are filled with new needs, such as (future) charging of electric cars. During load shifting, the top of the load profile is moved to the valleys, for example, by running a process at another time.

Load shifting is most common and is usually synonymous with load management. Through load management it is possible to reduce possible peaks in the energy demand. Thus, any costs for the expansion of grids, power charges, or costs for increasing boiler capacity can be reduced or completely avoided.

Load management may be achieved in several ways:

- Through production planning, that is, planning the production so that energy-intensive processes do not operate simultaneously

Introduction to Industrial Energy Efficiency. DOI: https://doi.org/10.1016/B978-0-12-817247-6.00012-2

- By storing products shorter time
- By storing energy shorter time (via batteries, flywheel, and so on)
- By temporarily changing the energy carrier (if it is possible to operate a process with different energy carriers)
- By using load prioritization systems, that is, use systems prioritizing processes to operate at temporary peaks

An added value of applying load management is that it gives an overview of how the plant works, which means that any deviations may be illustrated.

12.1.2 Energy storage

The use of energy storage is an approach for the company to increase energy efficiency. Energy storage occurs in various forms in industry, for example, ice water baths and accumulator tanks.

In addition to *chemical storage*, there are two principal ways of storing heat and cold: *sensitive* and *latent storage*.

- *Sensitive storage* means that heat or cold are stored by changing the temperature in the stored medium. *Sensitive storage* is the most common type of heat storage, and occurs in, for example, district heating systems where the temperature in the medium is raised before heat is delivered to the customer.
- During *latent storage*, a phase conversion takes place in the stored medium. One example is ice storage, where a large cooling effect is obtained during ice melting while heat is taken from the surroundings. Another example is (molten) salt energy storage, where heat is released when the salt reacts with water.
- *Chemical storage* means that chemical reactions occur, for example, via absorption and adsorption. In a reversible process, energy is released again.

There are several different types of heat storage. The most common heat storages are (Thollander et al., 2019):

- *Accumulator tanks* primarily store hot water. The most common example of an *accumulator tank* is the water heater. *Accumulator tanks* can either be pressurized or without pressure.
- *Pit storages* store hot water in a hole in the rock. A *pit storage* consists of a buried or excavated hole in rock or soil with an insulating cover, under which hot water is stored.
- *Pool storage* is essentially similar to pit storages, except that it is located indoors.
- *CTES* (cavern or mine thermal energy storage) is used seasonally. Hot water is supplied to the storage, in order to be used later when needed.
- *ATESs* (aquifer thermal energy storages) are groundwater reservoirs. Since an aquifer is natural, the cost is lower for constructing the storage than for other types of storages.
- *Borehole storage* mean that heat in the form of hot water is pumped down into a borehole to store it for later use. This type of storage can either be open or closed.
- *Salt hydrate technology* or simply *salt energy storage* utilizes phase conversion from water vapor to water and vice versa.

12.2 Energy deliveries – Industrial excess heat and internal electricity generation

12.2.1 Industrial excess heat

Excess heat is also called waste heat, surplus heat, residual heat, low-grade heat, and secondary heat, and these terms are often used interchangeably. There are some attempts to make a definition of the concept of industrial excess heat. According to one definition, so-called real industrial excess heat is the "heat that is left over" when an industrial process has been optimized (thermodynamically) (SEA, 2008). A weakness with this definition is the difficulty of knowing when a process is optimal. According to another definition, excess heat is the heat "... that cannot be used directly in the industrial process" (SEA, 2008). In the latter definition, it is instead possible to question whether there is a restriction on how inefficient a process may be. In an IEA project concerning excess heat (IEA, 2019), IEA suggests a consistent definition of excess heat as the heat content of all streams (gas, water, and air) which in any given moment is discharged from an industrial process. In addition to this definition IEA adds information regarding what can be (internally or externally) usable heat (technically and economically).

Despite difficulties with defining industrial excess heat, the concept of excess heat is apparently used in various applications. Industrial excess heat basically has three uses: (1) it can be used internally in the excess heat-generating process, (2) it can be used internally in the plant or externally, and (3) it can be used for electricity production. If the excess heat cannot be used internally at the company where it is produced, it is sometimes called true excess heat, or unavoidable excess heat. If the excess heat cannot be used externally, it can also be called waste heat, which thus gives a distinction between the concepts of excess heat and waste heat (IEA, 2019).

There are several examples of practical uses of excess heat, such as direct heating of internal processes, heating of processes in other industries, heating of greenhouses, heating of biogas plants, supply of heat to district heating networks, production of district cooling via absorption cooling machines, and electricity production. In the case of electricity production, Rankine cycle[1] may be used provided the excess heat has a sufficiently high temperature. Other power generation possibilities, for excess heat of lower temperature, include, for example, Organic Rankine Cycle (ORC), thermophotovoltaic (TPV), and phase change material (PCM) (Broberg Viklund and Johansson, 2014).

In Sweden, for example, there are plenty of collaborations of excess heat exchanges between industrial companies and district heating companies. However, there are examples of places where cooperation has not been realized, despite large

[1] The Rankine cycle, also called the steam cycle, is common for electricity production when burning fuels, for example, biomass and coal. It describes the process when steam is formed by water in a steam boiler (where fuel is burned), and where the steam is then expanded in a steam turbine (which rotates an electric generator which in turn produces electricity) and further to a condenser where the steam condenses. The condensate is then passed through a feedwater pump, which pressurizes the water (condensate), and this water is then vaporized in the boiler, and the cycle is closed.

excess heat amounts. The reasons are several (see e.g., Broberg Viklund, 2015). The existence of people with real ambition can be a crucial driver for cooperation to occur as well as the existence of (municipality owned) energy companies that may have a holistic approach, that is, not only focus on benefits for its own organization. Possible obstacles to cooperation may be, for example, that the company's flexibility decreases as the company become constrained by contracts, that there are cultural differences between the companies, that excess heat is a peripheral activity for the industry but core activities of the district heating company, that district heating companies want their own facilities, and that the excess heat is not available at the right time. Furthermore, there is always a risk that the excess heat generating company will be closed down, and then the district heating company will be without heat deliveries.

In some cases, the excess heat can be used directly and is then called primary excess heat. In other cases, the excess heat temperature must be raised before use and is called secondary excess heat. To make the excess heat available, heat exchangers, heat pumps (to increase the temperature), and radiation collectors are required. In some cases, it may also be necessary to store the excess heat due to the fact that needs and access do not match in time.

There is no reliable statistics on the total use of excess heat in Sweden. The proportion of delivered industrial excess heat to district heating networks, on the other hand, is documented and it amounts to about 10% of the district heating deliveries (about 5 TWh of primary excess heat annually). There are also no reliable potential estimates. Some estimate the industrial annual excess heat potential to between 6 and 8 TWh/year (primary), while another estimates the annual potential to between 20 and 50 TWh/year (primary and secondary) for temperatures that could be used in district heating networks. As the figures indicate, there are excess heat exchanges between industrial companies and district heating companies. The first collaboration in Sweden started in 1974 in the city of Helsingborg and today there are between 60 and 70 collaborations in cities (including, for example, Gävle, Borlänge, Gothenburg, and Varberg).

The use of excess heat reduces the use of other fuels. It is often assumed that the use of excess heat does not yield any CO_2 emissions, that is, the excess heat is considered as a by-product of industrial production. However, this approach can be questioned, especially if the industrial energy use increases to meet excess heat deliveries, or if the industrial process is not optimized. If the extra fuel required to deliver heat comes from the burning of fossil fuels, it may be difficult to argue for this approach. Excess heat can also be divided into subcategories, for example, green excess heat where the heat originates from biomass, which takes this problem into account. Other examples of subcategories are technically and economically useful heat (internally and externally).

Sometimes gases are included in the concept of excess heat, which are residual products from industrial processes (e.g., from the iron and steel industry, refineries, and petrochemical industry). In this case, the gases are not thermal excess heat directly, since it is not the heat content of the gas used, but the heat produced when the gas is burned. In some potential assessments, especially continental Europeans, heat from condensation plants that produce electricity, for example, coal condensing plants, are also considered as excess heat. Thus, other excess heat is also covered except the excess heat that is the result from industrial processes. In countries where district heating systems are well-developed, heat is generally used after electricity has been produced, in so-called cogeneration plants. This is true in, for example, Sweden where there are many cogeneration plants.

12.2.2 Internal electricity generation

Excess heat can be used to generate electricity. If the excess heat temperature is sufficiently high, the Rankine cycle (the steam cycle) can be applied. According to this, water is used as a working medium, which is pressurized in a feedwater pump and then delivered to a steam boiler where the water is heated and evaporated. The heating and evaporation are usually done by burning a fuel in a boiler, but also excess heat of sufficient high temperature can be used. The steam expands in a turbine, which in turn drives an electric generator. After passing the turbine, the working fluid is cooled in a condenser and transferred to the liquid phase, after which it is passed on to the feedwater pump, and the working cycle restarts. The condenser can in some cases be used as a heat source for, for example, district heating. This is called cogeneration or combined heat and power generation (CHP) (for deliveries to district heating networks) and back pressure [(steam) turbine] generation (for deliveries within the industry). In these cases, the expansion of the steam in the turbine must stop at a slightly higher pressure, which means that the electricity production will be somewhat lower. However, the overall efficiency is higher. In some companies, there is a need for steam of different pressures, which means that the turbine is designed to release steam at these pressures. This also leads to lower electricity production.

If the excess heat temperature is lower, the so-called ORC may be used. The principle is the same as for the Rankine cycle, but the working medium is not water but an organic medium that evaporates at lower temperatures. Various organic working media with different evaporation temperatures can be used. ORC has a lower efficiency than the Rankine cycle, which is because the temperature is lower. However, more excess heat sources with lower temperature can be used.

It is also possible to produce electricity from excess heat by using a Stirling engine, where the working medium is a gas which alternatively get heated and cooled. This is carried out with, for example, a two-cylinder engine with a cold and a hot cylinder. The working medium shrinks in the cold cylinder and expands in the hot, resulting in a movement of the gas. This in turn leads to the cylinder pistons moving, and since they are mechanically connected to an electric generator, electricity can be produced. Another method of producing electricity from excess heat is to use PCMs. In this case, a paraffin mix is used as the working medium. Solid paraffin is heated with excess heat to form liquid and then expands under high pressure. Liquid paraffin is then cooled to solid form, and at the same time shrinking. The mechanical energy created by expansion and shrinkage is utilized in a hydraulic system and then converted into electricity in an electric generator. Another method for electricity production of excess heat is to heat one of the ends of a semiconductor and at the same time cool the other end, where the ends have different so-called Seebeck coefficients (thermoelectric potential differences). This method, called thermoelectric generator (TEG), causes electricity to be produced.

Furthermore, solar cells can be used to produce electricity from radiation heat originating from industrial processes and products, so-called TPVs. By installing solar cells on the plant's roof or putting up solar cell parks, solar radiation can be utilized. One reason to implement such measures is to improve the company's goodwill or green image. Another reason mentioned by companies that installed solar cells is to gain control of the electricity costs.

References

Broberg Viklund, S., 2015. Energy efficiency through industrial excess heat recovery: policy impacts. Energy Efficiency 8, 19. Available from: https://doi.org/10.1007/s12053-014-9277-3.

Broberg Viklund, S., Johansson, M.T., 2014. Technologies for utilization of industrial excess heat: Potentials for energy recovery and CO2 emission reduction. Energy Convers. Manag. 77, 369–379. Available from: https://doi.org/10.1016/j.enconman.2013.09.052.

IEA (International Energy Agency), 2019. IEA IETS, Annex XV: Industrial Excess Heat Recovery – Technologies and Applications, Supported by Austria, Canada, Denmark, Germany, Italy, Norway, Portugal, and Sweden, Final Report – Task 2.

Pirkey, D., Eto, J.H., 1993. Demand-Side Management Data Collection Activities: The Role of A National Government. LBNL, Berkeley, CA.

SEA (Swedish Energy Agency), 2008. Analys av metoder för att öka incitament för spillvärme-samarbeten. ER2008:16, Statens energimyndighet. [In Swedish].

Thollander, P., Karlsson, M., Rohdin, P., Rosenqvist, J., et al., 2019. Energieffektivisering - Energikartläggning, energiledning och styrmedel (Energy Efficiency – energy audits, energy management and policies. Studentlitteratur, Lund [in Swedish].

Part II

Energy management gap

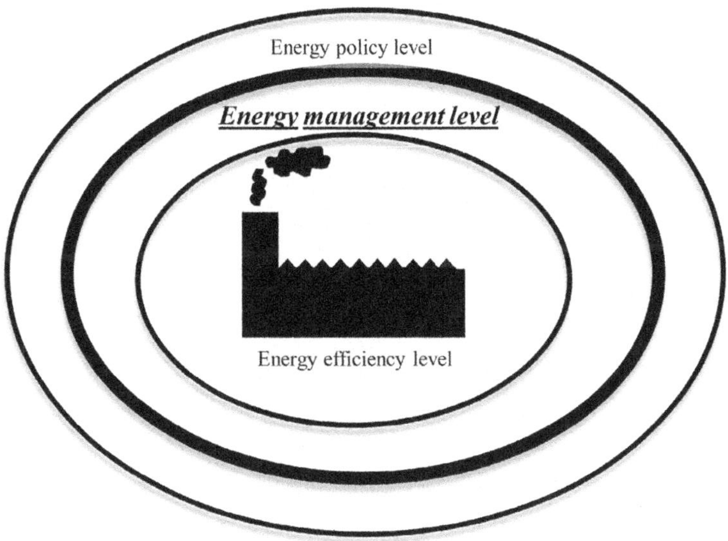

By continuous energy management practices in a company, and not just implementing new technology, the rate of energy efficiency is further improved, for example, energy efficiency potential is higher if energy management, organization, and behavior are included. This additional degree of energy efficiency is called the energy management gap. Part II focuses on energy management and what prevents improved energy efficiency in industry and is based on the so-called energy management gap.

Energy management

13

Chapter Outline

13.1 Energy management in industry[1]

Previous research has shown that barriers to energy efficiency in the industrial sector such as cost and risk of production disruptions, lack of access to capital/budget funding, lack of time and other priorities, other priorities for capital investments, slim organization, and lack of submetering, play an important role in explaining why energy efficiency investments that are cost-effective are not implemented (Thollander and Ottosson, 2008; Rohdin and Thollander, 2006), please see Chapter 15 for a further elaboration on barriers to energy efficiency. Energy management can be seen as an important means to overcome barriers to energy efficiency.

Research by Caffal (1996) has shown that industrial companies who adopt energy management practices may improve energy efficiency by up to 40% of their

[1] The text in this subchapter was originally published in Thollander and Ottosson (2008).

Introduction to Industrial Energy Efficiency. DOI: https://doi.org/10.1016/B978-0-12-817247-6.00013-4

total energy use. Both top management's wholehearted support and a strategic approach are of utmost importance if an energy management program is to succeed.

It is important to note that the goals in an industrial energy management system are on a lower organizational level than business energy strategies. While a strategy deals with how company leaders try and establish a direction for the organization and include predetermined courses of action and goals, improving energy efficiency and reduce energy costs using industrial energy management could be one of many goals within such a strategy (Mintzberg, 1987).

Reviews of the scientific publications on the topic of energy management in industry have been published by Schulze et al. (2016) as well as Trianni et al. (2019), and for Swedish industry, in particular, by Johansson and Thollander (2018).

13.2 Energy management gap—energy management provides improved energy efficiency

Energy management in a company involves the implementation of energy-efficient technology but also the implementation of processes on how technology and processes are used at the company. By initiating energy management in a company, which thus not only regards new technology implementation, the potential for energy efficiency will increase, see Fig. 13.1. This additional degree of energy efficiency is called the energy management gap.

Work towards more efficient end-use of energy focuses mainly on new technology and new system solutions. However, what is not emphasized to the same extent

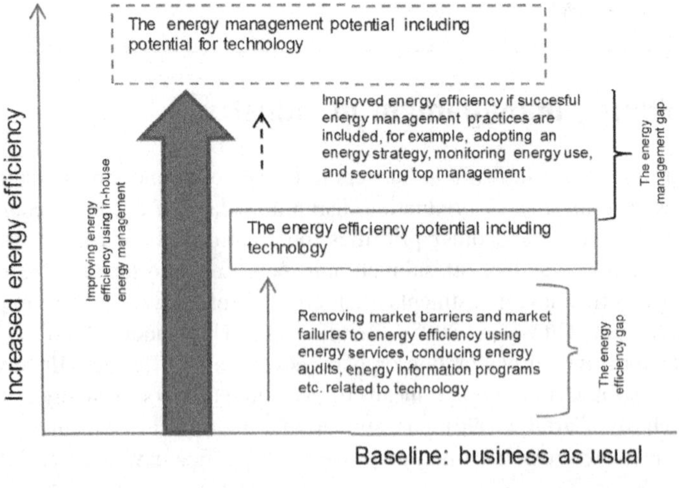

Figure 13.1 Energy efficiency potential if energy management is also included in the company (Thollander et al., 2013).

is management and organizational issues related to new technology, new processes as well as new system solutions. Early international research on energy management in industry shows that for manufacturing companies, deployment of in-house energy management practices can improve energy efficiency by between 4% and 40%, a result that can be attributed to both technology and management operations, that is, the in-house energy efficiency work should preferably include both technology and organization-related actions (Caffal, 1996).

13.3 Model for energy management

Since 2011 there is an international ISO standard, ISO 50001, for energy management. A revised standard came in 2017. Since the standard for energy management is general, the approaches to how energy management can be carried out depend on the specific company's characteristics. An energy-intensive integrated pulp mill with continuous pulp production and energy use of several TWh per year cannot be compared with a medium-sized non-energy-intensive engineering industry with discontinuous batch production and annual energy end-use of less than 1 GWh per year. The conditions are simply diametrically different. In both cases, it is essential to have knowledge of the internal energy-using processes, but more so for the pulp mill as the production is way more complex and also due to that energy efficiency improvements in relation to process knowledge can deliver very large cost savings for the pulp mill. Fig. 13.2 describes how the approaches can differ depending on how the organization accepts knowledge (the y-axis, absorption capacity) of the internal energy system and the production processes.

As shown in Fig. 13.2, an energy audit with suggested energy efficiency improvement measures can be an adequate approach for companies where the technology is not so closely intertwined with other processes and therefore can easily

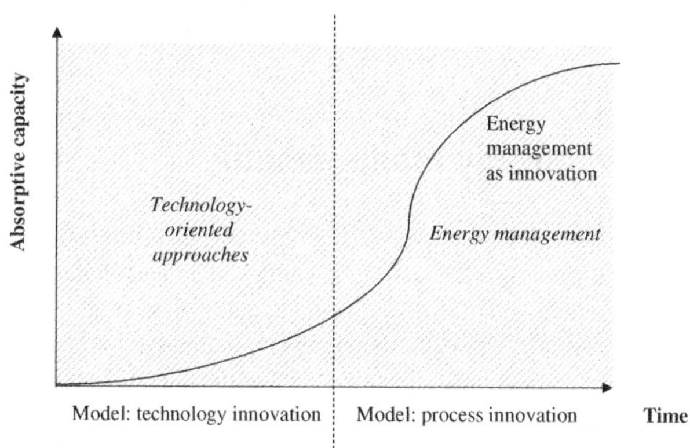

Figure 13.2 Model for energy management (Paramonova and Thollander, 2019).

be replaced. In these cases, the left part of the figure—technology innovation—is a good start.

For energy-intensive companies with complex production and support processes that are closely interconnected, collaboration between people and departments are often required to identify more energy-efficient ways of production operations. In these cases, the right-hand part of the figure—process innovation—is a desirable approach and a good target for the companies that have begun working within the framework of technology innovation.

13.3.1 Technologies and processes' complexity

One advantage—and risk—of investing in new energy-efficient technology is that the focus is on the technology, such as more energy-efficient electric motors and fans, that is, different types of products with a clearly defined system boundary. The risk is that the energy system as a whole is overlooked, that is, the focus on the new technology's efficiency improvements leads to suboptimization of the system as a whole. One example of this is air compressors which, given a procurement of heat exchangers to utilize excess heat from the cooling of the compressor, may create a technology lock-in. A new installation of this type of heat exchanger can lead to the incorrect perception that the compressed air system is optimal from the point of view of energy efficiency. Heat exchange, however, is the last step to be taken in the energy management work of an industrial compressed air system, see Chapter 9 on compressed air.

If the system boundary is widened in connection with the procurement, the chances of the investment becoming more successful from an energy efficiency point of view increase, but it also becomes more complex, and demands more thorough process knowledge. Increased demands are thus placed on the ability to understand and comprehend the company's energy systems and production processes and their interactions.

Procurement and implementation of industrial production systems and processes require careful planning. Installation of less complex types of technology, such as the installation of an electric motor, may be considerably easier to procure.

13.4 Effective energy management

As energy use, just like other factors, varies between different companies, it is necessary to adapt the approaches for efficient energy management, taking the specific company's characteristics into consideration. It cannot be emphasized enough that the potential for successful energy management depends, among other things, on the energy use of the process and whether it is continuous or batch-wise.

There are two fundamentally different ways of reducing energy costs for a manufacturing industry: more efficient supply of energy and more efficient use of energy:

- Focus on more cost-effective energy supply, for example, by negotiating the cost with the energy supplier, via load management, through investments with respect to supply, or by delivering excess heat or electricity.
- Focus on more cost-effective use of energy, including both technology and management measures.

On the part of use of energy, there are, as previously mentioned, four principally different means to reduce energy costs, namely:

- Operation changes, for example, more efficient operating and maintenance routines and changed behavior
- Investments in energy-efficient technologies
- Time control, for example load management
- Conversion from one energy carrier to another, for example, from electricity to district heating

The energy cost perspective is the perspective that is mainly considered in companies. Another perspective is the energy use perspective. In many cases, energy use (not the cost) is a limitation for a company's production. For example, an industry can be limited by an environmental permit from the government in relation to a maximum level of energy use per unit produced. There may also be limitations on emissions or on total number of produced products. Another perspective, the climate change perspective, can be a driving force for efficiency improvement. A company, for example, in its quest to limit its production's negative climate impact can also drive energy efficiency improvements.

13.5 In-house energy management

A company can use either an energy management system or an environmental management system in its internal energy management activities. Energy management is a way of working structure with the energy issue and a standardized energy or environmental management system is a tool that can be applied. However, it is important to note that a standardized management system as a tool, in order to be successful, requires anchoring in the organization. Research among industrial companies has shown that there are a number of general success factors that affect whether the company's energy management program will succeed or not (Johansson and Thollander, 2018). These success factors are as follows:

1. Full top-management support of the in-house energy management activities
2. The existence of a long-term energy strategy, preferably with quantified goals
3. A two-step energy plan: one covering 1-year, and one covering multiyear periods
4. A clear energy manager position, but not necessarily a full-time position
5. Real energy cost allocation based on submetering, not just energy costs allocated per square meter or number of employees
6. Clear KPIs (key performance indicators) which enable follow-up of results
7. Energy controllers at floor-level position, that is, one person per shift responsible for energy-efficient operation

8. Continuous energy efficiency education for employees
9. Visualization of energy efficiency progress at company level as well as at division level
10. Energy competition between divisions within the same company, encouraging improved energy efficiency

It is important to note that a well-implemented energy audit is a first very important step for initiation of energy management practices for the majority of companies.

13.5.1 Example: Swedish foundry industry

An energy audit in a company led to the company's CEO forming an energy group that met once a month. As the CEO made sure to always be present at these monthly meetings (and thus signaled that the energy issue was a high priority for him), all other managers also attended the meetings. This, in turn, led to the fact that the energy issue, for example, via the department heads, were also anchored at lower levels in the organization. As a result, electricity consumption in iron smelting decreased by 10%. Because the furnaces were charged immediately after emptying, the temperature of the iron that would be melted increased by about 100° thanks to the hot lining. Furthermore, LPG consumption decreased by 30% thanks to using the ladle preheating only when needed, that is, using a routine to preheat ladles sometime before the melting furnace was ready to be emptied (Thollander et al., 2008).

13.5.2 Methods for energy management

There are many different methods that all intend to promote energy management and more efficient energy use. Table 13.1 lists some of these methods with a connection to which main obstacles to energy efficiency as the methods mainly contributes to overcome.

13.5.2.1 Energy management standard and energy audit standards

The international standard for energy management, ISO 50001 is described later in this chapter, while the international standard for energy audit, ISO 50002, and the European energy audit standard EN 16247 are described in Chapter 5.

13.5.2.2 Energy audit software

Nordenaudit, a software for industrial energy auditing, launched in 2012, uses the unit process concept to create structure when the energy balance in the audit mapping is established (See Chapter 4 for information regarding the unit process concept). Fig. 13.3 shows an example of Nordenaudit's user interface.

Table 13.1 Tools for successful in-house energy management. The tools are selected based on a Swedish context (Johansson and Thollander, 2018).

Type of tool	Name	Could reduce barrier category	Reference
Energy management standard	ISO 50001	Organizational, competence	International and national standardization bodies
Energy audit standards	EN 16247; ISO 50002	Information, competence	International and national standardization bodies
Energy audit software	Nordenaudit	Information	Nordic Energy Audit (2016), Thollander et al. (2012)
Energy monitoring systems	Many actors on the market	Information	
Pinch analysis tools	ProPi; EINSTEIN	Information	Chalmers Industrial Technology (2016), Intelligent Energy Europe (2016)
Industrial energy system optimization	reMIND, MARKAL, TIMES, etc.	Information	Karlsson (2011), IEA (2011), Krzemień (2013)
Manufacturing simulation	Various software available	Information	SIMPLAN (2016), Solding (2008)
Exergy analysis		Information	Grip et al. (2011), Grip et al. (2012)
Investment calculation tools	Many actors on the market	Information	
Energy services	Various type of contracts offered by national ESCOs	Organizational, information, economic, competence	Apeaning and Thollander (2013), Christoffersen et al. (2006), Kindström et al. (2016), Brunke et al. (2014), Thollander and Ottosson (2008), Thollander et al. (2013)
Lean	Integrating energy efficiency in Lean	Organizational, behavioral	Johansson et al. (2011)
Energy efficiency network management systems	LEEN	Organizational, competence	Koewener et al. (2011), Paramonova et al. (2014)
Database of real energy efficiency measures	IAC, DEFRAM, NEAD	Information	IAC (2016), Blomqvist and Thollander (2015), NEAD (2018)
Education and training	Many actors on the market	Information, competence	

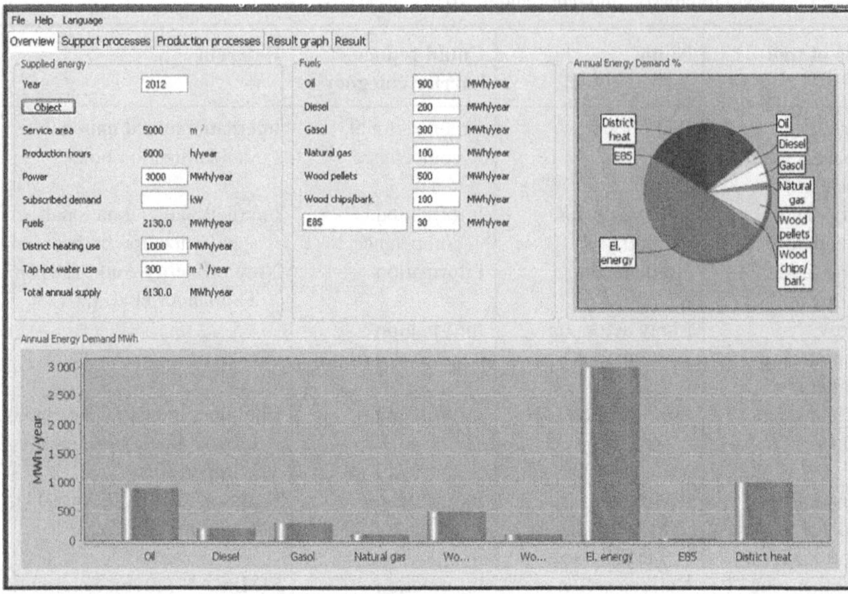

Figure 13.3 The industrial energy audit software Nordenaudit.

13.5.2.3 Energy monitoring system

There are several energy monitoring systems on the market, ranging from monitoring systems that relate to a specific installation to systems that monitor the company's entire operation and energy use and systems where the energy end-use of a company or group of companies can be analyzed and compared at regular intervals.

13.5.2.4 Process integration

An industrial process is usually a complex system consisting of different units, for example, reactors, separators, heat exchangers, and boilers which are connected to each other and interact during operation. Analyzing such a system in order to reduce energy use is complicated, and consequently, systematic methods are needed. Process integration consists of several methods that have been developed for this purpose, focusing on system solutions of processes and interaction between different subprocesses. Process integration, according to IEA's definition is: "Systematic and general methods for designing integrated production systems, ranging from individual processes to total sites, with special emphasis on the efficient use of energy and reducing environmental effects" (Gundersen, 2000).

Three methods are usually included in process integration: pinch analysis, mathematical programming (optimization), and exergy analysis. These methods correspond to level III audit according to ASHRAE's and the EN 50002 categorization, which means an in-depth study, for example, energy system optimization where data from a level II energy audit is used.

There are several reasons for using process integration methods, such as:

1. That an improvement in a subprocess does not always lead to an improvement of the entire system.
2. That the more complex a system is, the greater the probability of finding new solutions by a systematic method.
3. To enable comparisons of solutions to achieve different goals and to make trade-offs between conflicting goals.
4. To seek answers to strategic issues, new processes, changes, investments, etc.

13.5.2.5 Pinch analysis

Pinch analysis is a systematic tool for investigating the possibilities of heat integration (process integration) between processes (Linnhoff et al., 1982) and is most common of the three process integration tools. In a pinch analysis, streams that need to be heated are coupled to streams that need to be cooled. Between these streams, heat exchangers are inserted, which forms a heat exchanger network of different streams. With pinch analysis, it is possible to calculate the theoretically minimum possible external heat requirement for a heat exchanger network (i.e., the amount of heat that must be added to the system via, e.g., external boilers) and the amount of heat that must be cooled off in the network (i.e., the amount of heat that must be cooled off in the system via, e.g., cooling towers). It is also possible to determine the amount of heat that can be exchanged internally. Different curves are developed from which the system can be analyzed in different ways. From the composite curve diagram (CC), minimum possible external heating and cooling requirements, theoretically, are obtained. The same information is available from so-called grand composite curves (GCC) and in GCC foreground and background analyses can be made to identify possible integration of processes in an existing system. There are also other, more advanced pinch curves (Nordman and Berntsson, 2005).

13.5.2.6 Industrial energy optimization

Optimization (or mathematical programming) (see e.g., Rardin, 1998) is a process integration method. When optimizing, the problem is expressed with an objective function and constraints. The objective function should either be maximized or minimized. This can be, for example, maximizing the company's profit, or minimizing costs or CO_2 emissions from energy use. It is also possible to use a multiobjective function, where different objectives are weighted against each other. The objective function includes variables such as flows, temperatures, or pressures. The values of the variables are searched for when optimizing and consequently they influence the value of the objective function. These variables (and others) are also included in the constraints. The constraints allow the variables to assume certain values but exclude others. The constraints may, for example, relate to maximum flow in a process and describe the need for energy use in the process relative to the material flow in the process. When an industrial energy system is analyzed, the number of constraints can amount to several thousand. The problem formulated can be linear or nonlinear.

If some constraints are nonlinear, they can be linearized using integers. The problem formulated (i.e., the objective function and the constraints) is solved by an algorithm, which is governed by whether the constraints of the problem are linear, nonlinear, or linearized with integers.

13.5.2.7 reMind—decision support tool

reMIND is a decision support tool whose software is based on the MIND method (method for analysis of industrial energy systems). The tool is based on mixed integer linear programming, which means that nonlinear functions are linearized using integers. So far, mainly industrial energy systems have been modeled, but other types of systems can also be modeled provided that mathematical relations can be formulated. The tool has been used for the following type issues: optimal operating strategy, improved structure of processes, and the influence of different boundary conditions. The results include operation of, for example, boilers, storages and turbines, system costs, materials and energy flows, and investments. In the tool the dynamics can be illustrated by dividing a specific time period into smaller parts. (Karlsson, 2011).

13.5.2.8 Exergy analysis

An exergy analysis aims to minimize the exergy losses and thereby maximize the amount of work produced in the system. Thus, the exergy analysis intends to identify and evaluate quantitative causes of thermodynamic deficiencies in the energy processes and thereby provide information regarding improvement potential. In an exergy analysis, exergy into the system is always greater than exergy out of the system, due to the system's irreversibility. Additional exergy losses occur, such as unused exergy, which are exergy flows to the environment. These exergy flows to the environment constitute the improvement potential and are therefore important to identify in order to improve the energy efficiency of the system. By examining the exergy at different points in a series of energy conversion steps, the efficiency of energy and losses can be determined (Gong, 2004).

13.5.2.9 Manufacturing simulation

A simulation is an imitation of real processes or systems over time. Manufacturing simulation offers greater understanding of the system as a whole and can, among other things, identify bottlenecks, reduce operating costs, reduce downtimes, and make quick operation schedules. Simulation is not included in the process integration toolbox.

13.5.2.10 Example: manufacturing simulation

The foundry industry in Sweden was among the first to use manufacturing simulation for improving energy efficiency. One of the foundries, which began using manufacturing simulation in the early 2000s, wanted more efficient production planning. The company perceived this early adoption of manufacturing simulation to be invaluable but also that result could be improved if a more detailed manufacturing simulation model could be developed.

The company perceived that the manufacturing simulation was very valuable because of the large number of items and due to that many downtime occurrences when changing items. However, a foundry with fewer items and fewer downtimes would not achieve savings in the same order of magnitude. The company stated that by using the manufacturing simulation model during production planning, the energy costs in the processes involved (melting, shaping, molding, and cooling) could be reduced by about 10%–15%. It is important to note that this at the same time may face the risk of increasing lead times in the process, which counteracts the principles of Lean.

13.5.2.11 Investment calculation tools

There are primarily four ways to calculate whether an investment is profitable or not:

- Straight payback
- Net present value (NPV)
- Internal rate of return (IRR) and,
- Life cycle cost (LCC)

These four methods are described in Chapter 7 of the book.

13.5.2.12 Energy services

One market-driven approach for promotion of improved energy efficiency in industry and something that has been recognized in most developed countries are energy services and energy service companies, sometimes abbreviated as ESCO (energy service company). Energy services is a collective name for various form of energy services, each of which is aimed toward making energy use more efficient. Energy services include both advisory services, so-called indirect energy services, and operational services, as well as services for financing investments in more energy-efficient technologies and processes, so-called direct energy services. Examples of indirect energy services are energy audits and analysis of energy invoices and hourly data for a company's electricity consumption, as well as identification of energy efficiency measures. Examples of direct energy services are operation and maintenance and hands-on support for implementation of energy efficiency measures (Bergmasth and Strid, 2004).

The energy service companies' revenues are often based on how the energy efficiency of the studied company has changed, but fixed remuneration also occurs.

Energy services have mainly been used in the public sector and on building energy efficiency improvements. Table 13.2 describes the various types of energy services.

Table 13.2 Various types of energy services (SEA, 2013; Kindström et al., 2016).

	Type of energy service	Explanation
Basic energy services	Direct help to improve energy efficiency	This form of service studies data on energy use over time
	Energy audits	An energy audit can be done of either an individual process, as well as an entire company or industrial group. Large companies require that an audit is made by a certified energy auditor
	Enery analysis and advice	Energy companies as well as consultants as by energy companies
Advanced energy services	Direct improvement of energy efficiency by the customer	Direct help to improve energy efficiency
	Financing of investments by the customer	Direct help in financing an investment
	Operations and maintenance	Agreement linked to operation and maintenance at customer
	Functional contracts (e.g., contracts agreeing on a set indoor temperature)	The most complex form of energy agreement that defines function and performance
Others	Energy declarations	According to Swedish law, energy declarations must be carried out for all buildings that are not industrial properties. An energy declaration must be made by a certified energy expert

13.5.2.13 Lean

Lean is a methodology and a working method inspired by the Toyota Production System (TPS), which in turn is based on the Toyota Way where the company's values are described. Lean means, among other things, that waste is to be eliminated, for example, inefficient energy use, lead times should decrease, and the company should work with continuous improvements. In the Toyota Way, key concepts are found such as constant improvements and to challenge today's working methods. "Go and see" means to embark on reality in order to study how things stand, that is, don't settle for talking about problems in a meeting room.

An overall goal is to minimize waste by focusing on what is value creation for the customer (end customer or next process), which requires systematic improvement work including problem solving.

At Toyota, management focuses on reducing process lead times. Shorter lead times mean that the problems are made visible by making the processes more sensitive to disturbances.

The underlying idea is that if one is sensitive to interference, then one also sees their problems. And if you see the problems there are prerequisites for solving them. If one solve the problems, one will always be able to improve.

Chapter 14, Energy management using Lean, is fully devoted to explore Lean in relation to Energy Management and how in practice this is carried out.

13.5.2.14 Energy efficiency networks

Energy efficiency networks is a way of conducting energy management in group form. For a period of 3 to 4 years, the company works in a structured way with improving energy efficiency, starting with the implementation of an energy audit. The model originates from the energy networks created in Switzerland in the 1980s. Later on, energy efficiency networks were initiated in Germany and also in Sweden. In Sweden, the network concept was inspired by a successful network approach of certification of standardized environmental and quality management systems in group form entitled the Hackefors-model, which was initiated during the mid-1990s (Altea, 2018).

One of the most important drivers in an energy efficiency network is that the cost of energy management is significantly reduced, as several companies share the cost of the energy efficiency network coordinator. An energy coordinator can if the energy efficiency network functions well, be seen as a form of "in-sourced" energy manager.

13.5.2.15 Database with real energy efficiency measures

A database of real energy efficiency measures from previously implemented energy audits can assist energy managers and energy auditors in the identification of improvement measures. A database can thus significantly reduce the cost of, for example, searching for information and assessing the various measures' profitability (so-called transaction costs) in connection with the energy audit. The database also makes it easier for energy auditors who are not familiar with the production processes in the company, helps to improve the quality of energy audit and ensures that proposed measures are reliable, which in turn leads to a more thorough energy audit report. The database is also valuable for customers, in order to evaluate, review, and compare (validate) offers from various energy service companies.

The NEAD database covers more than half of the Swedish industrial energy use and includes the 10 years industrial policy programs since 2005, the Voluntary Agreement Program PFE (Program for improving energy efficiency in electricity-intensive industry) as well as the energy audit program. PFE's actions from the first and second program periods (between 2005 and 2014) cover approximately 2300 industrial energy efficiency measures. In addition to the PFE, the energy efficiency measures that were reported from the energy audit policy program between 2010 and 2015 were also included covering about 6200 measures in total.

From the perspective of the national energy audit policy programs, tools for energy audits are seen as key components in obtaining comparable results that can then be used to evaluate, analyze, and improve the program. However, it should be noted that a tool such as a database of real energy efficiency measures can never fully replace an experienced energy auditor in proposing improvement measures and find the most energy-efficient solution.

13.5.2.16 Education and training

Training of personnel is an important part of the company's competence development. Training in improved energy efficiency is advantageously done in two or more steps. Since the cost of educating everyone in the company can be high, a first step is often to train a selected part of the company's staff, that is, those that are primarily affected by the energy issue. In a second round, a larger group of employees are trained (Caffal, 1996).

13.5.3 Energy management standard

An energy management system, like other management system standards, is based on the PDCA (plan-do-check-act) model. Since 2011, there is an international standard for energy management, ISO 50001. It is largely based on the European standard SS EN 16001, which in turn was largely based on the Swedish standard SS 62 77 50.

The first standard for energy management was the Danish energy management standard. In the early 2000s, a new energy management standard, SS 62 77 50, was developed in Sweden. It was partly based on the Danish standard but was significantly changed to fit in with the Swedish national industrial policy program PFE.

13.5.3.1 Other standards related to energy efficiency in the industry

There are a number of standards that relate to energy efficiency in the industry. Below is a (not complete) list of such standards:

1. EN 16212:2012 Energy Efficiency and Savings Calculation—Top-Down and Bottom-Up Methods
2. EN 13187 Thermal Performance of Buildings—Qualitative Detection of Thermal Irregularities in Building Envelopes—Infrared Method
3. EN 15603:2008 Energy Performance of Buildings—Overall Energy Use and Definition of Energy Ratings
4. EN 15193:2007 Energy Performance of Buildings—Energy Requirements for Lighting
5. EN 12464-1:2011 Light and Lighting—Lighting of Work Places—Part 1: Indoor Work Places
6. EN 16231:2012 Energy Efficiency Benchmarking Methodology
7. ISO 50003:2014 Energy management systems—Requirements for bodies providing audit and ceritification of energy management systems
8. ISO 50006:2014 Energy management systems—Measuring energy performance using energy baselines (EnB) and energy performance indicators (EnPI).

13.6 Method management versus result management

There are two principal ways of leading an internal energy management work: result and method management. Normally, a successful energy management work is a combination of these two.

With result management, clear goals are formulated for how energy efficiency can be improved, which is an important success factor in the process of change. However, this type of intervention risks being heavily person-dependent, and many times successful change work goes back to business-as-usual if the responsible person changes position.

With method management, employees are instead given a greater freedom to develop routines and measures themselves in relation to energy efficiency improvement activities. This means that the staff will be personally involved and most likely more engaged in the actual change activities. Although the person responsible for the energy management changes position, the routines remain largely in the organization.

Result and method management can be likened to getting from point A to B on a sandy beach. In result management, the movement can, of course, go fast, it is simply to run from point A to point B. However, the waves quickly flush away the tracks in the sand and a second person who is to make the same attempt will not be helped by the previous runner. In method management, instead, a path is built that makes it easier for subsequent persons to move from points A to B. This naturally is way more time-consuming. When initiating and adopting in-house energy management in a company, it is proposed that both methods are used. The company's culture, past experience of streamlining work and the knowledge of the leader and the group, may then help steward the balance between the two approaches.

13.7 Monitoring progress of energy efficiency improvement

The most common way of benchmarking progress of energy efficiency improvements is by the energy KPI (e-KPI) specific energy consumption (SEC) (for more information on energy KPIs see Chapter 4). As noted previously in this book, the term itself is thermodynamically incorrect as "energy" cannot be "consumed" but nevertheless provides management with solid e-KPI to monitor progress. SEC is defined as using primary energy-use divided by number of products over a given time period.

Lawrence et al. (2019) in a review of the scientific literature on SEC found that the term has not been properly defined which in turn leads to very large variations in the actual magnitude of SEC. Fig. 13.4 outlines the major conclusion from Lawrence et al. (2019) and the information in the following subchapters are discussed in more detail in the same reference.

Examples of factors affecting the amount of energy used:

- Primary Energy factor
- System boundaries, e.g.,energy used by the main equipment
- Specifications of equipment, e.g., age
- Specifications of equipment, e.g., production rate
- Environmental conditions
- Availability and quality of information and data
- Assumptions

$$SEC = \frac{\text{Energy used}}{\text{Products}}$$

Examples of factors affecting the amount of products:

- Sold or produced products
- Partioning of products
- Availability and quality of information and data
- Assumptions

Figure 13.4 Major factors affecting specific energy consumption (SEC) (Lawrence et al., 2019).

13.7.1 Primary energy factor

What primary energy factor that is denoted to various energy carriers also affect the primary energy use.

13.7.2 System boundaries

Which system boundaries that one uses also affect SEC, for example, how are sold energy like district heating or electricity dealt with (inside or outside the system boundary line) and what losses are the production of district heating and electricity accounted for.

13.7.3 Specifications of production

If the production rate increases or decreases, the processes that are not directly related with production of products, so-called support processes, still remains at the same level of energy use. That means that an energy efficiency improvement can simply be increased production rate.

13.7.4 Environmental conditions

The environmental conditions have an effect on the energy use in the company. For example, a paper mill located in a colder climate will need energy to un-freeze the lodges entering the production during the colder season naturally affecting overall energy use.

13.7.5 Sold products or produced products

One common deviation in SEC between companies from the same industry is related to divergence in if sold products or produced products is used. If the latter is used, it complicates the calculations as yield is not taken into account. Due to that, sold or delivered products are the more appropriate term.

13.7.6 Partitioning of products

When partitioning of products is done, it is often very difficult to allocate the correct energy use to the different products. For example, how much compressed air, process ventilation, etc. was used for the specific product versus another product.

13.7.7 Availability and quality of information and data

The availability and quality of data have an effect, not on the actual energy use or production of products, but on the actual figures. Normally, electricity is metered and is both available and comes with a good data quality but energy carriers such as oil and wood chips are not always easy to distinguish between years as these are delivered in bulk to companies.

Data on sold or delivered products are normally both available and contains good data quality.

13.7.8 Assumptions

The uncertain factors presented above sometimes need to be assumed when calculating SEC. Due to that, it is important to clearly describe all the assumptions and calculations, for example, what PEFs that are used, preferably with reference and clear motivation.

References

Altea, 2018. The Hackefors-model. https://altea.se/en/the-hackefors-model/ (accessed 02.01.19.).

Apeaning, R.W., Thollander, P., 2013. Barriers to and driving forces for industrial energy efficiency improvements in African industries: a case study of Ghana's largest industrial area. J. Clean. Prod. 53, 204–213.

Bergmasth, M., Strid, M., 2004. Energitjänster på en avreglerad marknad—för en effektivare energianvändning? (Energy Services in a deregulated market—for a more efficient use of energy?), Doctoral thesis. School of Economics and Commercial Law, 2004. Gothenburg University, Gothenburg, Sweden [in Swedish].

Blomqvist, E., Thollander, P., 2015. An integrated dataset of energy efficiency measures published as linked open data. Energy Efficiency 8, 1125–1147.

Brunke, J.C., Johansson, M., Thollander, P., 2014. Empirical investigation of barriers and drivers to the adoption of energy conservation measures, energy management practices and energy services in the Swedish iron and steel industry. J. Clean. Prod. 84, 509–525.

Caffal, C., 1996. Energy Management in Industry. Centre for the Analysis and Dissemination of Demonstrated Energy Technologies (CADDET), Sittard, The Netherlands, Analysis Series 17.

Chalmers Industriteknik; 2016.

Christoffersen, L.B., Larsen, A., Togeby, M., 2006. Empirical analysis of energy management in Danish industry. J. Clean. Prod. 14, 516–526.

Gong M., 2004. Using exergy an optimization models to improve industrial energy systems towards sustainability, Linköping Studies in Science and Technology, Dissertation No. 868, ISBN 91-7373-933-2.

Grip C.-E., Elfgren E., Söderström M., Thollander P., 2011. Possibilities and problems in using exergy expressions in process integration. In: Proceedings of the World Renewable Energy Congress 2011 (WREC 2011), 9–13 May. Electronic Press, Linköping.

Grip C.-E., Thollander P., Söderström M., 2012. Stakeholder study on barriers to exergy analysis. In: Proceedings of the International Conference on Applied Energy ICAE. Suzhou, China; July 5–8.

Gundersen, T., 2000. A Process Integration Primer e Implementing Agreement on Process Integration. International Energy Agency Report, SINTEF Energy Research, Trondheim, Norway.

IAC, 2016. Energy and cost saving assessments for small and medium-sized US manufacturers – database.

IEA (International Energy Agency), 2011. Energy Technology Network. Energy Technology Systems Analysis Programme.

Intelligent Energy Europe, 2016. Einstein Software Tool.

Johansson, M., Thollander, P., 2018. A review of barriers to and driving forces for improved energy efficiency in Swedish industry: recommendations for successful in-house energy management. Renew. Sustain. Energy Rev. 82, 618–628.

Johansson, P.-E.-, Thollander, P., Moshfegh, B., 2011. Towards increased energy efficiency in industry: a manager's perspective. In: Proceedings of the World Renewable Energy Congress 2011. Linköping; 2011.

Karlsson, M., 2011. The MIND method: a decision support for optimization of industrial energy systems: principles and case studies. Appl. Energy 88, 577–589.

Kindström, D., Ottosson, M., Thollander, P., 2016. Driving forces for and barriers to providing energy services: a study of local and regional energy companies in Sweden. Energy Efficiency 1–19.

Koewener D., Mielicke U., Jochem E., 2011. Energy efficiency networks for companies - concept, achievements and prospects. In: Proceedings of ECEEE 2011 Summer Study, Energy Efficiency First, the Foundation of a Low-Carbon Society. pp. 725–733.

Krzemień, J., 2013. Application of Markal model generator in optimizing energy systems. J. Sustain. Mining 2, 35–39.

Lawrence, A., Thollander, P., Andrei, M., Karlsson, M., 2019. Specific energy consumption/use (SEC) in energy management for improving energy efficiency in industry: meaning, usage and differences. Energies 12 (247), 1–22. Available from: https://doi.org/10.3390/en12020247.

Linnhoff, B., Townsend, D.W., Boland, D., Hewitt, G.F., Thomas, B.E.A., Guy, A.R., et al., 1982. A User Guide on Process Integration for the Efficient Use of Energy. IChemE, Rugby, UK, Revised edition published in 1994.

MARKAL.

Mintzberg, H., 1987. The strategy concept II: another look at why organizations need strategies. Calif. Manag. Rev. 30, 25–32.

NEAD (Nordic Energy Audit Data), 2018. Database for industrial energy efficiency measures.

Nordman, R., Berntsson, T., 2005. Advanced pinch technology based composite curves for evaluating the usable excess heat potential. The 18th International Conference on Efficiency, Cost, Optimization, Simulation and Environmental Impact of Energy Systems. Trondheim, Norway, June 22–22.

Paramonova, S., Thollander, P., 2019. Technological change or process innovation: an empirical study of implemented energy efficiency measures from a Swedish industrial voluntary agreements program. Energy Policy (under review).

Paramonova, S., Ivner, J., Thollander, P., 2014. Outsourcing industrial energy management: industrial energy efficiency networks provided as an energy service. In: Deering, A. (Ed.), Outsourcing. Nova Science Publishers, Inc, pp. 71–98. Available at ⟨www.novapublishers.com/catalog/product_info.php?products_id = 53026&osCsid = c7e30ff61a1eb42eb3317a2532206f59⟩.

Rardin, R.L., 1998. Optimization in Operations Research. Prentice-Hall Inc, Upper Saddle River, NJ.

Rohdin, P., Thollander, P., 2006. Barriers to and driving forces for energy efficiency in the non-energy intensive manufacturing industry in Sweden. Energy 31 (12), 1836–1844.

Schulze, M., Nehler, H., Ottosson, M., Thollander, P., 2016. Energy management in industry: a systematic review of previous findings and an integrative conceptual framework. J. Clean. Prod. 112 (5), 3692–3708.

SEA (Swedish Energy Agency), 2013. Energitjänster i Sverige - Statusrapport för tjänster för energieffektivisering (Energy services in Sweden. status report for energy services). ER, 2013, 22 [in Swedish].

SIMPLAN, 2016. It has to be the right software.

Solding, P., 2008. Increased Energy Efficiency in Manufacturing Systems Using Discrete Event Simulation: Applied Studies on the Swedish Foundry Industry (Ph.D. dissertation). De Montfort University, Leicester.

Thollander, P., Ottosson, M., 2008. An energy-efficient Swedish pulp and paper industry: exploring barriers to and driving forces for cost-effective energy efficiency investments. Energy Efficiency 1 (1), 21–34.

Thollander, P., Rohdin, P., Karlsson, M., Rosenqvist, J., Söderström., M., 2012. A standardized energy audit tool for improved energy efficiency in industrial SMEs. In Proceedings of the ECEEE Industry Summer Study, 11–14 September, 2012.

Thollander, P., Backlund, S., Trianni, A., Cagno, E., 2013. Beyond barriers: a study on driving forces for improved energy efficiency in the European foundry industry. Appl. Energy 111, 636–643.

Trianni, A., Cagno, E., Bertolotti, M., Thollander, P., Andersson, E., 2019. Energy management: a practice-based assessment model. Appl. Energy 235 (1), 1614–1636.

Further reading

Nordenaudit, 2018. Energy audit software for industry. Nordic Energy Audit.

Ohno, T., 1988. Toyota Production System. Productivity Press.

Womack, J., Jones, D., Roos, D., 2007. The Machine That Changed the World: The Story of Lean Production – Toyota's Secret Weapon in the Global Car Wars That Is Now Revolutionizing World Industry. Free Press.

Energy management using Lean

14

Chapter Outline

Introduction to Industrial Energy Efficiency. DOI: https://doi.org/10.1016/B978-0-12-817247-6.00014-6

14.1 Reasons and benefits for energy management

Let's first clarify why an organization or corporation would want to work with energy management and to apply Lean energy thinking. There are several reasons and some are more abstract than others (in order of abstraction level starting with the most obvious):

1. Reduce energy cost
2. Contribute to protecting the environment
3. Positive impact from nonenergy benefits (NEB)
4. Train the organization in continuous improvements

14.1.1 *Reduce energy cost*

A corporation which has not previously worked in a structured way around energy efficiency is very likely to be able to make significant reductions of their energy bills. Reducing energy bills can be subdivided into several approaches where the most obvious is to reduce the amount of energy which is used. There are also in many cases saving potential from a pure commercial approach by re-negotiating existing supply and distribution contracts using purchasing best practice and getting several quotations. Further, the energy cost can potentially be reduced by switching from a high-cost energy source toward a low-cost energy source, paying attention to unwanted negative effects like increased CO_2 emissions, operational risk, and so on.

14.1.2 *Contribute to protecting the environment*

Energy management will contribute to protecting the environment by using less energy or at least improve energy efficiency and hence combat climate change by reducing CO_2 emissions. Even a corporation that is already, for example, buying green electricity will help to contribute if they reduce their electricity use as the electricity supplier will then have more green electricity available for other customers who currently buy electricity produced from fossil fuels. The environmental improvement factor is a particularly strong factor due to the urgency to reduce global warming and high awareness in the public and hence also among the employees. This high awareness among employees can be important when communicating about the energy management initiative and in trying to get improvement suggestions coming in.

14.1.3 Positive impact from nonenergy benefits (NEBs)

Positive impact from NEBs are numerous and is described in further detail in Part I, Chapter 7, of this book. Empirical research has indicated that the NEBs are financially of the scale 2.5 times higher than the energy cost reduction and hence significantly higher than the direct energy bill reduction (Hall and Roth, 2003). There are however difficulties in quantifying all the NEBs and hence it is difficult to directly see these in the financial business results. To ensure that energy saving and efficiency investments take place and that the right investment decisions are made it is important to try to include the financial details of NEBs to get the correct payback time. Some examples of NEBs are:

Safety can be improved for example by changing to energy-efficient LED lights. The additional benefit is that their life expectancy is significantly longer than traditional fluorescent lights and hence maintenance staff need to go up on height to replace them more seldom and hence lower risk exposure to working at height.

Quality can be improved in for example machining operations by doing tool dynamics optimization using modal analysis. This means that the optimized tool settings allow higher material removal rates and hence total time needed for machining is reduced and the machining center can be shut down and start saving energy. The side benefit is that the optimized tool now have lower levels of vibrations as it is set in a more stable process window. The lower levels of vibration results in lower surface roughness and hence potentially higher quality (depending on engineering requirements).

Unplanned breakdowns can be reduced as a nonenergy benefit from working with energy management. One way to reduce idle losses (energy used during nonproduction time) is to turn machines and equipment off during nonproduction time. Sometimes there is a resistance from either management or the workforce to turn some machines off as issues are anticipated when the machines are turned on again. Care must be taken so that no production losses occur as a result of energy management work but at the same time the "difficult" machines should be challenged. A first step could be to only turn off support systems (auxiliary systems, hydraulic pumps, and so on) and leave the main control system up and running. The issues should then be analyzed more in-depth and improvements should be implemented to be able to turn the machines fully off in the future. Do not forget to make secure back-ups of all programs and other information which might be needed to get the machines up and running again. Working systematically with "difficult" machines generally results in less unplanned breakdowns as we get to know the machines better and the consequence of the breakdown is less severe as we are used to working with the machines.

Improve productivity. For production stations, cells, and lines, it can very often be beneficial to use an energy meter to log energy use down to millisecond level (depending on equipment cycle time). This energy trace can then be compared to the programmable logic controller (PLC) signals or similar and with the value-adding process steps. Based on this comparison a significant amount can be learned about the current state such as when are we using energy, are there any unnecessary idle periods, can we clearly identify the bottleneck operation? Based on this analysis, a deeper understanding can often be built-up and productivity can be improved. This will allow the production volume to be produced in a shorter time and hence energy saving can start but also other nonenergy benefits due to increased productivity (free up capacity, reduce shift forms, insource parts which were not possible to produce before due to capacity constraints, and so on).

Reduced maintenance cost is probably the most significant NEB. The above example with longer lifetime of LED lights requires less maintenance interventions. A similar situation can be found in changing to more efficient brushless electrical motors which require significantly less maintenance. It is not unheard of that the whole business case for an energy-saving investment is based on reduced future maintenance needs.

Consumables is an area where nonenergy benefits can be relatively tangible. In the example with LED lights versus fluorescent light the LED lights will last significantly longer and hence a lower consumable cost after the initial investment. In addition, the deposit cost of burned-out fluorescent light tubes are also avoided which should be included in the investment calculation. Another example of reduced consumable cost is oil separators used in machining operations. Historically, these have been running during the whole production time whereas it is possible to connect them so that they only run when they are needed, which is during machining operation taking place (spindle rotating). This means that the filter change frequency is extended and fewer new filters need to be purchased.

Soft benefits such as increased employee engagement because the corporation is working on topics which are aligned with current topics such as climate change is also an NEB which should not be neglected. By achieving good results in energy management work we can see increased motivation among employees to work with continuous improvements in other areas, see Section 7.3, Chapter 7 for further elaboration on these soft benefits.

14.1.4 Train the organization in continuous improvements

Normally, improved energy efficiency is thought of as installations of new equipment. Training the organization in continuous improvements is often the least obvious reason for working with energy management. By using Lean thinking as a guiding principle when approaching energy management work, the underlying principles and methods are trained, which the organization can benefit from when working with continuous improvements in other areas such as safety, quality, and productivity.

14.2 Lean

Lean is the western world's copy of Toyota Production System (TPS). Lean as a term was first coined by Womack et al. in 1990 in their book "The machine that changed the world" (Womack et al., 1990). Womack et al. clearly showed the applicability of TPS even outside Japan and its effectiveness in cultures other than the Japanese. This section will provide a brief introduction to TPS to enable the reader to better understand Lean from its origins. This section is very brief as there is a significant amount of literature around TPS and Lean, already available if an interested reader wants to learn more. Womack et al. followed up with "Lean Thinking" (Womack and Jones, 1996) and there have been several books written by Taiichi Ohno which explains TPS (Ohno, 1988). To get an introduction to the evolution of TPS, Takahiro Fujimoto provides an in-depth overview (Fujimoto, 1999).

14.2.1 Origins of Toyota Production System (TPS)

TPS is a result of work by several key people as well as the whole of Toyota Motor Corporation. The story starts with the founding father of the Toyoda industrial empire, Sakichi Toyoda. Sakichi Toyoda was an inventor and industrialist who as a child watched his mother using a manual weaving loom. To support his family he started to improve the weaving loom, amongst other things, and his most famous invention is the automatic power loom and specifically the Type G model which featured autonomous automation known as Jidoka in Japanese. This invention meant that if a warp tread in the weaving loom broke the loom would stop automatically. The consequence of this invention is that it was impossible to produce defect fabric. Equally important, it also meant that an operator did not have to watch over the loom to visually make sure all the treads were intact. So, instead of one operator per loom, a single operator could supervise several machines. The results are fabric with better quality and significantly increased productivity, that is, reduced cost. Jidoka would later become the first main pillar of Toyota Production System. The patents for the Type G automatic loom was sold in 1929 to the Pratt Brothers in the United Kingdom for 1 Million Pounds Sterling. This was the starting capital for Toyota Motor Corporation with the aim to produce passenger cars.

It was the son of Sakichi, Kichiiro Toyoda, who was tasked with setting up car production. Kiichiro had traveled to Europe and the United States to study and benchmark automotive production as part of the preparations. In the initial days of Toyota and in the time post World War II there was a scarcity of capital, raw material, and parts to produce cars. This inspired Kichiiro Toyoda to establish the precursor of Just in Time (JIT). JIT would later become the second main pillar of Toyota Production System. JIT means that one should only produce the part which is needed, at the time when it is needed and in the quantity that is needed.

Whilst Jidoka and JIT are the two main principles in the Toyota Production System, it still needed to be put into a context and systematically applied to qualify as a production system. The main architect of Toyota Production System is Taiich Ohno. Ohno was working as a production engineer in the automatic loom works and was transferred to Toyota Motor Company to support the improvement work of the production. Whilst benchmarking, Toyota understood that Japanese productivity was significantly lagging behind the automotive industry in the United States. Apparently, it required nine Japanese workers to accomplish the same amount of work as one American worker! Ohno set out to close this gap and for more than 30 years worked to create the Toyota Production System with the full support of the senior management.

14.2.2 The Toyota Way

To explain the Toyota Production System there is a need to first explain the Toyota Way which are the values on which the production system is based (Toyota Institute, 2001). These values, as seen in Fig. 14.1, are very much lived by everyone in the company and enable the successful use of the Toyota Production System.

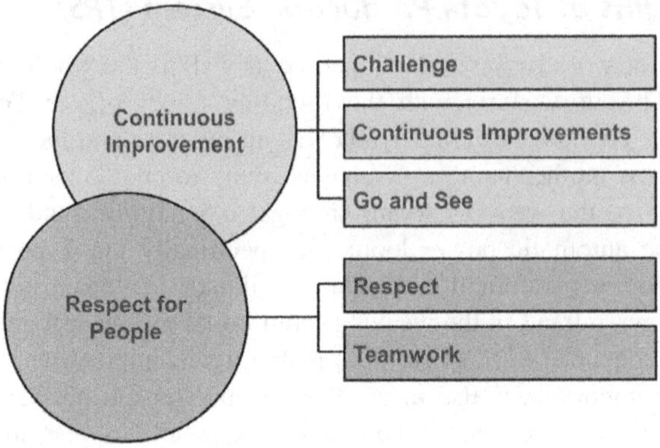

Figure 14.1 The Toyota Way.

The Toyota Way consists of two main components: Continuous Improvement and Respect for People. Continuous improvement can be further broken down into Challenge, Continuous Improvements, and Go and See. Respect for people can be broken down into respect and team work.

Challenge means that one should always challenge the current status in order to drive improvements. Why are we getting the results we are getting? Why is the current process design/functioning/used the way it is? Why is the layout/set-up/configuration the way it is? Why are we doing the job elements which we are doing and why in this sequence? Everything should be questioned. This way of thinking also reflects on the follow-up of activities and results. In Toyota it is better to fail to achieve a challenging target than to achieve a target which was not challenging enough. This is very different from most western companies where the main objective is to show ok conditions toward management.

Continuous improvements means that one should always strive to improve. "Always a better way" is a slogan which is used frequently by Toyota. When talking about continuous improvements the Japanese word Kaizen is often used. Kaizen means change (kai) good (zen) and is here used interchangeably with continuous improvements. When teaching kaizen it is also emphasized to focus on the many small improvements rather than the few large improvements. The reason for that is that the many small improvements are easy to find/come up with, normally simple and cheap to implement and comes with a low risk. If one focus on the large improvement, then it might take a long time to develop during which time there are no improvements taking place. A large improvement is potentially also much more complicated than a more simple operational change, it is more expensive to implement and thus entails a higher risk. If the large improvement does not work or do not give the results anticipated, then the step back is larger than if a small improvement does not work. Employees in Toyota all have four different job tasks: do their daily tasks, improve their daily tasks, improve themselves, and improve their team. This only works if there is an expectation from leadership that everyone should always be involved in continuous improvement.

Go and See. The idea behind Go and See is that management should be present where the value-adding process is being undertaken. In production this means that managers are not hiding in their offices but rather are out on the shop floor to fully try to understand the current conditions and to try and provide support in terms of problem-solving and challenging the current best level to drive continuous improvements. If a quality problem occurs, then Go and See would mean that you do not only sit in a meeting room to discuss the defect and potential root causes. Instead the team should go out on the shop floor to understand in detail by looking on the defect part/parts as well as where the defect was detected and where it was created to get a common and full understanding and thereby accelerate problem solving.

Teamwork is the emphasis on working as a team. The background, knowledge, and different perspectives from each team member are used to ensure that the contribution of the team is greater than the sum of its individual members. This applies to all functions and should drive strong cross-functional collaboration. For the Toyota Production System to function in a production environment, a significant amount of expectation is put on what they call the team leader (TL). In general, each team leader has approximately seven team members (TM) in his or her team but this can vary depending on the level of current issues, the level of absence among the TMs, wanted rate of improvement, and so on. The TL is fully trained on all processes in the team. The role of the TL is to support the team in achieving the daily production target. This can be done by a variety of different actions such as working on the assembly line to cover for absence of a TM, support individual TM's (so-called Andon response) when they are experiencing an abnormality, supporting in replacing consumables and if needed support with material replenishment. Based on the abnormalities such as quality defects, parts missing, and so on, the TL will start problem solving to ensure that the root cause is understood, and that same problem will not reoccur. To find improvement opportunities, the TL will on a regular basis conduct a process study of an individual production process within the team. Following the process study, the TL will have an improvement dialog with the TM in order to jointly find improvement opportunities and solutions.

Respect is probably the value of Toyota that is the most misunderstood and ignored when other corporations and organizations try to implement Lean. Respect means of course to treat each other respectfully and to build trust. However, it also means that the relevant people are involved in improvement activities. This is particularly true for the team member working on a process which is the subject of the improvement activity, after all he/she will be the person who understands the process the best and will hence have many relevant ideas. The team member will also continue to work on the process after the improvement activity, so it is only natural that he/she partake in the improvement activity to later sustain and further improve the process.

Furthermore, respect in Toyota also means giving people challenging and demanding jobs and tasks as a sign of respect of their capability. If someone's performance does not meet the expectations then as a sign of respect this feedback is provided. It would be considered disrespectful to not give feedback on poor performance because it would not enable the person to improve. So, respect is far more than being nice to each other and is probably one of the more difficult aspects of Lean to understand and also the part where the connections with Japanese culture and society is the strongest. Soft topics like respect are inherently more difficult to implement compared to, for example, hard Lean tools.

Successful Lean implementation requires building a culture which reflects the Toyota Way values. Without them, the right Lean thinking and culture will not be established and instead there will be a focus on Lean tools, which in time will be difficult/impossible to sustain.

14.2.3 Toyota Production System

The Toyota Production System (TPS) consists of three main principles: Standardized Work, JIT, and Jidoka as seen in Fig. 14.2.

Before going into the details about these principles we need to remember for which purpose and context in which TPS was developed: limited availability of capital and the need to reduce cost by closing the productivity gap toward US car production. This means that the purpose of TPS is cost reduction. The mechanism to achieve this is through problem sensitivity. In other words, at Toyota they want to be sensitive to problems. This may sound very counter-intuitive. One might think that we do not want to be sensitive to problems and that even if problem occurs, the production can still continue to operate. The thinking at Toyota is that if we are sensitive to problems then we will have an issue when a problem occurs and this issue will allow us to see the problem. If we can see our problems then we can also solve our problems. If we solve our problems, we will gradually become better. When Toyota is talking about working with continuous improvements it is not just because they know it is good to work with continuous improvements but rather that they are operating within a production system where problem-solving and continuous improvements are the only possibility. The role of management (at all levels) is to continuously make production more and more sensitive to problems to drive more continuous improvements.

When trying to apply TPS it is also important to understand the values, principles, methods, and tools and the connection between them. When visiting Toyota factories and reading about Lean, it is relatively easy to grasp the application of different Lean tools. There is then the need to appropriate this tool for the specific application. It might, for example, be counter-productive to apply a Lean tool from mass-production of cars to a specialist health care application. Instead it is important to understand the values, principles, and methods which Toyota tried to achieve to make it work outside Toyota.

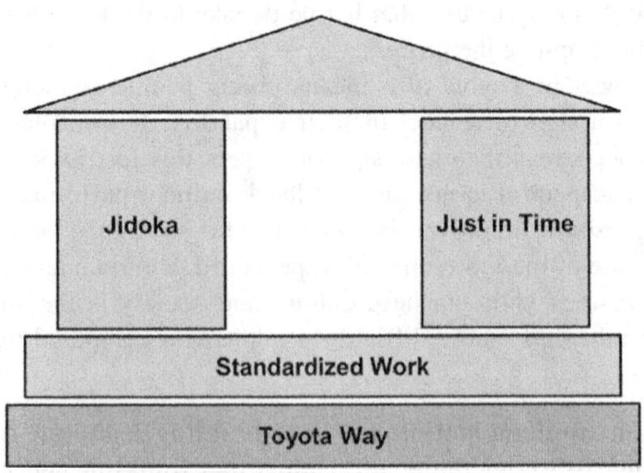

Figure 14.2 The Toyota Production System.

It should be emphasized that leadership plays a very important role in any Lean transformation. Leadership must have the right understanding, way of thinking, commitment as well as persistency. Failure to establish a Lean culture and behaviors based on the correct values and failure to appropriate different tools to specific context are the main reasons why many Lean initiatives fail.

14.2.4 Standardized Work

When applied correctly, Standardized Work is centered around human motion and describes the safe way to produce with the right quality and at the right time. Standardized Work is describing the currently best-known methods to conduct work. There is a saying within Toyota that "*without standard there can be no kaizen,*" meaning that without standard, continuous improvements will fail. The reason for this saying is that if the improvements that you make in a process are not documented and shared with all the team members doing the process, then how can we ensure that all team members will do the same sequence and not slide back to a less efficient way of working? This means that if there are no standards then there is no point in making any improvements as condition will most likely anyway deteriorate. Rather the first step in the improvement cycle should be to create a standard which can be used as a baseline upon which the improvements can build upon.

Standardized Work is often criticized for taking the initiative and creativity away from the team member and turning them into "robots." The standard should ideally be made by the production team member, if needed, in collaboration with the team leader. The standard should be available where the work is conducted, in the immediate vicinity, and used when training new team members. Experienced team members do not need to refer to the standard of each operational cycle as they know it by heart, but the team leader will use the standard when making a job observation to confirm that the correct sequence is being followed and that the time for each job element is correct. If one is unfamiliar with a process and watch someone doing that process then it is difficult to understand if the team member is doing the job correctly and in the correct time. However, if you have a standard in your hand you can clearly see if the correct sequence is followed. If the sequence in the standard is not followed, or if there is any gap in the timing that indicates that there is a problem somewhere, the standard helps in achieving problem sensitivity, which is the main underlying mechanism in TPS. If the sequence is not followed, that could be due to many reasons such as a team member not being correctly trained but also due to that the team member has found a better sequence of doing the work, in which case the standard should be updated. If the job takes longer time than outlined in the standard, it could be due to many things and can indicate for example, lack of detail in the work standard where a knack point/skill which some team members have might not have been captured in the standard and not systematically taught to all team members.

14.2.5 Jidoka

Jidoka, also known as "built-in-quality," was the first pillar of TPS and was one of the main inventions from Sakichi Toyoda. Jidoka means automation with a human touch in Japanese and the term autonomation is also sometimes used. The basic concept is that when a problem occurs the machine should automatically stop and signal that there is an issue. This means that no defect parts will be produced which will help to keep scrap cost down and reduce the risk of mixing good and bad parts. Additional benefit is that we do not need to have someone to supervise the machine and hence productivity can be improved. A common example from the textile weaving time was that before Jidoka, they used to have one operator looking after one weaving machine and with implementation of Jidoka, one operator could supervise several weaving looms. The benefit of stopping and signaling when a problem occurs is that root cause analysis can start immediately and all the conditions which caused the stop are still "fresh" which makes root cause analysis easier. To come back to a machine where a defect occurred a week ago makes root cause analysis much more difficult as all evidence might be gone. Other Lean tools like Andon also belong to the Jidoka pillar. Andon on an assembly line is a simple thin rope above the assembly line. When the rope is pulled a yellow light switches on showing to the team leader that one of the team members needs support. Andon in this case is Jidoka for a manual process.

14.2.6 Just in Time (JIT)

JIT is one of the founding principles along with Jidoka. JIT was introduced by the founder of Toyota Motor Corporation, Kichiiro Toyoda, for very pragmatic reasons. At the start of Toyota Motor Corporation, they could simply not afford to have large amounts of capital tied up in inventory and work in progress. JIT focuses on three key items: continuous flow, pull system, and takt time. The level of JIT is measured through lead time which in turn can be broken down in processing time and stagnation time. Lead time is the total time it takes for a product to go through the whole production process. To establish the lead time, one needs to measure the processing time of the different production steps and add the stagnation time of all the buffers. Stagnation time of buffers is calculated by multiplying the buffer size with the takt time. To reduce the lead-time we should first address the different types of waste which are creating the stagnation time. With short lead times we will be able to see problems easier. This will then force thorough root cause analysis to make sure the same problem will not happen again. When we are talking about JIT we especially mean JIT manufacturing, to deliver JIT is easy by building and stocking parts/products close to the customer.

Continuous flow is about keeping the flow of products (or services) flowing in a steady state. By keeping a flow, stagnation is avoided, and stagnation will increase lead time through excess inventory and work in progress. There are several basic methods to keep continuous flow such as streamlining the production flow, using multiprocess handling rather than multimachine handling and one-piece production.

Pull system is about connecting different production processes together, so that they only produce what is really needed by the downstream process. To achieve this, the concept of Kanban has been successfully used and widely publicized. An important but sometimes overlooked factor of pull system is that we can have inventory between different processes, as long as it is in a clearly defined area, with clear and scientifically calculated minimum and maximum levels. The levels depend on the fluctuations in the demand from the next process, reliability of the downstream process, and so on. To have zero inventory is a misunderstanding of TPS.

Takt time (TT) is about translating the customer demand to production demand. It is determined by how many parts the customer needs and how long the production process is available for production. In a simple example the customer needs 10 parts per day and the production runs 400 minutes per day meaning TT is 40 minutes. TT is always calculated and not something which is measured with a stop-watch and should not be confused with cycle time of the production process. Cycle time is the time the job takes to complete in a specific work process and therefore cycle time needs to be shorter than TT. Understanding the TT and to produce as close to TT as possible is key to reduce stagnation time as well as achieving a problem sensitive environment.

Leveling (Heijunka in Japanese) is a prerequisite to be able to apply JIT manufacturing. Leveling refers to leveling out both the production quantity and variety over time.

14.2.7 Waste elimination and 3M (Muri, Mura, and Muda)

When applying the Toyota Way values and Toyota Production System the outcome will be that one starts to see the problems better and can hence see all the waste which currently exists in the process. When talking about waste (or Muda in Japanese) it is important to first mention overburden (Muri) and unevenness (Mura).

Overburden (Muri) should be the first topic for improvement. Overburden is not necessarily physical overburden but can also be psychological. To overburden a team member is showing poor respect and from a practical perspective there is a significantly increased risk for injury as well as that the efforts are not sustainable over time. Overburden applies to people as well as machines.

Unevenness (Mura) should be the second topic for improvement. Unevenness can be physical unevenness but it can also be in time where workload is high in the morning and low in the afternoon. Or high workload in the beginning of the week and low toward the end of the week.

Waste (Muda) is the third topic for improvement when overburden and unevenness have been reduced or eliminated. There are seven main types of waste with an eight waste being added in recent times relating to un-used creativity of employees:

1. Over-production—Producing too much or starting too early
2. Inventory—A result from over-production which also has the effect of hiding problems
3. Defects and rework—Producing defect products and having to rework them
4. Motion—Unnecessary movement for completing the production process

5. Over-processing—Processing the product to a higher level than required
6. Waiting—Waiting for machines or material
7. Transport—Transporting products and material unnecessarily

When looking for waste it is useful to keep the method of "value-add" (VA) and "nonvalue add" (NVA) in mind. In any production process, the majority of the work done can be classified as NVA, as illustrated in Fig. 14.3 and this is what should be improved first by reducing or removing waste. Only after that should one start looking at improving the VA. An example from production could be a process which is tightening nuts, there is no point in trying to improve the speed of the nutrunner if the operator involved needs to walk a long distance to pick up the nuts. First, one should reduce the motion waste and hence reducing the NVA. It is important to remember that it is the customer who defines value. A useful question to ask is: "would the end customer be prepared to pay for it if he/she knew we did it?"

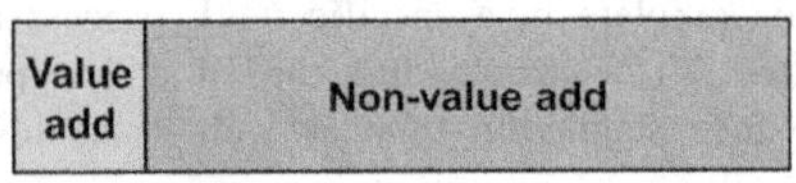

Figure 14.3 Value add and nonvalue add.

14.3 Lean energy

Lean energy is the application of Lean thinking to energy management by the application of the Toyota Way values and the principles in the Toyota Production System. One is then challenging and questioning why a process is using the amount of energy that it does and why it is using energy during nonproduction hours. One strives to continuously improve and reduce the energy used by implementing improvement ideas. We do not satisfy ourselves by sitting in a meeting room to discuss data but go out in the factory, office, field, or wherever the energy use takes place to see for ourselves, to ensure that we fully understand and that everyone in the team has the same view or sees the same picture.

14.3.1 Set a standard

As we could see in the Toyota Production System house, Standardized Work is the base. To be able to do Standardized Work we need a standard and hence this is a natural starting point. Everything should have a clear standard. We should know for each piece of equipment how much energy that equipment should be using under normal conditions. We should also know how much each support system (lighting, ventilation, and so on) is using. This is then our standard against which we can visualize actual/current conditions. If one has a gap between actual energy use and the standard, one can start activities to try and close this gap. One should prioritize to make sure that work with the largest gaps are in focus, so that one can focus attention and resources to the major potential improvements. One then starts to implement improvement ideas to close the gaps. We are not only satisfied to close a gap, but we also continue to challenge the standard.

The current standard level is a reflection of the technical nature and definition of the equipments and processes which we are currently using. Furthermore, it is a reflection of our culture and behavior around energy management. One can map out equipment in a factory and determine their current energy use and how it is being used. To make this easier and more visual, we can use a labeling of level 6 to level 1 as illustrated in Fig. 14.4. As shown in Fig. 14.4 we can make the analogy to a water tap in your kitchen at home. Level 6 is when the water tap is always open, and water is running into the sink regardless if you are at home or not. Level 5 is a first improvement where the tap is only open when you are at home, but even if you are sleeping it is still on. Level 4 is an improvement where the tap is only open when you are at home and awake. Level 3 is that the tap is only open when you are in kitchen. Level 2 is that the tap is only open when you need water but you do not know exactly how much water you need so there is still some wasted water. Level 1 is that you only open the tap when you need water and you only take exactly the amount of water that you need. The ideal condition is that we achieve level 1 for all equipment/systems. Even if level 1 cannot be achieved for all equipment, then improvement activities should be run to take it from level 6 to level 5 and so on. It is very fascinating to do this exercise in an organization, which has previously not worked with energy management, and to see the amount of level 6 and 5 equipment/systems that are in fact present. It is very common to find that ventilation systems are running 24 hours per day and even on weekends because nobody ever reflected on its use. If we use the ventilation example to illustrate level 1 in a meeting room, then ventilation should only be running when the meeting room is used and the flow rate should reflect the number of people in the room.

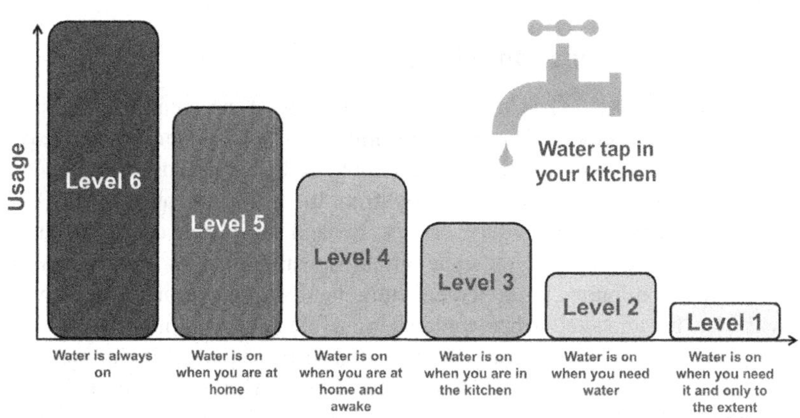

Figure 14.4 Energy use levels with water tap analogy.

14.3.2 Think in the right way

Fig. 14.5 shows an optimization example where we have an existing machine which is using 1 kW during 1 hour meaning the energy use is 1 kWh. We have two improvement opportunities on this machine. We can either reduce the power to 0.75 kW and still use it for 1 hour and hence use 0.75 kWh, or we can increase the power to 1.5 kW but then we only need to use it during 30 minutes meaning energy use is 0.75 kWh. The conclusion is that both improvement suggestions provide the same energy use: 0.75 kWh. Does it matter which one we choose? The benefit of improvement 1 would be that the maximum power outtake from the factory might be reduced which could impact distribution cost. However, the Lean way of thinking is improvement 2 where time is reduced during which we use the process. We run the process harder using more power but can turn it off earlier and can thus start saving energy from the process but also from all the surrounding support systems such as lights and ventilation, for the 30 minutes that we no longer need to operate. The overall total energy saving is thus likely to be larger with improvement 2.

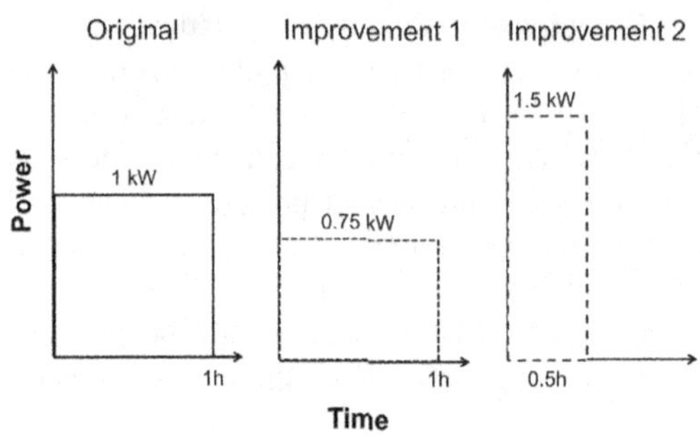

Figure 14.5 Optimization challenge power versus time.

14.3.3 Management commitment

To drive any change it is key to have, at the very least, acceptance and preferably support from management. In an organization which is not energy-intensive the energy costs are normally relatively small compared to other costs and hence there might be a lack of motivation or incentive from the management to start to address energy management and to reduce energy use and its related costs. If that is the case, there might be a need to do some initial education to show how energy can be used to drive a continuous improvement culture and that numerous nonenergy benefits will also be impacted when undertaking systematic continuous improvement work around energy. To showcase the potential to management it can be very useful to do some initial improvement work where high potential and low-cost improvements are implemented. This will enable you to use real cases in order to get management commitment. It is important that the management understands all aspects of why a company should work with energy management as outlined in the beginning of this chapter.

14.3.4 Energy manager

Someone needs to take initiatives on working level and organize and coordinate the activities. This person can be called energy manager or energy coordinator and can belong to any function in the organization. The role of the energy manager is to lead the working team by showing the right continuous improvement thinking, as well as motivating and sharing enthusiasm of the task ahead. In-depth energy competence is not actually required as this is more of a leadership role during the change management, but of course energy competence is beneficial.

14.3.5 Energy team

The energy team should ideally be cross functional in nature and involve representatives from all major functions. In a factory this could include production, maintenance, real estate/property as well as manufacturing engineering and other support functions. The team should meet together on a regular basis. This regular basis should be determined by the team or the main stakeholders but a suitable initial frequency could be weekly. Having a weekly meeting means that even if a team member misses one meeting he/she will at most be 2 weeks without feedback, a lower frequency could result in an improvement action being unaddressed for up to 1 month or more. A higher meeting frequency than weekly could be used in the start-up phase of an activity or if there are critical actions that require close monitoring.

14.3.6 Regular management reviews

When the energy manager/energy coordinator is in place, regular energy reviews with management can begin. These reviews should occur with a fixed frequency, preferably weekly but minimum monthly. The reviews should, in a visual way, highlight the current energy use versus target.

14.3.7 Open issue list/to do list

The energy team should together build an open issues list or to do list to guide and prioritize the activities. The base of this list is basic PDCA thinking (plan, do, check, and act). During the plan phase, different improvement ideas are collected from inside and outside the team. These are then prioritized and implemented. The check phase means that the improvement is followed-up to confirm that the saving has been achieved and that by measurement this can be quantified. The act phase should include updated standards to solidify the changes as well as horizontal expansion and implementation of the same improvement where applicable. An example of horizontal implementation could be an electrical cabinet where the cooling temperature could be changed from 25°C to 35°C, this should then also be implemented in the remaining electrical cabinets in the factory if applicable.

14.3.8 Energy metrics and understanding of current situation

One of the first tasks for the energy team is to understand the current situation and establish the baseline. The baseline is the level against which we will improve from and against which we will compare future improved levels. An example could be that we use the total energy use in 2019 as the baseline and then we set the target to improve with 2% annually. By doing this we have a clear baseline (energy use 2019) to compare with to see if we are doing well or not. To understand the current situation is very important to ensure the right focus. Part of this work should be focusing on which metrics that are important and relevant to measure and track as well as how these data are tracked. There are many metrics which could be measured and tracked such as total energy cost, total energy use, total electricity use, idle electricity use, total natural gas use, and so on. It is also useful to track production volumes so that energy use per produced unit can be visualized. More metrics are not necessarily better, but rather the focus should be on relevant metrics which gives an overview of the energy use and the rate of improvement/reduction. Each metric should have a definition document where it outlines how the metric is defined including such information as during which conditions and times the metric is taken, any possible calculations done to calculate the metric, and so on.

When doing improvement work it can be useful to set long term targets so that the improvement strategy can be kept stable. There is no need to rediscuss target levels each year and to allow long term investments to be incorporated and planned well in advance.

When tracking the energy improvement metrics it is also important to decide if they should be tracked according to a top-down or bottom-up approach and if it is static or dynamic measure.

> Bottom-up improvement metrics means that if energy saving is the metric, then each improvement action should be quantified. For example, if energy saving is due to LED light fixtures then there should be a measurement of the old lamp and the new LED lamp and the difference between them is the energy saving (multiplied by number of light fixtures). The equivalent top down approach would be to do a total electricity measurement before installation and then after installation and that would be the improvement metric. In the interest of being rigorous and scientific, a bottom-up approach is preferable.
> The second consideration is if these metrics are static or dynamic. In the example of the above LED light fixtures, if the energy improvement is calculated in a situation where the lights are turned on 8 hours per day and if production demand increases and overtime work is assigned, this means that the lights will now be on 10 hours per day. The question then arise, do we now then need to recalculate the saving for 10 hours (dynamic)? An alternative approach is that we keep the previous saving which was calculated during 8 hours (static) even though the actual saving with overtime work is larger. To reduce work load, a static approach is preferable, if possible, in order not to have to recalculate savings as production volumes change.

14.3.9 Energy measurement system

For a factory to start working with energy improvements it is beneficial to have some sort of measurement system so that the result of the actions can be seen and that implemented actions can be confirmed to be beneficial. However, there are many improvements which can be implemented even without a measurement system. For example, if you reduce the indoor temperature with one degree Celsius in the winter time (e.g., from 23°C to 22°C) when it is cold outside you will know that you have contributed to saving energy in most cases, that is, taken that the space heating does not fully emanate from excess heat in the building.

If a factory does not have access to its own measurement system, then perhaps data from the energy supplier can be used instead. The energy provider keeps track of how much energy the company uses as they send invoices to their customers in accordance with its logged use of energy. Most energy providers today have very sophisticated websites where one can log in and extract for example electricity use with 15 minutes resolution.

14.3.10 Targets and target break-down

To progress in the continuous improvement work the leadership should put some improvement targets on the organization. These targets can be different depending on the context and situation of the organization. In some cases perhaps only one overall target is formulated such as reducing energy cost with 5% per year for the next 5 years. In other cases, several targets can be set driving different objectives where one target drives an overall energy use reduction and another one drives reduction in energy use during nonproduction hours. In the latter case, the target during nonproduction hours is more likely to require more focus on culture and behavior rather than technical actions.

An important step relating to energy targets is to make a relevant target breakdown. The purpose of the breakdown is to make each department and team responsible for achieving their objectives. Let's illustrate this by a factory consisting of three production areas: weld, paint, and assembly. The overall objective is to reduce 10 GWhs of energy use over 5 years. Weld and paint is using 40 GWh each annually and assembly 20 GWh. This means that weld should save 0.8 GWh per year (40% of the total objective is 4 GWh divided over the 5 years). If we then have five different production teams in the weld department all using the same amount of energy they should each then save 0.16 GWh of energy annually. By doing this target breakdown we have gone from a very large objective of 10 GWh over 5 years, which might seem difficult and unattainable, down to a 0.16 GWh per year per weld team objective on which we can engage the team around and spend the coming year in working toward achieving this goal.

14.3.11 Communication

Data should be collected and consolidated to be able to communicate a clear picture. One approach is to consolidate the required information so that it can be understood at a glance. There is a difference between data being visual and visualized. This requires three things: (1) the measured data, (2) the target curve, and (3) a status symbol. Status symbol can be whatever the factories are currently being used to (green/red dots, smiley/sad face, and so on). In Lean thinking fashion, one would use the circle, triangle, and cross symbols. The circle, triangle, and cross symbols can be colored green, yellow, and red to further enhance the visualization. The advantage of the circle, triangle, and cross compared to colored dots is that even color-blind people can understand the status at a glance. Circle, triangle, and cross is very common in Japan, for example, train status is displayed using these symbols at train stations in Japan (see e.g., Fig. 14.6).

One way to enhance the communication around energy is for the energy manager to put together a brief newsletter where the main ongoing activities and change points are discussed. This newsletter can be sent out on a monthly basis. This can be very useful to drive awareness around energy improvements, to trigger ideas and proposals by people as well as making everyone aware of the different change points related to energy use. One example of good use of change point awareness could be if washing detergent is changed to allow lower water temperature in a component washing process. If everyone around this process is aware of this change, they are more likely to notice any adverse effects of it and hence to help in securing the quality of the process.

In order to be able to communicate with all stakeholders it might also be beneficial to create a summary document where the overall background and purpose as well as ongoing activities and results/status are consolidated. One example of this could be to use a visual dashboard as in Fig. 14.6. This example is based on A3 thinking from Toyota. An A3 at Toyota is a structured problem-solving method with eight clear steps. It is called an A3 because of the A3 paper format normally used (11″ × 17″ in the United States). In the example in Fig. 14.6, this has been modified to become a visual dashboard to give a brief summary of the overall current status and for the factories that are involved in this activity.

14.3.12 Energy corner/energy board

To ensure broad communication and awareness as well as commitment it can be beneficial to start posting information regarding the energy activities on a display board located in a strategic location which many of the employees pass by. On the board, data and information around the energy improvement activities can be visualized such as current energy use over time as well as the status (above or below target). There can be forward-looking information to inform about upcoming changes so that as many people as possible are aware of the change points in the factory. An empty space on the board can be provided for improvement suggestions related to energy use.

14.3.13 Feedback

To get engagement we need to show that we want to get input and a key to achieve this is to provide feedback to the different production teams. Ideally each production team should at any time be able to see how they are performing with regards to their energy target. This means that they need to be provided with information on energy use, that they have a target and that we can give a status based on the two

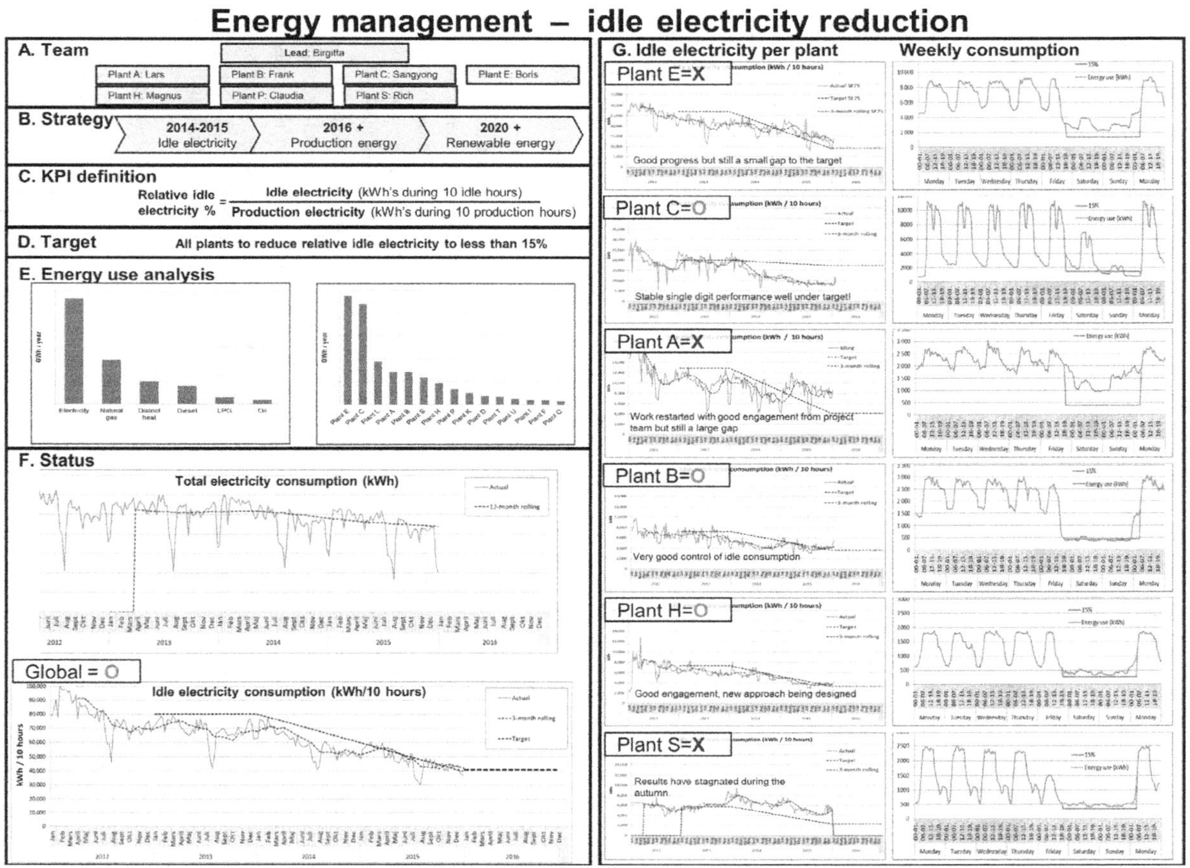

Figure 14.6 Visual dashboard for multi factory activity (Sannö et al., 2019).

previous data points (energy use above target equals red status and vice versa). The frequency of this information is ideally real time, but this is in many cases difficult and costly to achieve so a reduced but fixed frequency can be used such as daily or weekly. This data can also be very useful in identifying abnormalities. An example is an air pipe in the compressed air system which broke underground between two buildings. Since there was no monitoring system and no target it took 6 months before this was detected and only found out by chance when a visitor saw air bubbles in a water puddle in the parking lot between the buildings!

14.3.14 Training

To start any activity, it can be useful to provide relevant training to the people who will be involved. If a factory starts a Lean energy journey, everyone can benefit from a short introduction training in Lean as well as in Energy management and overall energy efficiency practices. The training can be very short and is ideally combined with some sort of practical exercise like a small shop floor activity to find energy-saving ideas in their work area such as a short energy kaizen event, which will be described below.

14.3.15 Energy kaizen events

The majority of the improvement ideas should ideally come out as a part of normal daily work by the team members on the shop floor, but it can be useful to accelerate this process by arranging different types of kaizen events. Here we describe two different kaizen events, one very short and one more elaborate.

> The short kaizen event can be a 2-hour event which is ideally run with all team members of a smaller production team (perhaps up to 20 members). The first 30 minutes are spent on introducing the event and its purpose as well as connecting it to the objectives of the factory and explaining why we are doing it. The following 60 minutes are spent on the shop floor in the production cell of that team where the team members are asked to write down energy-saving ideas on post-it notes as part of a brainstorming session. It can be useful to subdivide the team into smaller teams and ask them to focus on different pieces of equipment. The final 30 minutes are spent summarizing and discussing the ideas and arranging them on a four-block chart according to estimated size of energy saving versus estimated effort to implement. The team is then asked to implement the ideas. Support is given by the energy team if required and the ideas are tracked on the open issues list until implementation is completed.
>
> The long kaizen event is a more elaborate event running over two to three days. This event normally includes people with competence in energy as well as the facility support but should also include production team members. Depending on the type of production process the set-up can be different but for a production facility which only runs production in the weekdays, a good starting time could be Sunday morning. The whole Sunday can then be spent focusing on idle losses, which is energy used when production is not running. To walk around in a factory which is not running makes it easier to identify equipment, lights, and so on which are on when perhaps they could be shut off. The exact location of compressed air leaks is also easier to identify compared to during noisy production hours.

The Monday is spent focusing on production energy use and all improvement ideas are estimated and quantified as far as possible. The Tuesday is spent wrapping up the investigations during the previous 2 days and the workshop does a report-out to the factory management team in the afternoon with a clear list of suggested next steps and actions as well as showing the improvements which were implemented over the last 2 days. This sort of longer kaizen event is sometimes known as "energy treasure hunt."

14.3.16 Energy equipment kit

A successful energy team will need some basic equipment to do their job in an effective way. One important piece of equipment is a portable energy logger which can be connected to a piece of equipment to log the energy use over time. To be able to measure production equipment and visualize the energy use and discuss the findings with the production team members can be very powerful in getting ideas on how to reduce energy use. It is also very motivating to the production team members when the saving results can be quantified and presented back to them.

Other useful equipment can be ultrasonic air leak detector for searching for compressed air or vacuum system leaks. An infrared or thermography camera to understand temperature profile of equipment as well as facility can highlight many improvement opportunities such as poor insulation or actual temperatures of process media exceeding specifications. A lux meter to measure light levels can determine if light levels are in line with specified standards or if there is potential for dimming or turning off some lights based on the requirements.

For a more thorough description of metering devices and how to meter, see Chapter 6, Energy auditing and measurements in practice.

14.3.17 Suggestion schemes/recognition

If the factory is operating a suggestion or improvement recognition scheme, energy-related improvements should be included. If there is no such scheme in place, that could potentially be something to consider. To stimulate teamwork the suggestion scheme could be team-based and rather than monetary rewards the company could instead offer the team an outing for a group activity.

14.3.18 Friendly competition

Based on the above comments regarding reward scheme, a friendly competition can also be organized between different production teams to see which team can find the most energy savings. The result could be presented in a leaderboard as a percentage reduction for each team with gold, silver, and bronze positions clearly highlighted. This sort of competition could be arranged over different time horizons with a longer-term competition followed by shorter ones where the team is challenged over a week or day to achieve the largest reduction. This could, for example, be aligned with global events such as WWF Earth hour and use that as a starting point.

14.3.19 Night walks

An easy starting point for an energy improvement project and to get management buy-in could be to conduct a so-called night walk. A night walk is an event where a team (or several teams in a large factory) walks through a factory during nonproduction hours. Nonproduction hours do not necessarily mean during nighttime but could be late afternoon on a Friday when the production has stopped, and the building control system should have gone into nonproduction mode. Walking around, preferably together with representatives from the management team, will most likely reveal a lot of energy waste such as lights left on, compressed air leakage, machines and equipment not turned off, and so on. Based on this, a list of all findings can be made and all the findings can be tagged which will allow an improvement dialogue the following working day. People can then be assigned as responsible to ensure that this equipment is turned off for the following weekend. A night walk is preferably repeated on minimum a monthly basis initially in the lean energy management journey and then with some set frequency such as once every quarter. A creative way of doing the night walks is to take the support from the security staff. After a short training they could go around and tag equipment that should not be on. Additional benefits are that this forces standardization of which equipment that should be on or not. The security guard cannot decide this for the organization. An important component of the night walk is thus not just to spot improvement areas/measures, but also have the team members to create standards on how to operate the installations in a more energy-efficient manner.

14.3.20 Calculating return of investment

Implementing energy savings will reduce the energy bills but to get a good business result from investments, some sort of return on investment needs to be calculated and judged against some criteria level. A commonly used criterion is internal rate of return (IRR), which was explained earlier in this book. If the IRR of a specific investment is higher than the criteria, then the investment can be done. To increase the number of investments that meet the criteria different techniques can be used. Obviously the investment should be thoroughly scrutinized to minimize the investment size, but the saving potential can also be increased by using an artificial energy cost increase. This can be motivated since the risk of future energy prices is real and unknown and a reduced energy use will minimize this risk. If the energy cost including inflation is estimated to increase with 3% per year, then an artificial energy cost increase of 5% could be used in the IRR calculations. The benefit to the environment can also be used to motivate such a policy. Further, including nonenergy benefits as described earlier in this book also further strengthens the investment proposals.

14.3.21 Standardize the energy management

The way a factory is running the energy management work should also be standardized and documented. This will be very useful if there are personnel changes to reduce the start-up time of new employees. Creating a simple "play book" to outline the energy management work process could be one way of achieving this. The more advanced version of this would be to implement a full energy management system such as ISO 50001.

14.3.22 Energy management in purchase of new equipment or facility projects

The requirements which we put on equipment in a factory should be developed and documented together with the manufacturing engineering and equipment purchase team. If this is not done correctly then existing equipment in a factory is optimized for minimum energy use but that work would then have to be repeated every time new equipment is installed into the factory. For example, if we have a requirement that all equipment should be possible to be completely turned off to avoid idle energy losses, this needs to be considered when new equipment is purchased. When undertaking facility projects, energy management should also be a point which is considered.

14.3.23 Long term plan for equipment replacement

Connected with above item is the need to have a long-term plan for replacement of equipment. The moment when existing equipment is replaced is an excellent opportunity to make energy savings with good return of investment. Imagine the business case where you must replace the light system in a factory due to end of technical life. Compare this to a business case where you make the same replacement but of a perfectly well functioning existing light system which is only a couple of years old. In the latter case you might still be depreciating the previous investment and, in addition, you will have the full installation cost. Taking the opportunity of changes driven by other parameters can help improve energy saving business cases. Doing this on a long-term basis is also beneficial in order to improve the level of renewable energy used in a factory where changes are often very costly due to large support systems impacted. Replacing an oil burning heating system will be easier to justify if the oil burner has reached its maximum technical life.

14.4 Case study Volvo Construction Equipment (Volvo CE)

14.4.1 Brief introduction

Volvo CE is a construction equipment producer which is a part of Volvo Group with approximately 14,000 employees operating 15 production factories globally.

14.4.2 Strategy approach to energy management

Volvo Group has had Environmental Care as one of its core values since 1972. To challenge themselves, Volvo Group reached an agreement with World Wildlife Foundation (WWF) to enter their Climate Savers program, primarily focusing on the fuel consumption of the products. This agreement was later expanded to include further aspects such as production factory energy efficiency. Based on this, Volvo CE updated its Environmental Care focus to take a broader energy management approach using Lean energy thinking.

The priority prior to this was to reduce CO_2 emissions from the factories. The new strategy was turned around to first focus on energy conservation. This was followed by improved energy efficiency and then finally renewable energy as seen in Fig. 14.7.

- *Energy conservation* refers to not using energy in the first place by, for example, turning off equipment which is not used, turning on equipment later at the start of production and turning it off earlier and generally using existing equipment in a smarter way to eliminate or reduce the amount of energy used.
- *Energy efficiency* is the second step which relates to investing in more energy-efficient equipment. For example, new electrical motors (such as replacing IE2 motors with IE5) but also investing in energy recovery and fundamentally when investing to invest in right size equipment (equipment found in factories are often over dimensioned due to unclear initial requirements and exaggerated safety limits).
- *Renewable energy* is the final step in the energy pyramid. This was initially just called CO_2 neutral but changed to renewable not to include nuclear energy and hence participating in creating a global demand for renewable energy. Adding this as a last step is actually misleading as several factories have been purchasing renewable electricity for many years already and there are ongoing activities to improve the level of renewable energy (50% of energy used in 2016 was renewable). To put it as a last step however signals the new strategy, which compared to the old one where CO_2 reduction was the priority, is different. It also highlights the need to start to focus first on energy conservation and then on energy efficiency rather than just purchasing solar panels and then wasting the majority of the electricity produced by these solar panels.

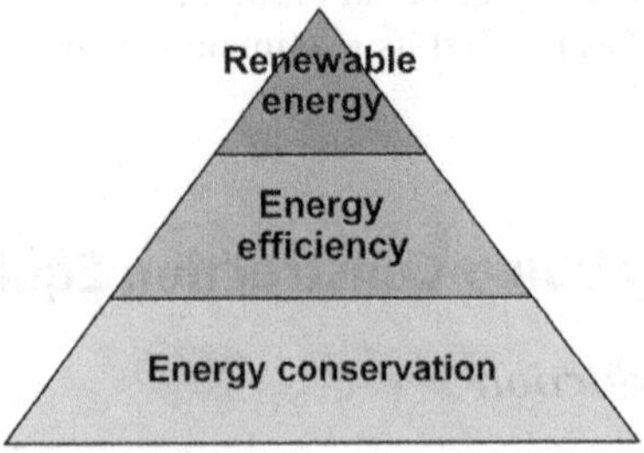

Figure 14.7 Energy pyramid.

Following the strategy outlined by the energy pyramid has several reasons and multiple benefits:

- *Build culture and behavior*: strong focus on energy conservation means involving everyone in the factory and hence starting to build the right behavior and culture around energy. The people working in the factory never see or have to pay the energy bills and hence the awareness is by default lower compared to a home-owner. In the factory, employees behave differently from when at home, for example, it would be considered normal to turn off the lights at home when you leave to go to work, the same is not obvious at work for many different reasons (awareness, not sure to be the last to leave, not knowing where the switch is, and so on). To build culture and behavior takes time and it is therefore important to start as soon as possible and to make sure to involve all employees and levels of management in the factory.
- *Low cost*: to focus on energy conservation does not cost much money as no significant investments are involved. The main cost is man-hours used to set up the approach and to train and involve everyone. This means that no capital expenditure or investment budget is needed to start which lowers the barrier and inertia to get started.
- *Self-funded*: not only is energy conservation low cost but it also enables the strategy to be self-funded. Through energy conservation the energy bills will be reduced and hence costs are reduced. These savings can then be utilized to invest in the following step, energy efficiency, which will reduce cost even more. This may in turn be utilized to improve the level of renewable energy which in some cases can be costly (e.g., replace factory heating system or paint shop oven heating source).
- *Reduce size of future investments*: is one benefit of energy conservation which is easily overlooked. By working with a system without making capital investments and just trying to do low cost waste elimination improvements one gets to know the system better and, for example, reduced flow rates will perhaps in a future investment enable the use of smaller and less expensive pumps and motors.

It is important to remember that energy conservation is a never-ending journey. In that way the visualization in the energy pyramid can be misleading. In order to keep a culture and behavior in place there needs to be some expectation from the organization and its leadership. If focus is lost, for example if significantly increasing production volumes happens, then there is a significant risk that performance and results will slide back.

Based on the revised strategy an energy management journey was started. Data analysis showed that Volvo CE is a low energy intensity company which was no surprise. Energy intensity being defined as annual energy cost divided with turnover was found to be 0.54%. The low level of energy intensity also helps to explain why energy use was not on the agenda of management. The data analysis further looked into the energy use by energy type as seen in Fig. 14.8 where it can be seen that electricity is by far the most used energy type. When electricity use was broken down per factory the Pareto in Fig. 14.9 could be made.

The outcome of the data analysis was that electricity was the prioritized energy type to address initially. To reduce the scope of the work it was also decided to initially only include the seven largest factories since that still captures 75% of the electricity use. The result of the data analysis is an example of prioritization and focus to get a stronger drive and achieve tangible results quicker. Based on a study

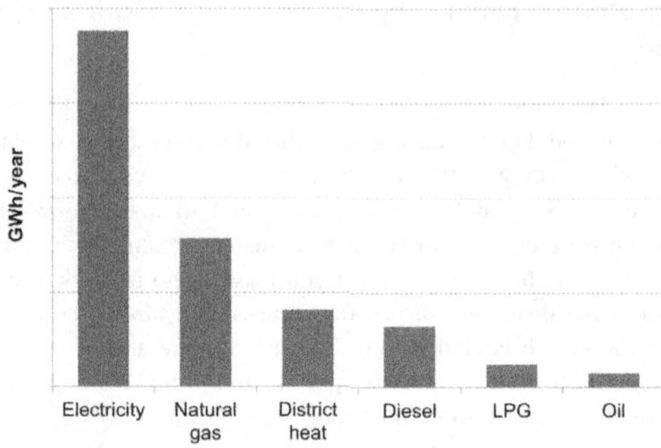

Figure 14.8 Energy use by type.

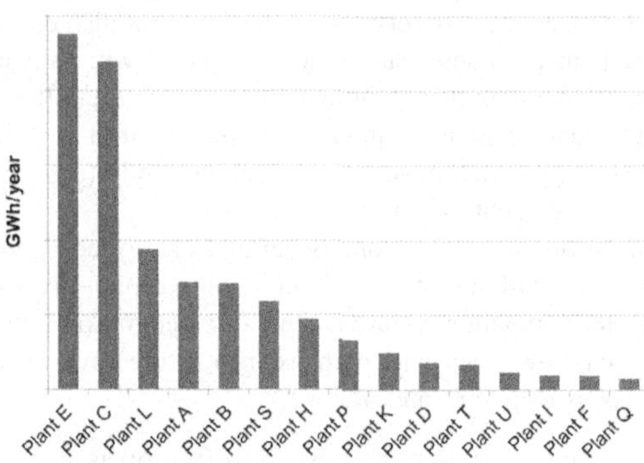

Figure 14.9 Electricity use per factory.

per factory it was estimated that more than 30% of the electricity used was actually wasted during nonproduction hours. Based on the strategy to start with energy conservation is was decided to start the energy management journey by focusing on electricity idle losses in the seven largest factories.

Prior to 2013, the only energy performance index which was followed on a global level in operations was energy use per produced machine (kWh/unit). This is a very limited measurement when driving a structured improvement work as it will mostly show if it was a warm or cold winter season and if production volume increased or decreased. So, the next challenge was to define the key performance index for idle electricity losses and to ensure that when comparing the seven factories with each other we would be comparing apples to apples. Idle electricity use was defined as ten consecutive hours in the weekend between 19.00 (7 pm) on Saturday evening and 05.00 (5 am) on Sunday morning. A second measurement was defined as production time electricity which was taken as 10 consecutive hours on the following working day (normally following Monday unless vacation/

holiday). If the idle electricity was divided with production electricity then a third measurement could be established which was called relative idle electricity as seen in Eq. (14.1). The relative idle electricity allowed comparison between the different factories despite difference in product, size, working hours, and so on. A KPI definition document was created to outline the exact details.

$$\textbf{Relative idle electricity \%} = \frac{\textbf{idle electricity} \text{ (kWh's during 10 idle hours)}}{\textbf{Production electricity} \text{ (kWh's during 10 production hours)}}$$

Equation 14.1 Relative idle electricity calculation.

Data were collected from all the factories and the relative idle electricity was calculated. This revealed that the best factory was at 17% and the worst at 42%. Based on this, a target was set for all factories to achieve less than 15% before the end of 2015. Benchmark studies showed that a good level of idle electricity for similar companies would be less than 10% which gives a potential for a more challenging target in the future. An issue which was highlighted in some factories was that they did not have an electricity measurement system installed, in those cases the data supplied by the electricity supplier was used.

14.4.3 Results

In this section we will briefly review the results which were achieved. The energy use per unit produced was reduced with 27% over the 5 years. Fig. 14.10 shows idle electricity use and shows that more than 50% decrease was achieved from the baseline year of 2012 until mid-2016. The lowest level occurred in September 2016 when the 15% target was achieved but since then the results have been going in the wrong direction. The main reason for this is significant production volume increase which has two effects, firstly, that some production cells are being used during the time window of idle and, secondly, that due to the high production volumes the attention from management on energy management topics is lower. The culture around energy management is still not solidified. As can be seen in the target curve, the ambition level is increasing gradually with a new target of 12% and later 10%.

Fig. 14.11 shows the total amount of energy savings which has been implemented. During the first 5 years period, a total of 44 GWhs worth of energy savings were implemented compared to the target of 31 GWh. The way this is measured is a static bottom-up measurement at the time of implementation, meaning that this is volume independent. The energy savings included initially the 10 largest factories and later the project was extended to include all factories. The savings are audited by an independent third party on an annual basis through conference call as well as factory visits. The target for the second 5 year period (2019–23) is 24 GWhs.

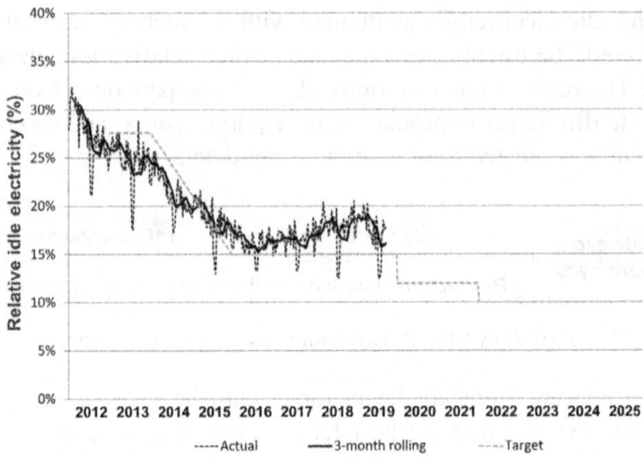

Figure 14.10 Idle electricity use.

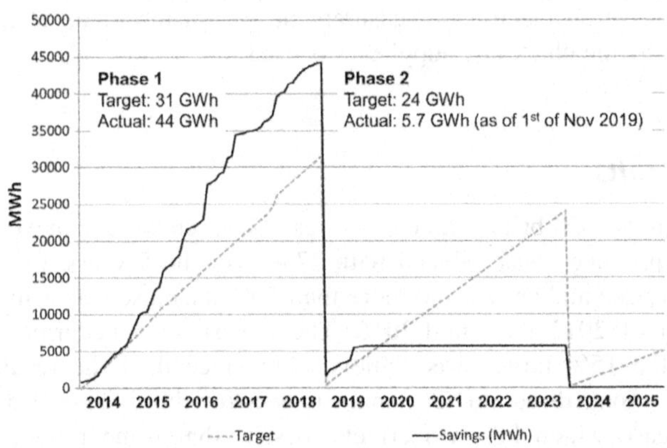

Figure 14.11 Energy savings implemented.

In summary, it has been shown that a large nonenergy intensive company can achieve considerable energy reductions if systematic improvement activity is deployed such as Lean Energy. There has been a significant reduction in total energy use and hence energy bills as well as contributing positively to the environment and harvesting numerous nonenergy benefits as well as training the whole organization in Lean and Continuous improvements. The cumulative financial impact is estimated to over 10 M EUR over the last 5 years.

References

Fujimoto, T., 1999. The Evolution of a Manufacturing System at Toyota. Oxford University Press, Oxford.

Hall, N.P., Roth, J., 2003. Nonenergy benefits from commercial and industrial energy efficiency programs: energy efficiency may not be the best story. In: Proceedings of the 2003 International Energy Program Evaluation Conference, Seattle, WA, USA, August 20–22, 2003; pp. 689–702.

Ohno, T., 1988. Toyota Production System: Beyond Large-Scale Production. Productivity Press, New York, NY.

Sannö, A., Johansson, M., Thollander, P., Wollin, J., Sjögren, B., 2019. Approaching sustainable energy management operations in a multinational industrial corporation. Sustainability 11 (3).

Toyota Institute, 2001. The Toyota Way 2001. Toyota Motor Corporation.

Womack, J.P., Jones, D.T., 1996. Lean Thinking. Free Press, New York, NY.

Womack, J.P., Jones, D.T., Roos, D., 1990. The Machine That Changed the World. Free Press, New York, NY.

Part III

Energy policy gap

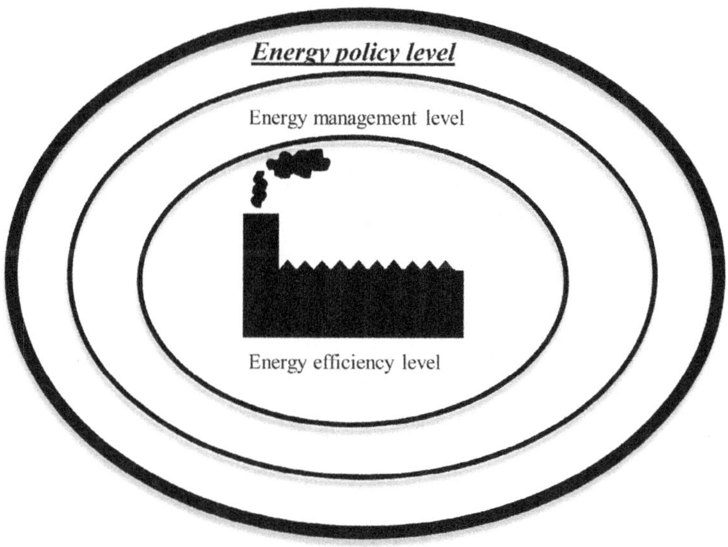

Improved energy efficiency from a governmental perspective can be achieved by adoption of different types of energy policy programs. Properly designed, these policy programs can create incentives for companies to improve both the energy efficiency and energy management gaps. The difference between the potential for energy efficiency and energy management by means of policy programs, and the policies, which currently are in place, is referred to as the energy policy gap. Part III focuses on policy programs for improved energy efficiency and how these can be designed, implemented, and evaluated, and originates from the energy policy gap.

Barriers to energy efficiency

15

Chapter Outline

15.1 Theoretical barriers to energy efficiency

There are many scientific publications that have shown the gap between potentially cost-effective energy-efficient measures and measures that are not being implemented, the so-called energy efficiency gap or the energy paradox (Rohdin and Thollander, 2006; Rohdin et al., 2007; Johansson and Thollander, 2018). There are several barriers to energy efficiency that can explain this energy efficiency gap. A barrier is defined in this context as an assumed mechanism that prevents improvement measures, even though these measures are both energy-efficient and perceived cost-effective.

An important distinction of barriers is that of so-called market-failure barriers or market imperfections and that of market barriers. Market failure or market imperfection barriers constitute a stricter form of barrier than market barriers. Market imperfections or market failures are a cause of a violation of the theoretical principles of well-functioning markets. The existence of such barriers may in turn lead to the government intervening with, for example, energy policy programs, to address the market failure barriers inhibiting improved energy efficiency. A market barrier, on the other hand, is a wider definition of a barrier to energy efficiency which may not solely be a cause of violations of the principles of well-functioning markets. For a more thorough explanation of this, see Chapter 16. Table 15.1 shows different theoretical explanations to the energy efficiency gap.

Introduction to Industrial Energy Efficiency. DOI: https://doi.org/10.1016/B978-0-12-817247-6.00015-8

Table 15.1 Classification of barriers to energy efficiency (Sorrell et al., 2000; Rohdin and Thollander, 2006; Rohdin et al., 2007; Thollander et al., 2010a,b).

Theoretical barriers	Comment
Imperfect information (Howarth and Andersson, 1993)	Lack of information may lead to cost-effective energy efficiency measures opportunities being missed
Adverse selection (Jaffe and Stavins, 1994)	If suppliers know more about the energy performance of goods than purchasers, the purchasers may select goods on the basis of visible aspects such as price
Principal–agent relationships (Jaffe and Stavins, 1994)	Strict monitoring and control by the principal, since he or she cannot see what the agent is doing, may result in energy efficiency measures being ignored
Split incentives (Jaffe and Stavins, 1994)	If a person or department cannot gain benefits from energy efficiency investment it is likely that implementation will be of less interest
Hidden costs (Jaffe and Stavins, 1994)	Examples of hidden costs are overhead costs, cost of collecting and analyzing information, production disruptions, inconvenience, and so on
Access to capital (Jaffe and Stavins, 1994)	Limited access to capital may prevent energy efficiency measures from being implemented
Risk (Jaffe and Stavins, 1994)	Risk aversion may be the reason why energy efficiency measures are constrained by short payback criteria
Heterogeneity (Jaffe and Stavins, 1994)	A technology or measure may be cost-effective in general, but not in all cases
Form of information (Stern and Aronsson, 1984)	Research has shown that the form of information is critical. Information should be specific, vivid, simple, and personal to increase its chances of being accepted
Credibility and trust (Stern and Aronsson, 1984)	The information source should be credible and trustworthy in order to successfully deliver information regarding energy efficiency measures. If these factors are lacking this will result in inefficient choices
Values (Stern, 1992)	Efficiency improvements are most likely to be successful if there are individuals with real ambition, preferably represented by a key individual within top management
Inertia (Stern and Aronsson, 1984)	Individuals who are opponents to change within an organization may result in overlooking energy efficiency measures that are cost-effective
Bounded rationality (DeCanio, 1993)	Instead of being based on perfect information, decisions are made by rule of thumb
Power (Sorrell et al., 2000)	Low status of energy management may lead to lower priority of energy issues within organizations
Culture (Sorrell et al., 2000)	Organizations may encourage energy efficiency investments by developing a culture characterized by environmental values

15.2 Design of a barrier study

When a study of barriers to energy efficiency is carried out, the choice can be made between two basic approaches. Either the study is made on the basis of so-called grounded theory, that is, the study is not based on any actual theory or previous research on how energy efficiency can be understood. Before such a study is carried out, a pilot study can be carried out, for example, by asking a smaller number of respondents what major barriers to energy efficiency they experience, and then proceed with designing the study. The person who carries out the study can also make additions with regard to barriers that are assumed to be important.

The second way of conducting a study on barriers to energy efficiency is to start with a certain theory formation, for example, that risk aversion means that a person is not willing to take a decision about an energy efficiency measure. With such an approach, a review is made of the primary theories that exist in the field, then a number of "indications" or "stated barriers" are chosen. In such a study, the ranked barriers can be related to the theory the study was based on. Risk of production disruptions is sorted under the theory of risk, that is, risk, as an explanation to why energy efficiency measures are not implemented, etc.

These two principal means of research, also denoted inductive and deductive remain a fundamental aspect of any good textbook in the field of social sciences theories wherefore we direct a reader who is interested to dig deeper into this, to the social science area of research.

When studying barriers, it is important to define which unit to study. In a study that focuses on energy efficiency, the energy efficiency measures that can be implemented constitute part of the definition. For the study not to become too extensive, the factor "profitable" or "cost-effective" should be included in the definition (to avoid the plethora of measures that are directly unprofitable and which seem relatively uninteresting to undertake research about if it relates to the energy efficiency gap). Finally, it is not the technology itself or the operational measure that is being studied, but the actual decision to implement such a measure or rather why this is not being done. This provides us with the following proposed definition of the analyzing unit in a barrier study: decisions on cost-effective energy-efficient measures.

The normal way of then proceeding with the study, be it inductive or deductive, is to either conduct an interview study or send out a questionnaire with an array of proposed barriers to energy efficiency that one asks the respondents to rank.

15.3 Empirical barriers to energy efficiency—the case of Sweden

Based on a deductive approach, Sorrell et al. (2000) created a questionnaire on barriers where each stated barrier in the questionnaire is related to a theoretical explanation. In an early work in Swedish industry on barriers by Rohdin and Thollander (2006), the authors in response to the ranking of real barriers provided by the respondents, then related back to theory, see Table 15.2

Table 15.2 Results from a barrier study (Rohdin and Thollander, 2006).

Rank (points)	Stated barrier	Comments	Main theoretical barrier
1 (12)	Cost of production disruption/hassle/inconvenience	Five out of eight respondents considered the cost of production disruption a major barrier to energy efficiency	Hidden costs
2 (12)	Lack of time or other priorities	Lack of time was something that the interviewees often perceived to be a large barrier	Hidden costs
3 (11)	Cost of obtaining information about the energy consumption of purchased equipment	Half of the respondents considered this to be a large barrier to energy efficiency, especially if the information comes from equipment suppliers	Imperfect information
4 (11)	Technical risk such as risk of production disruptions	The risk of production disruptions was considered a barrier by half of the respondents	Risk
5 (11)	Other priorities for capital investments	Generally, production-related investments had a higher priority among the studied companies than investments in energy-efficient technology	Hidden costs
6 (10)	Technology is inappropriate at this site	This was considered a large barrier by two of the respondents	Heterogeneity
7 (9)	Lack of staff awareness	Staffs are not aware of or do not pay attention to energy efficiency and its potential	Imperfect information
8 (9)	Lack of technical skills	Some of the respondents felt that lack of technical skills within their organization constitutes a barrier to energy efficiency	Imperfect information

(*Continued*)

Table 15.2 (Continued)

Rank (points)	**Stated barrier**	**Comments**	**Main theoretical barrier**
9 (9)	Access to capital	Access to capital may constitute a barrier. However, the respondents claimed that it is not a major barrier as long as the investment criteria are fulfilled	Access to capital
10 (8)	Poor information quality regarding energy efficiency opportunities	Even though energy audits had been carried out, six out of eight respondents considered this to be a barrier to energy efficiency. Only two considered it to be major barrier, however	Imperfect information
11 (8)	Possible poor performance of equipment	Six out of eight considered this to be a barrier, two considered it to be large	Risk
12 (7)	Cost of identifying opportunities, analyzing cost-effectiveness and tendering	This was considered a substantial barrier by two of the respondents	Hidden costs

In a review of the carried out research in Swedish industry on barriers to energy efficiency, the studies carried out between 2004 and 2018 were summarized. Table 15.3 shows this summary of undertaken barrier studies in Swedish industry.

The table shows that the barriers vary considerably between different industries. Among the highest ranked barriers are, for example, the difficulty of obtaining accurate information on the energy use of procured equipment (imperfect information) and lack of submetering (split incentives). Furthermore, lack of time, other priorities, and other priorities for capital investment can be due to split incentives and the principal–agent relationship. Lack of capital has been ranked high in a few of the industries.

Table 15.3 Largest ranked barriers to improved energy efficiency in Swedish industry identified in empirical studies (Johansson and Thollander, 2018).

Stated barriers / Sector	Non-energy intensive manufacturing industry	Foundary industry	Pulp and paper industry	Manufacturing SMEs	Largest industrial companies in five municipalities	Foundary industry (Sweden and Europé)	Iron and steel industry	Industrial companies that participated in the Swedish Energy Audit Programme
	1. Cost of production disruption/ hassle/ inconvenience	1. Lack of access to capital	1. Technical risk such as risk of production disruption	1. Lack of time or other priorities	1. Lack of time or other priorities	1. Lack of budget funding	1. Technical risk such as risk of production disruption	1. Lack of time or other priorities
	2. Lack of time or other priorities	2. Technical risk such as risk of production disruption	2. Cost of production disruption/ hassle/ inconvenience	2. Other priorities for capital investments	2. Other priorities for capital investments	2. Other priorities for capital investments	2. Access to capital	2. Access to capital
	3. Cost of obtaining information on the energy use of purchased equipment	3. Lack of budget funding	3. Technology is inappropriate at this site	3. Lack of access to capital	3. Long decision chain	2. Lack of access to capital	3. Other priorities for capital investments	3. Slim organization
	4. Technical risk such as risk of production disruption	4. Difficulty of obtaining information on the energy use of purchased equipment	4. Lack of time or other priorities	4. Cost of production disruption/ hassle/ inconvenience	4. Energy objectives not integrated into operating, maintenance or purchasing procedures	2. Lack of time or other priorities	4. Lack of time or other priorities	4. Lack of technical skills
	5. Other priorities for capital investments	5. Other priorities for capital investments	5. Lack of access to capital	5. Lack of budget funding	5. Technical risk such as risk of production disruption	3. Lack of technical skills	5. No available cost-effective technical measures	5. Costs of identifying and analyzing opportunities

	6. Technology is inappropriate at this site	6. Possible poor performance of equipment	6. Slim organization	6. Lack of submetering	6. Lack of submetering	4. Department or workers not accountable for energy costs	6. Uncertainty about future energy prices	6. Lack of staff awareness
	7. Lack of staff awareness	7. Lack of submetering	7. Possible poor performance of equipment	7. Difficulty of obtaining information on the energy use of purchased equipment	7. Slim organization	5. Cost of production disruption/hassle/inconvenience	7. Limited authority of energy manager	7. Difficulties in obtaining information about the energy use of purchased equipment
	8. Lack of technical skills	8. Poor information quality regarding energy efficiency opportunities	8. Lack of budget funding	8. Lack of technical skills	8. Lack of budget funding	5. Slim organization	8. Insufficient top management support	8. Lack of budget funding
	9. Lack of access to capital	9. Cost of identifying opportunities, analyzing cost-effectiveness and tendering	9. Other priorities for capital investment	9. Low priority given to energy management	9. Lack of staff awareness	6. Technical risk such as risk of production disruption	9. Poor information quality regarding energy efficiency opportunities	9. Energy objectives not integrated into operation, maintenance or purchasing procedures
	10. Poor information quality regarding energy efficiency opportunities	10. Low priority given to energy management	10. Lack of staff awareness	10. Lack of staff awareness	10. Cost of identifying opportunities, analyzing cost-effectiveness and tendering	7. Energy objectives not integrated into operation, maintenance or purchasing procedures	10. Lack of technical skills	10. Other priorities for capital investments
	11. Possible performance of energy efficiency opportunities	11. Lack of time or other priorities	11. Long decision chain	11. Technical risk such as risk of production disruption	11. Lack of access to capital	8. Low priority given to energy management	11. Lack of staff awareness or motivation	11. Technical risk such as risk of production disruption

(*Continued*)

Table 15.3 (Continued)

Stated barriers \ Sector	Non-energy intensive manufacturing industry	Foundary industry	Pulp and paper industry	Manufacturing SMEs	Largest industrial companies in five municipalities	Foundary industry (Sweden and Europé)	Iron and steel industry	Industrial companies that participated in the Swedish Energy Audit Programme
	12. Cost of identifying opportunities, analyzing cost-effectiveness and tendering	12. Lack of technical skills	12. Cost of identifying opportunities, analyzing cost-effectiveness and tendering	12. Slim organization	12. Cost of production disruption/hassle/inconvenience	8. Energy manager lacks influence	12. Uncertainty regarding hidden costs	12. Poor information quality regarding energy efficiency opportunities
# of companies in the case study (and year)	8 (2004)	27 (2005)	40 (2006)	47 (2006)	21 (2009)	20 (2012)	23 (2012)	105 (2013)
Method	Questionnaire and semistructured interviews	Questionnaire	Questionnaire	Questionnaire and semistructured interviews	Questionnaire and semistructured interviews	Questionnaire and semistructured interviews	Questionnaire and semistructured interviews	Questionnaire
Reference	Rohdin and Thollander (2006)	Rohdin et al. (2007)	Thollander and Ottosson (2008)	Thollander et al. (2007)	Trygg et al. (2010)	Trianni et al. (2013)	Brunke et al. (2015)	Backlund (2014)

SME, Small and medium-sized enterprises.

15.4 Empirical barriers to energy efficiency—emerging economies

In a novel attempt to explore barriers to industrial energy efficiency in Africa, industrial companies in Ghana's largest industrial part were studied (Wentemi Apeaning and Thollander, 2013). A selection of papers related to barriers to industrial energy efficiency from emerging economies includes but are not limited to:

- Barriers to and driving forces for industrial energy efficiency improvements in African industries (Wentemi Apeaning and Thollander, 2013);
- Barriers to energy efficiency in the Bangladesh iron and steel industry (Hasan et al., 2018);
- Barriers to energy efficiency in the Bangladesh textile industry (Hasan et al., 2019a); and
- Barriers to energy efficiency in the Bangladesh non-energy intensive industry (Hasan et al., 2019b).

The abovementioned studies' results are presented in Table 15.4.

15.5 Characteristics of energy efficiency

Research shows that barriers differ between industries, between the type of production, and between different size of companies, which region the company is located in, etc. In a novel attempt, Fleiter et al. (2012) addressed energy efficiency improvement measures and various characteristics of energy efficiency measures arguing that this may in fact be directly related to the degree of adoption. The characteristics outlined were (Fleiter et al., 2012):

- Internal rate of return;
- Payback period;
- Initial expenditure;
- Nonenergy benefits;
- Type of modification;
- Distance to core process;
- Scope of impact;
- Lifetime;
- Transaction costs;
- Knowledge of planning/implementation;
- Diffusion progress; and
- Support process or production process measure.

In a later paper, Trianni et al. (2014) elaborated on this further and inspired by Fleiter et al. (2012) outlined 17 characteristics that impact deployment of an energy efficiency measure:

- Waste reduction;
- Payback time;
- Implementation costs;
- Productivity;

Table 15.4 Largest ranked barriers to improved energy efficiency in studies from Ghana and Bangladesh (Wentemi Apeaning and Thollander, 2013; Hasan et al., 2018; Hasan et al., 2019a; Hasan et al., 2019b).

Sector / Barriers	African industries	Bangladesh iron and steel industry	Bangladesh textile industry	Bangladesh non-energy intensive industry
	Lack of budget funding	No available cost-effective technical measures	Inadequate capital expenditure	Inadequate capital expenditure
	Access to capital	High perceived risk due to uncertainty about future energy prices	Inadequate technical cost-effective measures	Other preferences for capital venture
	Other priorities for capital investments	Poor information quality regarding energy efficiency opportunities	Poor research and development	Time limitation or other significant work
	Technology is inappropriate at this site	Uncertainty regarding hidden costs	Inadequate financial incentives	Unpredictability nature of *energy parameters*
	Technical risks such as risk of production disruptions	Technical risk	Inadequate financial incentives	Unpredictability nature of energy parameters
	Lack of technical skills	Other priorities for capital investment	Other preferences for capital venture	Inadequate financial policy
	Conflicts of interest within the company	Lack of information about allocation of energy costs	Unpredictability nature of energy parameters	Insufficient attention from government
	Cost of production disruption/hassle/inconvenience	Insufficient top management support	Inadequate financial policy	Inadequate technical experts
	Cost of identifying opportunities, analyzing cost-effectiveness and tendering	Lack of staff awareness or motivation	Inadequate information about allotment of energy expenditure	Inadequate financial incentives
Reference	Wentemi Apeaning and Thollander (2013)	Hasan et al. (2018)	Hasan et al. (2019a)	Hasan et al. (2019b)

- Operation and maintenance;
- Working environment;
- Saving strategy;
- Activity type;
- Ease of implementation;
- Likelihood of success/acceptance;
- Corporate involvement;
- Distance to core processes;
- Checkup frequency; and
- Indirect effects.

15.6 Barriers to energy efficiency in compressed air systems

The vast array of research papers on barriers to energy efficiency treat energy efficiency measures and barriers to implementation in general terms. One of the few studies of technology-specific barriers is a study by Nehler et al. (2018) on barriers to energy efficiency in compressed air systems. The major barriers to compressed air energy efficiency are presented in Table 15.5.

Table 15.5 Largest ranked barriers to improved energy efficiency in compressed air systems (Nehler et al., 2018).

Technology / Stated barriers	Compressed air
	Other priorities for capital investments
	Access to capital
	Lack of budget funding
	Energy objectives not integrated into operating, maintenance or purchasing procedures
	Costs of identifying and analyzing opportunities
	Costs of production disruption/hassle/inconvenience
	Department or workers not accountable for energy costs
	Technical risk such as risk of production disruption
	Lack of time or other priorities
References	Nehler et al. (2018)

15.7 Barriers to energy efficiency, social regimes, and management theory

Research has shown that the occurrence and, not least, the importance of various barriers can be linked to so-called social regimes and the type of management strategy applied (Palm and Thollander, 2010). Regarding social regimes, there is a risk that groups are formed in a company or industry that is relatively closed, where people in between different groups do not communicate with each other. One example is the energy manager who often sits far down in an organization and rarely or never gets the opportunity to communicate with the manager. Research on management strategies shows that in companies with a strong foundation in core business or core competence risk strengthening, certain barriers to energy efficiency as the resources are mainly allocated to just core competence and core business (Thollander and Ottosson, 2010). As efficiency improvements of internal energy use in a manufacturing industry per definition cannot be a core competence or core business (since it does not generate revenue but only a saving), energy efficiency faces the risk of becoming a low-priority issue.

15.8 Energy service companies as a means of overcoming barriers to energy efficiency

One solution may be to in-source an energy service company (ESCO), which core business is energy and energy efficiency. Major barriers which an ESCO can help overcome are presented in Table 15.6.

Table 15.6 Implications for designing energy services based on the barrier theory (Thollander et al., 2010a,b).

Barrier	Implication for energy services
Heterogeneity	– This barrier may be effectively avoided by having successfully adopted cases to rely upon and skilled ESCO staff
Hidden costs	– Having an ESCO involved in the initial phase of an energy efficiency investment may greatly reduce hidden costs as the ESCO staff are specialists in their field and know where to find information
Access to capital	– An ESCO providing third party financing enables this barrier to be fully eliminated
Risk	– The industry's risk may be considerably reduced by having "specialists" involved, which greatly enhances the investment's stability. The industry's risk of entering into a business agreement with the ESCO and vice versa, however, is not as easily reduced, calling for a complex agreement to be set up which in turn may increase the magnitude of the two hidden cost barriers

(*Continued*)

Table 15.6 (Continued)

Barrier	Implication for energy services
Imperfect information	– Service opportunities exist where the information for customers/users is imperfect – Providing information on energy consumption for economically efficient decisions may contribute to service provision – Keeping information on how to increase energy efficiency within a provider may turn out to be a source of services
Asymmetric information	– Building up an agent responsible for both costs and benefits may be a key to introduce energy-efficient solutions (split incentives). – Visualizing and guaranteeing quality of products/services may lead to purchase of better solutions (adverse selection). – Transferring (and/or translating) the information at the agent level up to the principal level may be effective (principal–agent relationships)
Bounded rationality	– Routines and everyday activities do not support energy efficiency. Establish routines that contribute to "right" decisions being embedded in everyday practices
Form of information	– Different regimes within a company need different kinds of information and information packages that relate to their needs and demands
Credibility and trust	– The industry's perception of ESCOs needs to be strong as regards their credibility and trust if an energy service is to be carried out
Values and Inertia	– Different value systems can exist in a company. Promote and support value systems that benefit energy efficiency
Power	– Identify different power arenas in a company to know where to target different kinds of information and measures
Culture	– Embedded knowledge and routines need to be identified to initiate reflection on how to change and improve them

However, solutions with ESCOs do not always work out smoothly because the split incentives risk being greatly strengthened and the transaction costs for the service become too high. Further, knowledge of the production and support processes, which is gained over time by the persons operating these, may not be scalable for smaller countries, that is, the cost for gaining unique information for one specific industrial process may not payoff if a country or region only have a very limited number of such production plants. Thus the potential national or regional energy service market is not big enough.

References

Backlund, S., 2014. Efficient improvement of energy efficiency in small and medium-sized Swedish firms. Linköping studies in science and technology [Dissertation No. 1634]. Linköping University.

Brunke, J.C., Johansson, M., Thollander, P., 2015. Empirical investigation of barriers and drivers to the adoption of energy conservation measures, energy management practices and energy services in the Swedish iron and steel industry. J. Clean. Prod. 84, 509–525.

DeCanio, S., 1993. Barriers within firms to energy efficient investments. Energy Policy 9 (1), 906–914.

Fleiter, T., Hirzel, S., Worrell, E., 2012. The characteristics of energy-efficiency measures: a neglected dimension. Energy Policy 51, 502–513.

Hasan, A.S.M.M., Hoq, M.T., Thollander, P., 2018. Energy management practices in Bangladesh's iron and steel industries. Energy Strategy Rev. 22, 230–236.

Hasan, M., Rokonuzzaman, M., Amin Tuhin, R., Salimullah, A., Ullah, M., Hasan Sakib, T., et al., 2019a. Drivers and barriers to industrial energy efficiency in textile industries of Bangladesh. Energies 12 (9).

Hasan, M., Amin Tuhin, R., Hasan Sakib, T., Thollander, P., 2019b. Empirical investigation of barriers and driving forces for efficient energy management practices in non-energy-intensive manufacturing industries of Bangladesh. Sustainability 2019 (11), 2671. Available from: https://doi.org/10.3390/su11092671.

Howarth, R., Andersson, B., 1993. Market barriers to energy efficiency. Energy Econ. 15 (4), 262–272.

Jaffe, A.B., Stavins, R.N., 1994. The energy efficiency gap: what does it mean? Energy Policy 22 (10), 60–71.

Johansson, M., Thollander, P., 2018. A review of barriers to and driving forces for improved energy efficiency in Swedish industry: recommendations for successful in-house energy management. Renew. Sustain. Energy Rev. 82, 618–628.

Nehler, T., Parra, R., Thollander, P., 2018. Implementation of energy efficiency measures in compressed air systems: barriers, drivers and non-energy benefits. Energy Efficiency 11 (5), 1281–1302.

Palm, J., Thollander, P., 2010. Interdisciplinary perspective on industrial energy efficiency. Applied Energy 87 (10), 3255–3261.

Rohdin, P., Thollander, P., 2006. Barriers to and driving forces for energy efficiency in the non-energy intensive manufacturing industry in Sweden. Energy 31 (12), 1836–1844.

Rohdin, P., Thollander, P., Solding, P., 2007. Barriers to and drivers for energy efficiency in the Swedish foundry industry. Energy Policy 35 (1), 672–677.

Sorrell, S., Schleich, J., Scott, S., O'Malley, E., Trace, F., Boede, E., et al., 2000. Reducing barriers to energy efficiency in public and private organizations. SPRU's (Science and Technology Policy Research).

Stern, P.C., 1992. What psychology knows about energy conservation. Am. Psych. 47 (10), 1224–1232.

Stern, P.C., Aronsson, E. (Eds.) 1984. Energy Use: The Human Dimension. The National Academies Press, Washington, DC.

Thollander, P., Ottosson, M., 2008. An energy-efficient Swedish pulp and paper industry: exploring barriers to and driving forces for cost-effective energy efficiency investments. Energy Efficiency 1 (1), 21–34.

Thollander, P., Ottosson, M., 2010. Energy management practices in Swedish energy-intensive industries. Journal of Cleaner Production 18 (12), 1125–1133.

Thollander, P., Rohdin, P., Danestig, M., 2007. Energy policies for increased industrial energy efficiency: evaluation of a local energy programme for manufacturing SMEs. Energy Policy 35 (11), 5774–5783.

Thollander, P., Palm, J., Rohdin, P., 2010a. Categorizing barriers to energy efficiency and interdisciplinary perspective. Invited publication in Energy Efficiency. InTech Publisher.

Thollander, P., Palm, J., Sakao, T., 2010b. Energy services in industry: an interdisciplinary approach with engineering- and social science aspects. In: Proceedings of the IPS2 Conference, pp. 51–56.

Trianni, A., Cagno, E., Thollander, P., Backlund, S., 2013. Barriers to industrial energy efficiency in foundries: a European comparison. J. Clean. Prod. 40, 161–176.

Trianni, A., Cagno, E., De Donatis, A., 2014. A framework to characterize energy efficiency measures. Appl. Energy 118, 207–220.

Trygg, L., Thollander, P., Broman, G., 2010. Evaluation of industrial energy audits in SMEs. In: Proceedings of the International Energy Program Evaluation Conference (IEPEC) 2010, Paris, France.

Wentemi Apeaning, R., Thollander, P., 2013. Barriers to and driving forces for industrial energy efficiency improvements in African industries: a case study of Ghana's largest industrial area. J. Clean. Prod. 53, 204–213.

Energy policy programs for improved energy efficiency

16

Chapter Outline

16.1 Barriers to energy efficiency—the political perspective

In 2012, the European Union (EU) adopted the so-called energy efficiency directive (EED). The directive aims at "... remove barriers in the energy market and overcome market failures that impede efficiency in the supply and use of energy ..." (EU, 2012). In particular, the directive highlighted the elimination of market barriers and market failures as central. Through this wording of the directive, the EU states that:

1. There is a so-called energy efficiency gap, and
2. An opportunity for each Member State to pursue an energy policy that not only aims at eliminating market failures but also provides the opportunity to work to eliminate market barriers.

The difference between a market barrier and a market failure is central to, in particular, European and Western World energy policy-making, where (Thollander and Palm, 2012).

Introduction to Industrial Energy Efficiency. DOI: https://doi.org/10.1016/B978-0-12-817247-6.00016-X

- Market failures (market imperfections) derives from a violation of the theoretical principles of well-functioning markets.
- Market barriers are defined as "*any factor that could give rise to an energy efficiency gap*," that is, any barrier including market failure (market imperfections).

16.1.1 Market failures

Market economic theory states, among other things that the following factors should be met in order for a market to be considered perfect:

- A complete set of well-defined property rights in the market, where buyers and sellers can freely exchange assets.
- Customers and sellers perform competitively by maximizing the benefits (utilities) and minimizing costs.
- Market prices are known to all consumers and businesses.
- Transaction costs, that is, the cost procurement is zero.

A violation of any of these principles gives rise to a so-called market failure. Market failures or market imperfections can be divided into four subgroups, namely

- Incomplete markets
- Imperfect competition
- Imperfect information
- Asymmetric information.

Among these, it is especially the latter two, information asymmetries and imperfections that are of particular importance in regard to barriers to energy efficiency. Asymmetric information can, in turn, be divided into split incentives, principal–agent relationship, and adverse selection.

The two first-stated market failures, incomplete markets and imperfect competition are alone barriers to a functioning market. However, these market failures cannot directly explain why apparently cost-effective energy efficiency measures remains unadopted are not implemented. Already without eliminating these first stated market failures, there is a profitable energy efficiency potential that is not implemented. Thus, if incomplete markets and imperfect competition were eliminated, the energy efficiency gap would be significantly greater. These market imperfections can thus not therefore directly explain the existence of the energy efficiency gap.

16.1.2 External effects

Even externalities, or so-called external effects, is normally sorted into the category of market failures. There are positive and negative external effects. A negative externality is a cost that is not included in the market price, often due to absence of ownership. Hence negative externalities are costs that the user does not pay for without market intervention.

Emissions of carbon dioxide and other greenhouse gases is an example of an external cost (negative environmental impact) but which the user does not pay for without policy measures. There are also numerous other external costs related with burning of fuels such as emissions of NO_x, SO_x, VOCs (volatile organic compounds), and so on.

One way to create more efficient markets with improved resource efficiency is to internalize, that is, distribute the external costs among the users. This can be achieved with different types of policies, such as a carbon dioxide fee or a carbon dioxide tax.

16.1.3 Market imperfections and industrial energy policy programs

As regards barriers to energy efficiency, a distinction is made between market barriers which has a wider definition (any factor accounting for the energy efficiency gap) and market imperfections, which is a narrower definition of barriers, where the existence of the latter constitutes a violation of market theory principles.

Among western world's market economies (regions or countries), the incidence of market imperfections is one of the major reasons for considering policy program interventions. The basic principle is that the utility of the proposed policy must be weighed toward its costs. This weighing is done with a so-called cost-benefit analysis.

With the European Energy Efficiency Directive, the EU has partially abandoned this principle and indicates that other types of barrier, that is, not just market imperfections, may be adequately addressed with governmental policy program intervention. Lack of access to capital, risk aversion, and hidden costs are all examples of market barriers that are not considered to be market imperfections.

16.1.4 Energy intensity

Energy intensity is defined as primary energy use divided by (industry) GDP (gross domestic product) measured at fixed prices. The numerator in the function—primary energy use—is calculated as (Thollander et al., 2013):

$$E_{tot,ind} = \sum_{i=1..I} (P_i x_i + S_i x_i) + S_0 \tag{16.1}$$

where,

$E_{tot,\ ind}$ = industry's annual primary energy use
P_i = production processes' specific primary energy use to produce a product
S_i = the production-related support processes specific primary energy use (e.g., pumping within a pulp mill) to produce a product
x_i = the amount of products produced, i

S_0 = base load, that is, the primary energy use of support processes which is required even if the production is currently on hold

The denominator—industry's GDP—can in turn primarily be calculated in three different ways, such as the sum of either:

- All expenses
- Value added
- Income of producers

These should theoretically give the same result. If the sum of value added is used, it can be defined as the value of a sold item or service subtracted with material and energy costs, transportation, and other production-related costs.

The standard SS-EN 16212: 2012, Energy Efficiency and Energy Saving Calculations—Top-Down and Bottom-Up Methods, describes how energy efficiency is calculated from a top-down perspective. Fig. 16.1 illustrates this.

The projection up to the level of point 4 implies an increased primary energy end-use per year, without regard to any of the factors listed in the figure. Structural effects, meaning that an industry in a region is knocked out for competitive reasons, reduces projected development by a certain percentage per year. If also autonomous

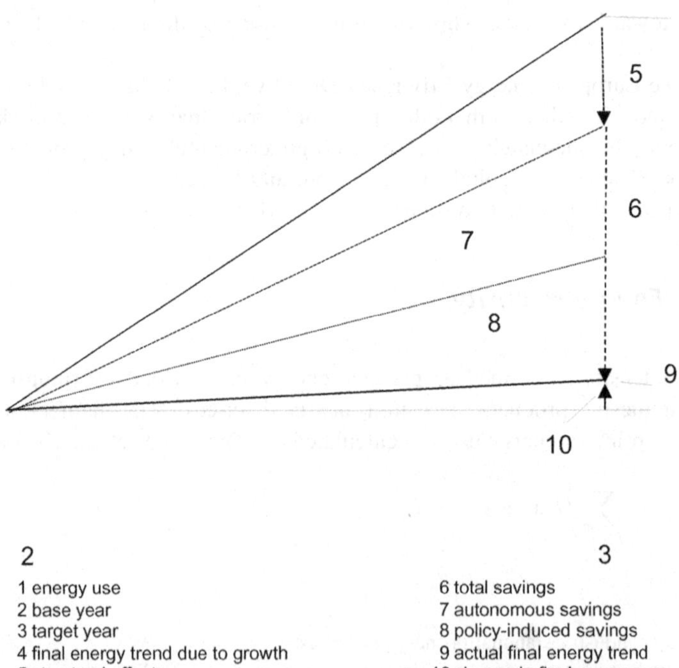

Figure 16.1 Model for calculating energy efficiency from a top-down perspective (EN 16212: 2012).

effects, that is, measures that are implemented as a result of, for example, rising energy prices, normal replacement of existing technologies and processes, and so on, are included in the calculation, reducing the primary energy end-use by a further percentage per year. If account is also taken of the effects of policy instruments, the primary energy end-use is assumed to decrease by a further percentage per year. This level corresponds to some increased primary energy end-use per year and is the level assumed to be achieved if no further measures are taken.

16.1.5 Varying energy efficiency potential

The potential for energy efficiency in a country, region, or sector varies depending on, among other things:

- Base year, that is, from which year you measure—meaning if the potential from one year to another is infinite, technology development and also production activity differ meaning that from which year you measure has an effect on the stated level of potential.
- Economic activity (e.g., structural effects), not just energy savings—if part of a sector is under heavy competition by other country's businesses, the total energy efficiency potential may be lower if already some companies have been out-ran by competitors.
- Previous energy price structure—low electricity prices in a country may have led to a higher relative electricity consumption (e.g., autonomous effects), which in turn provides a higher potential for electricity energy efficiency.
- Energy prices—rising energy prices means that more measures will be profitable (e.g., autonomous effects). On the contrary falling energy prices lead to investments in energy efficiency becoming less profitable.
- The definition of "profitable energy efficiency measure"—constitutionally, in a market economy the definition of what is perceived as "cost-effective" is defined by the market actors. In practice, there is however also governmental legislation in the form of various environmental acts that may force actors to undertake investments. Despite so, the definition of what is cost-effective varies from company to company meaning that the stated potential for improved energy efficiency, for example, 1 year straight pay-off, 50% internal rate of return or 3-year straight pay-off, and so on.
- Effect from previous policies.

16.1.6 Example: EU 2020 targets

In October 2006, the so-called the Energy Efficiency Action Plan was adopted, describing the EU energy-saving target of 20% by 2020. The goal was to reduce primary energy end-use by 20% up until 2020 compared to a projected level from 2005. Fig. 16.2 illustrates the stated target for 2020 in relation to projected values.

A development without regard to any of the factors listed in Fig. 16.2 results in 2.3% increased primary energy use per year. If structural effects are included, primary energy end-use decreases by 0.6% per year. If autonomous measures are also included, primary energy use decreases by an additional 0.85% per year. If so-called previous policies are also included, the primary energy use is assumed

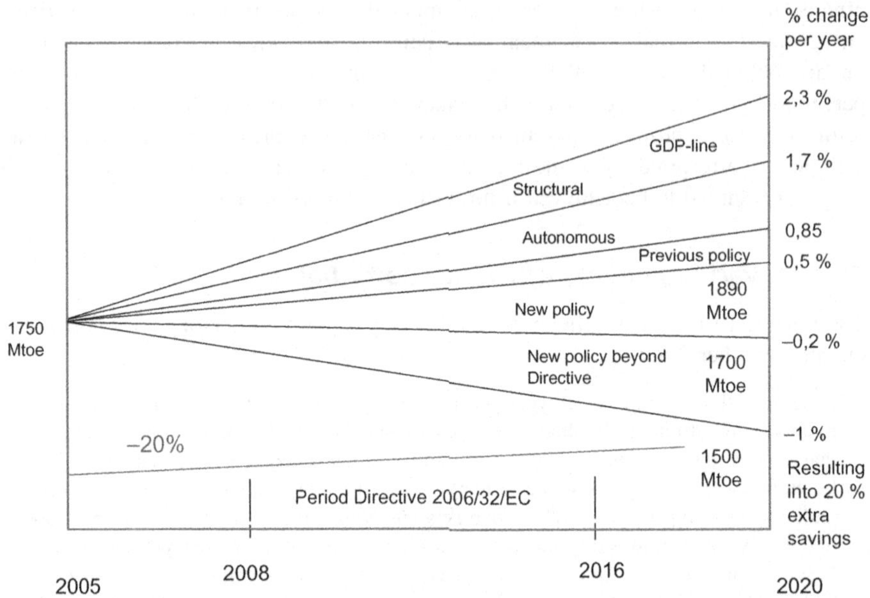

Figure 16.2 Annual energy intensity change (EC, 2006) (Mtoe = million tons oil equivalent, 1 toe = 11.6 MWh).

to decrease by an additional 0.35% per year. This level, that is, 0.5% increased primary energy use per year, the level is assumed to be achieved if no further measures are taken.

The EU Action Plan for 2020 states that 1.5% improved efficiency of primary energy use will be achieved, in addition to the current trend of 1.8% improved energy efficiency per year, which is impact on the effectiveness of previous EU community legislation (0.35% per year) and other effects (0.6% due to structural changes, such as industrial development, and 0.85% due to "autonomous improvement," such as normal replacement of existing technologies and process, per year). Together, this means a reduction of energy intensity by 3.3% per year, if energy end-use increases by an average of 0.5% per year and if a GDP growth of 2.3% per year is assumed (EC, 2006).

An increase in primary energy use by an average of 0.5% per year and a GDP increase corresponding to 2.3% per year means a real reduced energy intensity corresponding to about 1.8% per year, that is, what the EU action plan intended to achieve as a result of the above-mentioned factors (structural, autonomous, and previous policies).

The bottom red curve in Fig. 16.2 is the baseline for the 2020 target for primary energy use. The curve starts from a 20% lower primary energy level in 2005. Thereafter, primary energy end-use is allowed to increase by 0.5% per year (due to economic growth) and in 2020 finally reach the actual level

(1500 Mtoe = 17,100 TWh), that is, the level obtained by the above-mentioned effects and policy means.

16.2 Categorization of energy policies for improved energy efficiency

Energy end-use efficiency policy has an important role to remove market barriers or market failures to energy efficiency. There are three approaches in such policy intervention:

- *Administrative or regulation directed policies,*
- *Information policies*, and
- *Economic policies.*

In addition to these three categories, there is also a fourth, *research and development*, the purpose of which is to drive technology development to be able to launch new products and services at the free market at a later stage. In practice, the policies are often a mix of these categories.

Whether a policy is a consequence of the existence of a market failure (imperfection) or not is of importance to analyze prior to the design of a new policy.

16.2.1 Economic policies

Economic policies include subsidies and tax. Often, polices are a combination of different types of, for example, energy- and carbon taxation, and subsidized energy audit program includes both an economic part, the subsidy, and an information part, the energy audit report delivered to the company. From the market failure perspective, a goal for something would rather be the means for well-functioning markets. Another example of market-based instrument is the European Emission Trading Scheme (EU ETS).

Economic policy instruments provide price signals and change economic incentives which affect the behavior of individuals and companies in a desirable direction.

16.2.2 Administrative policies

Administrative or regulative polices include laws, technology standards, performance standards, and management standards. The last type, which is also called enforced self-regulation or management-based regulation, is a way of regulation which requires firms to have management systems, and the firms determine the details of the management systems by themselves (Ayres and Braithwaite, 1992; Gunningham and Grabosky, 1998). The Japanese Energy Conservation Law on firms is a good example of administratively based regulation (Kimura and Noda, 2010; Kimura and Noda, 2014).

This category also includes quantitative restrictions on resource use or emissions, as well as rules for the design of industrial plants and production methods, as well as the authorities' supervision of the application of the rules.

Requirements for education and certain equipment are other examples of administrative policy instruments. Administrative or regulatory policies often form the baseline for policies in a country.

The fundamental difference between administrative and economic (or incentive-creating) policy instruments is that administrative policies indicate how a particular goal is to be achieved, while economic instruments indicate that the goal is to be achieved, but do not state how.

The European Industrial Emission Directive is an example of an administrative policy where the policy, for example, states BAT (best available technology)-documents for various sectors.

16.2.3 Information policies

Information can be used as a direct policy for influencing knowledge creation, attitudes, and user behavior. Information can also be used as a necessary complement to financial and administrative policies. For example, in order for energy services to have an effect, it is necessary for the companies to be informed of the conditions and rules that apply. *Information policies* include voluntary guidelines, manuals, and training. Nonmandatory energy audit programs also belong to this category.

16.3 International industrial energy efficiency policy programs[1]

A review of about 300 industrial energy policy programs in the IEA countries was conducted by Tanaka (2011), stating that many of the programs did not target SMEs. Tanaka (2011) explicitly outlined some programs toward industrial SMEs such as the German Special Fund for Energy Efficiency in SME, Japanese, UK, and French low interest rate loans for investments. For energy-intensive industry, two of the foremost policies are energy audits and so-called long-term or voluntary agreements programs (VAPs), a policy where an agreement on improved levels of energy efficiency between the government and either a sector organization or individual companies is set (Bertoldi, 2001). A review of 23 VAPs in 18 countries was made by Price (2005) who made a categorization of three different types of VAPs. Price (2005) stated that the VAPs that are completely voluntary show lower participation and weaker results while VAPs with regulatory or economic components showed higher participation rates and improved results. Rezessy and Bertoldi

[1] The introductory part of this sub-chapter was originally published in Thollander et al. (2015).

(2011) reviewed EU VAPs and found that VAPs, though historically focused on process-industry, now also have started to include medium-sized companies and outline eight EU Member States that include industrial SMEs in their VAPs. Cornelis (2012) evaluated the Belgium auditing VAPs in Flanders, comprising medium-sized companies and showed that the economic strength of the company had impact on the adoption of measures. Most VAPs, like the one in Flanders, include energy audits as one major component. Lindén and Carlsson-Kanyama (2002) and Uggla and Avasoo (2001) presented an evaluation of the Swedish VAPs EKO-Energi and stated that the energy audit within the VAPs was one of the most appreciated gains according to the participating companies. Price and Lu (2011) presents a review of 22 industrial energy audit program from 15 countries, and Johansson et al. (2007) shows an overview of the current Swedish industrial energy policy mix, presenting an overview of current demand and supply policies toward the Swedish industry. Thollander et al. (2007) evaluated a regional Swedish energy audit program and also presents an overview of Swedish industrial energy-end use programs. Thollander and Dotzauer (2010) presented the logics behind the Swedish energy audit program. Ottosson and Peterson (2007) presented the major functions of the Swedish VAPs PFE. Stenqvist and Nilsson (2012) and Henriksson and Söderholm (2009) evaluates the Swedish PFE program showing a gross impact of more than 3% improved electrical efficiency where 30% of participating companies are SMEs.

Kimura and Noda (2010) and Kimura and Noda (2014) present an evaluation of the effectiveness of regulation, the Japanese Energy Conservation Law and showed that it indeed is a successful policy toward medium-sized companies. Johansson et al. (2007) in their presentation of industrial energy end-use policies in Sweden outlined that the Swedish law enforcement, the Swedish Environmental Code is a not an extensively used policy even though it holds juridical power.

Harris et al. (2000) evaluated the Australian EEAP, primarily targeting larger companies, and found that a deployment rate of five out of six proposed measures from the energy audits were implemented. Anderson and Newell (2004) and Corbett et al. (2009) present evaluations of the American IAC, the world's largest energy audit program in terms of participating companies. They showed that about half of the suggested measures were implemented, where measures with low investment costs were preferred (Anderson and Newell, 2004; Corbett et al., 2009). Fleiter et al. (2012) presents an evaluation of the German energy audit program and show that participating SMEs adopted 1.7–2.9 of the measures. Farla and Blok (1995) found that the investment subsidy, the Dutch Energy Bonus, suffered from a large degree of free-rider effects. Gruber and Brand (1991) presented results for a German grant for energy audits in SMEs and found that the German subsidy for energy audits were not known to even exist by most SMEs, and some that were aware of the subsidy did not know that they could apply for it. Yet another type of policy is benchmarking schemes, for example, USEPA who developed plant energy benchmarks known as Energy Star Energy Performance Indicators for more than 10 industrial sectors (Boyd, 2013). Also, a more recent type of policy that has only recently been scientifically evaluated are energy networks or learning energy efficiency networks, emerging form of policy that have proven to be successful in Switzerland and Germany, especially for medium-sized companies (Koewener et al., 2011). Energy networks show a double impact compared with stand-alone energy audit (Koewener et al., 2011).

Xu et al. (2014) investigated the energy conservation and CO_2 emission reductions in China's 11th Five-Year Plan from a top-down perspective, showing that economic growth is the biggest driver for increased energy use and GHG emissions. The paper concludes that renewable energy and promoting economic restructuring to limit the increase of energy-intensive sectors are the main challenges to overcome (Xu et al., 2014). Li et al. (2014) summarizes the Chinese energy policies for energy-intensive industry and evaluate these using indicators. Wang and Chang (2014) examined the development of Chinese low-carbon policy instruments and suggest more rigid law enforcement toward in particular the industrial sector to improve efficiency and reduce GHG emissions. In another study of the Chinese automotive industry, Wachtmeister (2013) stated that regulative measures using command-and-control are the most effective climate policy measures but states that market oriented measures are underrepresented in the policy mix compared with Europe. Lo (2014) states that industry uses 60%–70% of the Chinese energy use and therefore is an important sector to focus on, in relation to renewable energy and energy efficiency policies and states that currently China has four strategies: energy efficiency obligation, forced closure, energy efficiency funds, and electricity pricing.

Brannlund et al. (2014) studied the Swedish carbon tax and found that the tax have had impact in decoupling economic output and GHG emissions. Ericsson et al. (2011) studied the strategic changes in the Swedish pulp- and paper industry and found that change was driven by both economic market conditions, for example, electricity price increase, as well as policies such as the EU ETS, the Swedish VA, the PFE, and the electricity certificate scheme. Backlund and Thollander (2015) studied the impact of the Swedish energy audit program and its cost-efficiency.

Napp et al. (2014) conducted a review of more energy efficient technologies for climate change mitigating in the industrial sector and made some concluding remarks on suitable policies for various cost-effective energy efficient technologies.

Wakabayashi (2013) presents results from three case studies of VAPs in Japanese industries and concludes that voluntary activities are more effective compared with alternative policy measures, provided a proper institutional framework exists.

Sugino et al. (2013) in their study of carbon pricing mechanisms found that such scheme would create an uneven burden on sectors which are energy-intensive and heavily dependent on international trade when applied to the Japanese economy. Mochizuki (2011) also conducted an ex post evaluation of Japanese voluntary emission trading systems and concludes that while these systems worked to some extent, they are still in a pilot phase and there remain substantial barriers for effective implementation.

Kuramochi (2015) presents a comprehensive review of recent developments of Japanese energy and climate policy and argues that the current national policy for climate mitigation in Japan is weak. His review, however, focuses on economic instruments such as emission trading schemes and carbon taxes as well as supply-side policy such as the feed-in tariff (FIT) scheme and does not cover demand-side energy efficiency policy.

Chou and Liou (2012) studied Taiwanese energy-intensive industries and its climate change policies and concluded that policies aimed toward industry, not give rise to absolute energy savings. Rather, energy use is increasing. They conclude that this may be a general challenge for newly industrialized countries (Chou and Liou, 2012). Their study is further supported by Wang et al. (2012) stating that industrial economic growth together with increased fossil dependency in Chinese industry is the main drivers for the increased carbon dioxide emission.

From a management perspective, firms are well motivated to behave responsibly, if proactive environmental practice has a positive impact on firm's financial performance. Both positive and negative relationship between firm's environmental performance and financial performance has been reported. Sariannidis et al. (2013) suggested that different features of investment behavior may account for this divergence; socially responsible firms are likely to be punished when CO_2 increases while conventional firms are not. Their argument implies that motivating mechanisms are more needed for SMEs than large firms, considering most SMEs lack financial incentives.

16.4 Swedish energy policy programs for improved energy efficiency – an example

Swedish industry has over the years been subject to various policies aimed toward a more efficient energy use. Since Sweden joined the European Union, Sweden is now subject to submit to the EU-related legislations. The most important instruments for the industrial sector in 2018 are as follows:

- Energy and carbon dioxide taxes
- EU Emissions Trading Scheme
- Electricity certificate system
- The Swedish environmental code
- Energy audit policy program for SMEs
- The law on energy audits for large companies
- Energy efficiency networks for SMEs

Another policy program, the program for energy efficiency in energy-intensive industry (PFE), was used between 2004 and 2014. The reason why this policy was terminated was that the program's design was partly not in line with the EU's new tax legislation. With its vast impact on energy-intensive industry and because PFE was the world's first policy program that adopted an energy management standard for industry, PFE is also described below.

16.4.1 Economic policies

16.4.1.1 Energy and carbon dioxide taxes

One of the most common general policy instruments is taxation. Through taxation, such as energy and carbon dioxide taxes, the cost of using energy as well as the cost of emissions are increased. This cost increase signals that industrial companies must act for more efficient energy use.

16.4.1.2 The EU emissions trading scheme

The EU emissions trading scheme, the EU ETS, was launched in 2005 as a policy for reducing greenhouse gas emissions. The EU ETS means that every company in the EU that has boilers, furnaces, and so on of a size exceeding 20 MW must have emission rights corresponding to the company's greenhouse gas emissions. For example, if a company is in need of more emission permits or does not need to use its emission permits, they can be procured or sold on the market for tradable emission rights. The EU annually reduces the number of emission allowances within the EU ETS.

16.4.1.3 The program for energy efficiency in energy-intensive industry (PFE)

The PFE was initiated in 2004 in connection with the introduction of an energy tax on electricity equivalent to approx. 5 EUR per MWh and ended in December of 2014. Through participation in PFE, the Swedish government offered a complete tax rebate on the energy tax. About 100 companies participated during PFE's two 5-year program periods.

During the first 2 years, participating companies would carry out an energy audit, introduce a certified energy management system, introduce procedures for purchasing, planning, and renovation that included life cycle cost. During the final 3 years, the detected energy efficiency measures with a pay-back less than 3 years was to be implemented.

The outcome of PFE's first program period was that measures corresponding to an annual energy saving of 1.45 TWh per year were reported in the form of 1,254 measures, with a total investment cost of just over SEK 700 million (Stenqvist and Nilsson, 2012).

16.4.1.4 The electricity certificate system

The electricity certificate system was launched in 2003 for the Swedish electricity market, but since 2012, it also includes Norway. The purpose of this policy is to increase the production of renewable electricity in a cost-effective manner.

For each MWh renewable electricity produced, the electricity producer receives an electricity certificate from the government. These certificates can then be sold on a tradable market, which gives electricity producers of renewable electricity extra revenue which is added to the sale of produced electricity. Buyers of electricity, on the other hand, have a so-called obligation quota[2]. In addition, the quota levels for renewable electricity are increasing every year.

16.4.2 The energy ladder

Companies that are included under the Act on law for energy audits for large companies can, within the energy ladder, apply for two types of support:

- Project grant
- Investment grants

The design support comprises a contribution to design a technical improvement proposal from the energy audit and the investment grant can be applied for costs that are seen as extra costs for an energy-efficient investment.

16.4.3 Administrative policy

16.4.3.1 The Swedish environmental code

The Swedish Environmental Code came into force in 1998 and addresses, among other things, energy efficiency as a key aspect. According to the code, the best available technology (BAT) should be implemented. This is due to the fact that the European Industrial Emissions Directive (IED) sets requirements for BAT. In the previous European regulations, the Industrial Pollution and Prevention Control (IPPC) was in force were more than 20 best reference documents (BREF) for a number of industries exist. Under the new directive, the IED, new documents are designed for different industries.

The impact of the application of the Code is difficult to evaluate since an actual ex post evaluation is lacking. However, the assessment is that the Code will have an increasingly important role in the future, but its application needs to be balanced against other policies, such as the Law for energy audits for large companies.

16.4.3.2 The law on energy audits for large companies

Based on the EU Energy Efficiency Directive (EED), the Swedish Energy Agency in 2014 launched regulations concerning the law for energy auditing for large companies. About 1 100 companies in Sweden are covered by the law,

[2] Obligation quota means that electricity suppliers are required to hold a certain amount of electricity certificates in relation to sales. The quota obligation is reported every year to the Swedish Energy Agency.

which stipulates, among other things, that the audits are carried out according to one of the two available energy audit standards ISO 50002 or EN 16247.

The law further state that the energy audit shall be carried out by a person with special competence, that is, a certified energy auditor. Companies that have a certified energy or environmental management system can carry out the audit themselves, provided that they have personnel who meet the competence requirements for certified energy auditors. The Swedish Energy Agency has produced guides for various industries that show how the energy audit should be carried out.

16.4.4 Informative polices

16.4.4.1 The energy audit policy program for SMEs

A governmental subsidy for the implementation of energy audits was introduced in 2010. In the first program period of the Swedish energy audit policy program, which ran between 2010 and 2014, approximately 1000 small and medium-sized enterprises (SMEs) applied for subsidy. The major aim with this policy is to provide industry with targeted and specific information on energy use and energy efficiency improvement measures (Paramonova and Thollander, 2016).

In 2015 the Swedish Energy Agency launched a second program period, which runs until 2019. Companies with an energy use exceeding 300 MWh per year can apply for financial support to carry out an energy audit. The support amounts to 50% of the audit cost, but a maximum of 5000 EUR (Paramonova and Thollander, 2016).

16.4.4.2 The energy efficiency networks program for SMEs

In 2015 the Swedish Energy Agency initiated a national energy network program for SMEs. The program comprises between 300 and 400 companies, which have been divided into about 40 unique networks, where groups of about 10 companies work with the support of a network coordinator and an energy expert. The network period begins with the companies conducting an energy audit and then the companies meet regularly for a period of 3 to 4 years. The company with an annual energy use of at least 1 GWh/year may participate (Carlén et al., 2016).

References

Anderson, S.T., Newell, R.G., 2004. Information programs for technology adoption: the case of energy-efficiency audits. Res. Energy Econ. 26, 27–50.

Ayres, I., Braithwaite, J., 1992. Responsive Regulation: Transcending the Deregulation Debate. Oxford University Press, Oxford, UK.

Backlund, S., Thollander, P., 2015. Impact after three years of the Swedish energy audit program. Energy.

Bertoldi, P., 2001. Effective policies and measures in energy efficiency in end-use equipment and industrial processes. In the 2001 Workshop on Good Practices in Policies and Measurers.

Boyd, G., 2013. Plant energy benchmarking: a ten year retrospective of the ENERGY STAR energy performance indicators (ES-EPI). ACEEE 2013 Industry Summer Study.

Brannlund, R., Lundgren, T., Marklund, P.O., 2014. Carbon intensity in production and the effects of climate policy: evidence from Swedish industry. Energy Policy 67, 844–857.

Carlén, A., Rosenqvist, M., Paramonova, S., Thollander, P., Municio, S., 2016. Energy effiency networks for small and medium sized enterprises-bosting the energy efficiency potential by joining forces. In: ECEEE Industry Summer Study, Berlin, 12–14 September.

Chou, K.T., Liou, H.M., 2012. Analysis on energy intensive industries under Taiwan's climate change policy. Renew. Sustain. Energy Rev. 16, 2631–2642.

Corbett, C., Muthulingam, S., Benartzi, S., Oppenheim, B., 2009. Adoption of profitable energy efficiency related process improvements in small and medium sized enterprises. In: Proceedings of the 5th European Conference on Economics and Management of Energy in Industry (ECEMEI-5).

Cornelis, E., 2012. Analysis of the differences of the energy saving measures proposed by medium-sized industrial companies in Flanders, Belgium in 2006 and 2009. In: Proceedings of the ECEEE Industry Summer Study, 11–14 September.

EC (European Commission), 2006. Announcement by the commission: action plan for energy efficiency. COM (2006) 545 final.

EC (European Commission), 2012. Directive 2012/27/EU on energy efficiency, amending Directives 2009/125/EC and 2010/30/EU and repealing Directives 2004/8/EC and 2006/32/EC. Vol. 4: 1–56.

Ericsson, K., Nilsson, L.J., Nilsson, M., 2011. New energy strategies in the Swedish pulp and paper industry: the role of national and EU climate and energy policies. Energy Policy 39, 1439–1449.

Farla, J., Blok, K., 1995. Energy conservation investment behaviour of firms: analys is of investments in energy efficiency in the Netherlands in the 1980s. In: Paper presented at the 1995 American Council for an Energy Efficient Economy (ACEEE) Summer Study on Energy Efficiency in Industry.

Fleiter, T., Gruber, E., Eichhammer, W., Worrell, E., 2012. The German energy audit program for firms: a cost-effective way to improve energy efficiency? Energy Efficiency 5, 447–469.

Gruber, E., Brand, M., 1991. Promoting energy conservation in small and medium-sized companies. Energy Policy 19, 279–287.

Gunningham, N., Grabosky, P., 1998. Smart Regulation: Designing Environmental Policy. Oxford University Press, Oxford, UK.

Harris, J., Anderson, J., Shafron, W., 2000. Investment in energy efficiency: a survey of Australian firms. Energy Policy 28 (12), 867–876.

Henriksson, E., Söderholm, P., 2009. The cost-effectiveness of voluntary energy efficiency programs. Energy Sust. Dev. 13, 235–243.

Johansson, B., Modig, G., Nilsson, L.J., 2007. Policy instruments and industrial responses - experiences from Sweden. In: Proceedings of the 2007 ECEEE Summer Study "Saving Energy - Just Do It," Panel 7:1413–1421.

Kimura, O., Noda, F., 2010. Effectiveness of regulations on firms by Japanese Energy Conservation Law. CRIEPI Report No. Y09010 2010. Central Research Institute of Electric Power Industry, Tokyo, Japan [in Japanese].

Kimura, O., Noda, F., 2014. Does regulation of energy management systems work? A case study of the Energy Conservation Law in Japan. In: Proceedings of ECEEE Industry Summer Study, Arnhem, 2–5 June.

Koewener, D., Mielicke, U., Jochem, E., 2011. Energy efficiency networks for companies – concept, achievements and prospects. In: Proceedings of ECEEE 2011 Summer Study. Energy Efficiency First. The Foundation of a Low-Carbon Society, pp. 725–733.

Kuramochi, T., 2015. Review of energy and climate policy developments in Japan before and after Fukushima. Renew. Sustain. Energy Rev. 43, 1320–1332.

Li, L., Jianjun, W., Zhongfu, T., Xinquan, G., Jian, Z., Xiaozhe, Y., 2014. Policies for eliminating low-efficiency production capacities and improving energy efficiency of energy-intensive industries in China. Renew. Sustain. Energy Rev. 39, 312–326.

Lindén, A.L., Carlsson-Kanyama, A.A., 2002. Voluntary agreements: a measure for energy-efficiency in industry? Lessons from a Swedish programme. Energy Policy 30, 897–905.

Lo, K., 2014. A critical review of China's rapidly developing renewable energy and energy efficiency policies. Renew. Sustain. Energy Rev. 29, 508–516.

Mochizuki, J., 2011. Assessing the designs and effectiveness of Japan's emissions trading scheme. Climate Policy 11, 1337–1349.

Napp, T.A., Gambhir, A., Hills, T.P., Florin, N., Fennell, P.S., 2014. A review of the technologies, economics and policy instruments for decarbonising energy-intensive manufacturing industries. Renew. Sustain. Energy Rev. 30, 616–640.

Ottosson, C., Peterson, K., 2007. First results from the Swedish LTA programme for energy efficiency in industry. In: Proceedings of the 2007 European Council for an Energy-Efficient Economy (ECEEE) Summer Study "Saving Ergy - Just Do It," Panel 7:1517–1525.

Paramonova, S., Thollander, P., 2016. Ex-post impact and process evaluation of the Swedish energy audit policy programme for small and medium-sized enterprises. J. Clean. Prod. 135, 932–949.

Price, L., 2005. Voluntary agreements for energy efficiency or GHG emissions reduction in industry: an assessment of programs around the world. In: Proceedings of the 2005 American Council for an Energy Efficient Economy Summer Study on Energy Efficiency in Industry, West Point, NY, USA, July.

Price, L., Lu, H., 2011. Industrial energy auditing and assessments: a survey of programs around the world. In: Proceedings of the ECEEE 2011 Summer Study, Energy Efficiency First: The Foundation of a Low-Carbon Society.

Rezessy, S., Bertoldi, P., 2011. Voluntary agreements in the field of energy efficiency and emission reduction: review and analysis of experiences in the European Union. Energy Policy 39, 7121–7129.

Sariannidis, N., Zafeiriou, E., Gianarakis, G., Arabatzis, G., 2013. CO_2 emissions and financial performance of SR firms: the empirical survey of DJSI with a non linear model. Business Strat. Environ. 22, 109–120.

Stenqvist, C., Nilsson, L.J., 2012. Energy efficiency in energy-intensive industries: an evaluation of the Swedish voluntary agreement PFE. Energy Efficiency 5 (2), 225–241.

Sugino, M., Arimura, T.H., Morgenstern, R.D., 2013. The effects of alternative carbon mitigation policies on Japanese industries. Energy Policy 62, 1254–1267.

Tanaka, K., 2011. Review of policies and measures for energy efficiency in industry sector. Energy Policy 39, 6532–6550.
Thollander, P., Dotzauer, E., 2010. An energy efficiency program for Swedish industrial small- and medium-sized enterprises. J. Clean. Prod. 18, 1339–1346.
Thollander, P., Kimura, O., Wakabayashi, M., Rohdin, P., 2015. A review of industrial energy and climate policies in Japan and Sweden with emphasis towards SMEs. Renew. Sustain. Energy Rev. 50, 504–512.
Thollander, P., Palm, J., 2012. Improving Energy Efficiency in Industrial Energy Systems: An Interdisciplinary Perspective on Barriers, Energy Audits, Energy Management, Policies & Programs. Springer.
Thollander, P., Rohdin, P., Danestig, M., 2007. Energy policies for increased industrial energy efficiency: evaluation of a local energy programme for manufacturing SMEs. Energy Policy 35, 5774–5783.
Thollander, P., Rohdin, P., Moshfegh, B., Karlsson, M., Soderstrom, M., Trygg, L., 2013. Energy in Swedish industry 2020: current status, policy instruments, and policy implications. J. Clean. Prod. 51, 109–117.
Uggla, U., Avasoo, D., 2001. EKO-Energi: successful voluntary agreements on energy efficiency and environmental control in Swedish industry. In: Proceedings of the 2001 ECEEE Summer Study "Further Than Ever from Kyoto? Rethinking Energy Efficiency Can Get Us There," vol. 1, pp. 252–260.
Wachtmeister, M., 2013. Overview and analysis of environmental and climate policies in China's automotive sector. J. Environ. Dev. 22, 284–312.
Wakabayashi, M., 2013. Voluntary business activities to mitigate climate change: case studies in Japan. Energy Policy 63, 1086–1090.
Wang, N., Chang, Y.C., 2014. The development of policy instruments in supporting low-carbon governance in China. Renew. Sustain. Energy Rev. 35, 126–135.
Wang, Z., Zhang, B., Yin, J., 2012. Determinants of the increased CO_2 emission and adaption strategy in Chinese energy-intensive industry. Natural Hazards 62, 17–30.
Xu, J.H., Fan, Y., Yu, S.M., 2014. Energy conservation and CO_2 emission reduction in China's 11th Five-Year Plan: a performance evaluation. Energy Econ. 46, 348–359.

Further reading

Saidur, R., Rahim, N.A., Ping, H.W., Jahirul, M.I., Mekhilef, S., Masjuki, H.H., 2009. Energy and emission analysis for industrial motors in Malaysia. Energy Policy 37 (9), 3650–3658.
Trygg, L., Karlsson, B., 2005. Industrial DSM in a deregulated European electricity market: a case study of 11 plants in Sweden. Energy Policy 33 (11), 1445–1459.

Further reading

Design, monitoring, and evaluation of industrial energy policy programs

17

Chapter Outline

17.1 Design

Energy audit programs and voluntary agreement programs (VAPs) have historically been the foremost policy programs for improved industrial energy efficiency. While

Introduction to Industrial Energy Efficiency. DOI: https://doi.org/10.1016/B978-0-12-817247-6.00017-1

energy audit programs is a policy means for, in particular, the nonenergy intensive and small and medium-sized industries, VAPs are more suitable for energy-intensive companies. This is mainly due to the fact that the capacity to work with energy efficiency is more limited for the previous category, which means that a resource-intensive VAP may not particularly suitable. For energy-intensive companies, however, a VAP may be appropriate. A VAP is an agreement on energy efficiency between the government and an industrial company or an industry trade association. VAPs are considered one of the most effective policies within the EU to enforce energy efficiency in the industrial sector, and for decades, the Japanese government have had VAPs in place. In addition to a focus on energy technology opportunities, VAPs also aim to introduce energy management components in the companies, for example, via certification of an energy management system.

Another form of policy is general information campaigns, where dissemination of information takes place through seminars, workshops, training, and so on. General dissemination of information in the form of education and seminars has been shown to increase knowledge of both the possibilities of energy efficiency opportunities as well as the environmental problems that can be related to energy use. However, this type of program shows a very scarce impact in terms of actual energy efficiency improvements (Stern and Aronsson, 1984). For example, the evaluation of a Swedish seminar series aimed toward SMEs did not show any impact in terms of implemented energy efficiency measures (Hardell, 2006).

Industrial energy policy programs for nonenergy-intensive and smaller and medium-sized industries can be divided into two different types:

1. Support for detecting more energy efficient technology improvements (through energy audits and technical support).
2. Investment subsidies or other forms of support to facilitate the implementation of identified improvement measures.

Combining these two types in an energy program often provides the most desired outcome. However, research on the Dutch program "Dutch Energy Bonus," where investment support was offered to introduce energy-efficient technologies, showed that at least 85% of the measures would have been implemented even without support, that is, constituted by so-called free-riders (Farla and Blok, 1995). Based on these experiences, a pure investment subsidy cannot, therefore, be recommended in the first place. However, some form of financial support may be appropriate after the energy audit has been carried out.

17.2 Important factors when implementing an industrial energy program

As stated previously, energy audits are one of the most widely used policies for disseminating information about the possibilities of reducing energy costs and facilitating implementation of more energy-efficient technologies for nonenergy-intensive and smaller and medium-sized industries. The outcome in terms of improved energy efficiency of energy audit depends on several factors, depending on what one wants to achieve. The type of information provided is of great importance, for example, if the information from the audit is only provided via a report or also presented in person, by the auditor. Energy audits where specific information is provided is more effective than general information campaigns. The latter, however, should not be underestimated as an important measure for raising the level of knowledge and awareness of energy efficiency in general and new policy programs in particular.

The importance of the energy auditor having in-depth knowledge of the specific industry's production processes is highlighted as another important factor. Thus, industry organization's staff that possess this technical and sector-specific competence can be well suited to carry out the audits. On the other hand, the vast majority of energy efficiency measures identified in auditing nonenergy-intensive and small and medium-sized industries, consist of nonproduction-related measures in so-called "nonproduction" support processes.

Credibility and trust in the source of information is of course also of great importance. An efficient energy audit includes a technically skilled auditor with good knowledge of possible technology and process solutions and good theoretical skills to carry out calculations. Despite a technically sound audit, there may sometimes be a need for additional support for the measures to be implemented. Implementing a number of audits in the same region has proved more effective than individual audits.

Good examples have also proved important. Furthermore, evaluations of energy audits have shown that it is not enough for a technically competent auditor. The auditor's social skills, or rather the lack thereof, can lead to the audit results being rejected. In the European Energy Audit Standard EN 16247-5, some general guidelines related to competence of an energy auditor has been addressed to facilitate excellence in regard to among other factors, the way the auditor behaves.

Another important factor is continuity in the energy audit policy programs. Regular contacts and energy audits (with a few years' intervals) are needed in order to promote the companies' continuous work on energy efficiency improvements. Simple methods, such as reminding companies of simple routines or requesting statistics on energy use, give good results in the long term. The need for continuity in the communication process shows that actors who work with energy efficiency, for example, engineering consultants, energy service companies, and authorities, with a close relationship with the companies, are more suited to providing support than, for example, government agencies with little or no relation to the companies.

When authorities must cooperate with companies, contextual differences must be taken into account, that is, a program that works in one country may be less appropriate in another country or region due to cultural differences, the structure of the energy program itself and past experience.

Studies of a Finnish energy audit program showed that energy audit programs can create entirely new business opportunities, which means that tax revenues increase and the costs of unemployment decrease (Väisänen, 2003). If these factors are included in the calculations, a program can in fact even give a positive state-financial outcome.

An energy audit program is usually very capital and public labor intensive in the first phase, that is, major costs arise in the policy start-up. Based on previous experience of energy audit programs, cost efficiency increases with the length of the program. There are therefore good reasons to consider a more long-term strategy for this type of policy program, in comparison with other governmental programs.

17.3 Design of industrial energy policy programs[1]

An energy audit program consists, according to Väisänen et al. (2003), of a number of different elements, namely:

- Program objectives
- Legal framework
- Marketing
- Subsidy rate
- Players
- Program structure
- Energy audit models
- Monitoring and evaluation
- Training of energy audits
- Certification of energy auditors
- Quality control
- Tools for energy audits

17.3.1 Program objectives

It is important to formulate annual targets for the number of audits, this in order to follow-up the policy progress in total number of audits planned to be conducted during the lifetime of the policy. Also, in order to plan budget-vise for the total cost of the policy, it is also important to consider that it takes between 2 to 4 years to come up in volume, so the goals should be set higher during the latter part of the program period. It is then also important to allocate resources for the fact that the latter part of the program period may need larger budget allocations than the first years (apart from the previously stated high start-up costs). The fact that the audit subsidy runs out in the middle of the program period is not a desirable situation as it contributes to uncertainty. A program with clear goals provides a structure for the entire program and is one of the keys to success (Väisänen et al., 2003).

17.3.2 Legal framework

Energy audit programs can be linked to existing laws on a voluntary basis. The basic question is whether current legislation is in line with the planned program. Changing laws are normally a slow and time-consuming process and because the

[1] Section 17.3 is inspired by Väisenen et al., 2003.

decisions must be very well-founded. If governmental funds are used to establish a program, the legal framework is often already in place in the form of regulations (Väisänen et al., 2003).

17.3.3 Marketing

The marketing of a policy program is very important in order to reach high number of audits to be carried out. Marketing of the industry consists of two parts: the marketing of the program itself and the marketing of the actual audit (Väisänen et al., 2003).

17.3.4 Subsidy rate

Governmental subsidy, as in the case of the Swedish energy audit program (Paramonova and Thollander, 2016), is very common among audit programs. A too small subsidy can mean that only the companies that are really interested are joining the program. However, a too excessive subsidy creates the risk of reducing the cost-effectiveness of the program. (Väisänen et al., 2003).

The companies' contribution to the program is, of course, dependent on the size of the subsidy, but with a 100% subsidy, on the other hand, there is a risk that the program is not fully appreciated since the companies do not have to pay anything for their participation. It is important to find a level that triggers the companies to want to take part of the subsidy, and which at the same time can ensure that they invest enough time themselves to want to manage the new specific information given from the energy audit report, that is, that the effort is large enough that the companies should not neglect, for example, information of the suggested improvement measures (Väisänen et al., 2003).

17.3.5 Players

Four main players can be distinguished in an energy audit policy program, namely (Väisänen et al., 2003):

- Program administrator
- Program operator
- Energy auditor
- Company

Energy auditors can be found from a number of stakeholder, for example, consultancy firms, energy utilities, and research institutes and universities.

The program administrator initiates the program itself, while the program operator is responsible for the program. Usually, but not always, the program operator is a neutral organization, which is governed by the program administrator (Väisänen et al., 2003).

An example of division of responsibility (Väisänen et al., 2003):

- The program administrator is responsible for monitoring, support for the program operator, financing of audits, and financing of the program operator.
- The program operator is responsible for education, authorization, quality control, monitoring, reporting to the program administrator, mentoring and support for auditors, and marketing.
- The energy auditor is responsible for energy audits, marketing, and assistance with the company paper work.
- The company is responsible for cofinancing of the audit, reporting to the program operator, and some level of quality control reporting.

For an energy efficiency network policy program, there will also be a fifth player, the network coordinator, having the role of coordinating each implementation network consisting of 5–12 companies per network.

As regards the structure of an energy efficiency network, models for how to operate such networks can be found by Koewener et al. (2011) related to the German LEEN (Learning Energy Efficiency Networks). Further, Paramonova et al. (2014) presented a general model on how to operate an energy efficiency networks program categorized into the following parts:

- network strategy and goal formulation,
- promotion and marketing,
- framework,
- administrative model,
- network operationalization,
- quality control, and
- network finalization.

17.3.6 Program structure

A large number of activities are carried out in an audit policy program, such as financing, reporting, marketing, databases, and help-desk services, which are distributed among the various actors within the program. It is important that the responsibility for the activities is distributed at an early stage of the planning process (Väisänen et al., 2003).

17.3.7 Monitoring and evaluation

In an energy audit policy program there should always be some form of monitoring present, although this can vary considerably between different policy programs. Next to the cost of the energy auditing, the development and operation of the monitoring system is normally the largest cost item. On the other hand, monitoring and evaluation is an absolute necessity; a program without feedback risks ending prematurely due to lack of support for continuation of the policy program. Planning and development of a monitoring system is a long process and should therefore be initiated at a very early stage. The monitoring is in turn linked to the choice of energy audit model. Available data are obtained mainly from the energy audit reports. It is thus of great importance that the energy audits are carried out and reported in a structured manner in order to keep down the costs of monitoring the policy program. Monitoring of the program mainly consists of eight parts, namely (Väisänen et al., 2003)

- Costs for the policy program
- Number of energy audits
- Energy saving potential
- Theoretical energy saving of implemented measures
- Measured energy savings at company level
- Verification of results
- Number of energy auditors
- Categorization of energy end-use processes.

The first three points do not require any specific methodology. However, for the consecutive three, there are some different methods to choose from. For follow-up of the theoretical and measured energy savings at company level, a questionnaire is a recommended tool to be used, this because of the extensive costs that would arise if all savings were to be followed up with measurements before and after. Of course, with regard to verification of the results of theoretical and measured energy savings, this cannot be done among all participating companies. As a result, this follow-up is often done among a small selection of companies (Väisänen et al., 2003; Paramonova and Thollander, 2016).

The last two parts, number of auditors and categorization of energy end-use is important to keep track of potentially under-performing energy auditors and as for the end-use, understand among which processes that the majority of the energy use is found and which processes that have shown to have the largest energy saving potential (Väisänen et al., 2003).

In a comparison of different national policy programs, it was shown that the way data were categorized differed making it basically impossible to compare results between countries (Thollander et al., 2015).

At a later stage of the program, evaluation is an important activity, not least to evaluate whether the program objectives have been achieved. The evaluation is very closely linked to the monitoring of the program. If the program is monitored continuously, sufficient information will be available to conduct a good evaluation, both a process as well as impact evaluation. If the evaluation is also conducted well, both the program administrator and the program operator are provided with very important feedback, which can be used to improve the policy program or future policy programs (Väisänen et al., 2003).

In order to be able to carry out sound monitoring and evaluation activities of the policy program, it is very important to establish a structured and well-thought-out database. Due to the slow increase in the number of energy audits over a policy program period, an evaluation of the objective fulfillment cannot in fact be adequately implemented until after 3 to 4 years (Blomqvist and Thollander, 2015).

Many public bodies use an online reporting tool for sending in the data of the results from the policy. Research shows however that the quality control of data oftentimes is foreseen leading to low-quality of the data reported, which in terms makes the time for analysis very time consuming at best, and at worst, makes the analysis and compilation of results basically impossible to undertake with thousands of companies each having 5–20 energy efficiency measures reported.

17.3.8 Energy audit models

Generally, there may be a need for different energy audit models or energy audit levels[2] depending on which sector the program is aimed toward. This range from an energy audit model which may be classified as walk-through audit to a level three audit with advanced optimization or simulation applications. In order to keep costs down for monitoring and evaluation, a model should be worked out for how the audits should be carried out and reported (Väisänen et al., 2003).

17.3.9 Training of energy audits

Some form of training is usually found in an energy audit policy program. Often, the program operator has few choices in the choice of energy auditors. A solid training (or certification) ensures, on the one hand, potentially very skilled energy auditors, but the cost of such training can be high. The first important decision that is to be taken is to decide who should carry out the energy audits. With this as a starting point, training can be designed where the teaching staff is made up of for example experienced energy auditors. It is important that the cost of the training ends up at the right level, that is, not too high, but sufficient to avoid uninterested participants. Of course, the training is very closely intertwined with the certification of energy auditors surveyor and the choice of energy audit model, all three elements with the aim of quality control the energy audits (Väisänen et al., 2003).

17.3.10 Certification of energy auditors

The certification is a license that allows the energy auditor to undertake energy audits. There are two ways to get certification. Either the persons who possess expertise are given a certification, or a minimum standard is set for the energy audits within the program. The former places demand on the program operator to offer a good training and the latter gives the program operator the opportunity to opt-out of energy auditors that do not meet the requirements for the set minimum standard for an audit. In Sweden, certification of energy auditors was handled within the Law for Energy Audits for Large Enterprises and the Law for Energy Declarations. Within the energy audit policy program for small and medium-sized companies, instead, minimum standards were made on the energy audit report, and no demand was placed on the energy auditor skills.

[2] Read more of the three levels of energy auditing in Chapter 5 of the book or in the ISO 50002-energy audit standard.

17.3.11 Quality control

As far as quality assurance is concerned, there is sometimes a too strong reliance upon training and certification of energy auditors. Without a continuous quality control of completed energy audits, the energy auditors face the risk of making the same mistake over and over again. Since the cost of carrying out energy audits is the by far largest cost of the whole policy program, low-quality energy audits may reduce both the degree of implementation and cost-effectiveness and downgrade the policy program's reputation (Väisänen et al., 2003).

Another important area is that procurement competence is ensured, that is, that the companies possess the right competence to procure an energy audit. Further, the part responsible for quality control should be objective, possess technical competence, and have the necessary resources to meet the deadlines set. (Väisänen et al., 2003).

17.3.12 Tools for energy audits

Tools and support for energy audits and energy audit programs can be documents, models, and applications. A software for the data collection for example serves as a tool to cut down public policy program cost. One such tool that exists was Nordenaudit[3]. The intention is that energy audit tools, such as manuals, files, websites, and so on, should serve as a support for the energy auditor, as well as for the program operator and program administrator. Below is a list of different types of support that can be valuable in energy audits. However, several of these overlap. It is therefore of great importance that long before the energy audit policy program starts, decide which tools need to be developed and used so that these are ready to be implemented when the policy program starts (Väisänen et al., 2003).

- Information/documentation regarding technical, industry-specific factors.
- Handbook/manual for energy auditing and energy management manual.
- Checklists for energy audits.
- Calculation methodology and energy audit software.
- Data collection form, preferably cloud based.
- Report template.
- Checklist for quality control of the audit report.
- Key figures and benchmarking.
- Databases with energy efficiency measures.

17.4 Designing energy policy programs for industrial SMEs

In an international research project under the IEA IETS revolving around improved energy efficiency in industrial SMEs, some general guidelines were presented on

[3] A software for industrial energy auditing developed at the division of Energy Systems, Department of Management and Engineering at Linköping University.

Table 17.1 Characteristics of various types of industrial SMEs (Thollander et al., 2014).

	Medium-sized SMEs and energy-intensive industrial SMEs	**Small-sized SMEs and nonenergy-intensive industrial SMEs**
Amount of energy used	Medium	Small
Human resource for energy management	Limited, but they usually have a couple of responsible engineers	Very limited, often without responsible personnel
Type of technology	Production and support[2] processes	Mainly support processes

suitability of various policy approaches presented in ranked form (Thollander et al., 2014). The characteristics of various types of industrial SMEs are presented in Table 17.1.

Based on the two primary types of industrial SMEs, general guidelines of suitable energy efficiency policy approaches were outlined:

EE-policies for medium-sized and energy-intensive industrial SMEs (Thollander et al., 2014)

1. Energy Conservation Law/LTA/VAP
2. Energy audit programs for industrial SMEs, preferably but not necessarily located regionally or locally
3. Energy networks (preferably locally or regionally anchored)
4. Investment subsidies mainly for investments in production-related technologies
5. Benchmarking
6. Sector guidelines

EE-policies for small-sized and nonenergy-intensive industrial SMEs (Thollander et al., 2014)

1. Energy audit program (preferably locally or regionally anchored)
2. Energy networks (preferably locally or regionally anchored)
3. Investment subsidy
4. Benchmarking
5. Sector guidelines

17.4.1 Method for developing an industrial SME energy policy program

In a review of the literature on industrial SMEs and energy policy programs, Johansson et al. (2019) presented a synthesis of major general steps that need to be taken when designing energy policy programs for industrial SMEs, see Table 17.2.

Table 17.2 Major steps when designing policies for industrial SMEs (Johansson et al., 2019).

Design step	Comment
Decide primary target sector group	The primary target sector group should be decided and also, preferably, should be as homogenous as possible, both in terms of sector(s) and company size
Map the annual energy	The annual energy demand and energy end-use is important and mapping the energy use and major technologies and processes are key steps. The share of energy use in support and production processes varies between sectors
Review the current energy policies	For the major energy end-using technologies and processes in the former step, it is important to assess which existing public policy programs are in place and thus already provide support for deployment
Make an energy efficiency potential estimation	Using available policy documents and scientific papers, an estimation should be made of the energy efficiency potential for the policy program. If no documents and studies are available, one can undertake a pilot ex-ante study where companies are asked about this
Review the barriers and drivers for energy efficiency	Review the scientific publications regarding barriers and drivers for the specific sector group targeted. If the information regarding barriers and/or drivers for the targeted sector group is scarce, this could be supplemented with interviews or questionnaires in a prephase study
Suggest appropriate policy	Consider the most appropriate policy or policy mix in relation to the findings in the previous steps
Evaluate the impact of the policy program	Make an evaluation of the policy program to evaluate the achieved energy efficiency and cost-effectiveness of the program. Include the expected implementation rate in the potential estimation. Anderson and Newell (2004) found that approximately half of the energy efficiency measures in the American IAC were implemented and evaluation of the Swedish Project Highland showed an implementation rate of above 40% (Thollander et al., 2007). A common estimation of the implementation rate can be 50%. Further, one should aim to include quantified nonenergy-benefits (NEBs) in the assessment of energy efficiency measures which could cut the payback time of measures (Finman and Laitner, 2001)

17.5 Method for developing a regional industrial energy policy program

In a study, Backlund et al. (2014) proposed a general method on how to develop a regional (county) industrial policy program:

1. Determine the primary energy factors and emission levels for the various energy carriers
2. Conduct an overview of major energy end-using industrial sectors in the county
3. Conduct a mapping of the major national energy end-use policies and its effect in the county
4. Conduct a study on industrial energy efficiency potential in the county
5. Based on 1–3, conduct a review of the scientific literature of the major technologies of interest to improve energy efficiency in the largest energy end-using sectors
6. Based on 1–4, create an action plan with strategic measures related to the county and the national environmental goals

17.6 Impact assessment of energy policy programs

17.6.1 *Ex ante and ex post policy program evaluation*

When undertaking a policy evaluation, one can either undertake the assessment or evaluation ex ante or ex post. Ex ante refers to an evaluation before the policy is finalized while ex post refers to an evaluation when the program has ended. Naturally, both ways have its pros and cons. While the ex post evaluation has a higher form of validity the ex ante evaluation can support the policy-maker in estimating various kinds of budget for the policy, goal-setting, cost-effectiveness of the policy, and so on.

17.6.2 *Formative and summative policy program evaluation*

When undertaking an evaluation, one can either undertake the evaluation in a summative manner meaning that the evaluation undertaken is summarized in a report or in another format. This way of undertaking a policy evaluation is often reflected in the available published scientific paper on industrial policy program evaluation.

Another way to undertake an evaluation is formative, meaning that the results from the evaluation are used to further develop and streamline the policy, that is, the result forms a partly different design.

One example of a formative ex ante evaluation was published by Carlén et al. (2016) where the main question was if the Swedish national energy efficiency networks program for SMEs should have a minimum annual allowed energy use by participating companies of 2 GWh/year as originally decided or if that figure should be lowered to 1 GWh/year. This study was made before, ex ante, the policy was finalized. Results showed that the cost-effectiveness of the policy program would not be affected to a high extent if the minimum energy use would be lowered to 1 GWh/year, so this evaluation allowed the policy to be reformed, that is, it was also a form of formative policy program evaluation (Carlén et al., 2016).

17.6.3 Process evaluation

In a process evaluation of a public policy program, the process of the policy is studied. This way, the policy operator or policy administrator can receive feed-back on the various processes involved in the policy program operationalization.

17.6.4 Example of part of a policy program evaluation, the case of the Swedish energy audit program

When the national energy audit program was launched some of the ideas and experiences from earlier programs were included. However, to improve the program, to make it easier to carry out, implement, evaluate, and draw conclusions from, some additional proposals for improving the program, are presented below. These include, for example, improvements in the program related to the disbursement of support, goal-setting, and involvement of actors.

17.6.4.1 Proposal for progressive disbursement of support[4]

It is up to the energy audit consultant to set the price for an energy audit and deliver the business offer to the company in competition with other technical consultants. However, many small companies have little time to spend on surveying the market and have no or minor experience of energy audits. In order to avoid excessive costs for the assistance of technical consultants who charge high costs for an energy audit a progressive disbursement of energy audit support is proposed. It is proposed that the support is based on the annual energy costs for the company. A suggestion is that the support offered is 0.5% of the annual energy cost as justified below.

An energy audit can cost from 0.5 to 1.0 EUR/m^2 or ranging way above, if a more detalied energy audit is needed (Thollander, 2010). Energy use per square meter (or feet) is a common figure in premises and buildings. This figure, however, varies greatly between industries. A reasonable figure of energy use per square meter for a manufacturing industry may be around 400 kWh/m^2 although both lower but notably higher figures are present (Thollander, 2010). This means that a company that uses approximately 5000 MWh annually has an estimated surface of 12,500 m^2. If one uses the figure of 400 kWh/m^2, this in turn means that the least cost for the energy audit of such an industry is approximately 6250 EUR. With an energy price of 70 EUR/MWh this would mean that the cost

[4] In order to compare figures, one Euro was assumed to be equivalent to 10 Swedish SEK in this chapter.

for the energy audit would be about 1.8% of the energy cost. A sensitivity analysis using 600 kWh/m^2 instead of 400 kWh/m^2 shows that this will give a minimum energy audit cost of approximately 4167 EUR or approximately 1.2% of the energy cost.

A proposal for the design of energy audit support could therefore be to put the subsidy in relation to the annual energy cost, for example, that the support amounts to half the energy audit cost that in the example above would be equal to 0.6% or 0.9% of its annual energy cost. However, setting a lowest figure of, for example, 1000 EUR and a maximum of 3000 EUR, which is the current maximum in the program, is highly suggested.

17.6.4.2 *The need for goal-setting*

It is of great importance, in designing an industrial energy program, to formulate targets that are quantified for the desired results in the form of reduced energy use and that the number of companies participating in the program is sufficiently large (Väisänen et al., 2003). This goal-setting will then help the program operator, in this case the Swedish Energy Agency, to annually verify the effectiveness of the policy. The absence of such targets will most likely cause a more project-based character of the policy, both in terms of resource prioritization at the Swedish Energy Agency but also in the view from the companies participating in the program. To prevent this and give the program a greater long-term approach it is proposed to set up goals in the program. In a study by Thollander and Dotzauer (2010) the energy audit program was evaluated ex ante, expecting to create savings of 0.7 to 1.4 TWh/year with about 1000 industrial companies participating in the program. The reason for the lack of such goals in Swedish energy efficiency programs in general may be explained by a strategy from the Swedish government to not provide goals for specific sectors, for example, industry, transport, and buildings (Thollander et al., 2012).

17.6.4.3 *Involvement of local and regional actors*

It is proposed that the local authority energy and climate consultants[5] are responsible for informing about and promoting the energy audit support to businesses through personal contact (by phone or personal visit), such as the model of the successful Project Highland. Several of the companies that participated in the regional policy program Project Highland stated that the energy audits have been the best the municipality ever had initiated (Thollander, 2010). This supports the idea of the importance of a locally anchored base of a policy, in this case to promote energy audits, although they will not carry out the actual energy audit. By having the local authority energy and climate consultants responsible for informing about and promoting energy audit support, companies will experience, based

[5] The local authority energy and climate consultants are working in the public municipality organization offering support and advices to private persons and businesses in areas concerning energy and climate.

on results from Project Highland, that the municipality (public sector) is committed to the ambitious energy and environmental issues and thus becomes a role model.

It is suggested that the Regional Energy Agencies[6] are responsible for the collection of the energy audit reports (within 2 months of support received) and are responsible for the collection of the annual assessments of the energy audits. This is motivated by at least three reasons. First, a regional impression is created of how the industry manages energy efficiency issues, which is important in the evaluation of regional climate change targets. Although the regional energy agencies are not the actor working explicitly with the issue, it is considered important that there is a regional focus provided. Secondly, this gives the companies a natural opportunity to address issues regarding the energy audits and the results from the audits to an independent regional player, which should drive the level of participation in the program and increase the rate of implementation of the measures. This has proven to be an important factor in other sectors where information from colleagues in the industry or organizations, such as trade associations, is valued higher than the information from the state (Rohdin et al., 2007). Thirdly, it also promotes the creation of local networks in the region.

17.6.4.4 Proposals for short training course of energy auditors

Some form of education is found in most major energy audit programs (Väisänen et al., 2003). Often, an operator of an energy program has few choices when it comes to the choice of energy auditors. A solid education gives on one hand very well-trained energy auditors, but on the other hand creates practical barriers such as cost of such education which often puts constraints. The cost of training for operators who want to carry out energy audits under the program should not be too high, but high enough to avoid participants that are not actually interested in carrying out professional energy audits of high quality. The participants themselves should pay for course costs, but if the price is too high only energy auditors from the major consulting firms and ESCOs will participate.

The proposed course is very closely linked to a possible authorization of energy auditors and the choice of energy audit model and also provides information about the actors that will carry out energy audits in the program. All three of these elements have the purpose to assure the quality of the energy audits.

Authorization is a type of license that gives the energy auditor an opportunity to carry out energy audits. The authorization may have different purposes: (1) give those who have expert competence, an authorization and (2) set a minimum standard for the energy audits within the program. The former put requirements to provide a top-ranked course while the latter provides the ability to remove energy auditors that do not fulfill the requirements. In a small country like

[6] A regional energy agency is a body which promotes improved energy efficiency and increased use of RES. There are 380 regional agencies in the EU and currently 13 under operation in Sweden.

Sweden the latter is perhaps not preferable because the final number of energy auditors may be small.

Based on positive experiences of education and authorization of energy auditors in Norway and Austria, a shorter course for energy auditors is proposed in the Swedish program. Minimum requirements for admission regarding, for example, previous knowledge of energy audits and the structure of the course must be thoroughly scrutinized. It is also desirable to establish a network for energy auditors for regular meetings where experiences can be exchanged.

To minimize cost for authorization it is suggested not to use an official accreditation agency. Instead, it is suggested that the authorization is simply approved by participation in the proposed course for energy auditors. This strategy gives the Swedish Energy Agency an overview of authorized energy auditors and gives the instrument to remove the authorization for an energy auditor when carrying out energy audits not meeting minimum requirement for an energy audit. Minimum requirement must be established by the agency, in consultation with experts in the area, in order to ensure quality of the energy audits.

The regional energy agencies together with the Swedish Energy Agency are proposed to be responsible for the course. However, the teachers themselves should be very experienced energy auditors.

17.6.4.5 Energy auditing tools

In general, there may be a need for different models depending on the sector targeted by the program. This is not a prerequisite in the Swedish program where the intended recipients of the energy audits may be considered relatively uniform, where the savings are particularly found in the so-called support processes (e.g., ventilation, lighting, and compressed air). To keep the costs as low as possible regarding monitoring and evaluation, a model should be developed for how the energy audits shall be conducted and reported.

17.6.5 Method for undertaking an impact assessment of an energy policy program

When undertaking an impact assessment also denoted impact evaluation, one can either undertake the assessment or evaluation ex ante or ex post. In an impact assessment that is made before the program is started, some key steps need to be taken. These are, but not limited to, the following steps shown in Table 17.3.

17.6.5.1 Example of an ex ante impact estimation

17.6.5.1.1 Energy efficiency

Assuming a 3-year national policy program consisting of energy audits free of charge for 1000 industrial SMEs with an average energy use of 800 MWh/year. The SMEs have an expected energy efficiency potential of 25%. An expected

Table 17.3 Major steps when undertaking an impact assessment (Andrei et al., 2019).

Major important steps	Comment
Decide primary energy factors for all relevant energy carriers	Primary energy factors for the major energy carriers should be set, for example, the European Energy Efficiency Directives have suggested that for, for example, electricity a primary energy factor of 2.5 is relevant if nothing else can be motivated (EU, 2012)
Decide emission factors for all relevant energy carriers	Emission factors vary between countries and regions and should preferably be used on a local or region level. Make sure to use credible sources such as a regional or national EPA or Energy Agency. This holds in particular for electricity use where figures can range from basically zero to almost 1 kg carbon dioxide per kWh of electricity if relying on a marginal price perspective (Trygg and Karlsson, 2005)
Decide the target group's approximate average annual energy for the relevant energy carriers	To make an estimation of the average target groups' company-specific annual energy use is of key performance. Not seldom are impact estimations provided way too high when these are not related to the total annual energy use. This holds in particular for SMEs where annual energy use can be way below 1 GWh/year while for a pulp mill, annual energy use of several TWh/year is normal
Decide the approximate total number of participating companies and the program's expected total annual energy use	When the average energy use per company is set, the total number of companies expected to be affected by the program should be set as well as the total expected annual energy use consisting of all participating companies
Decide how the primary energy use equation looks like and where the major energy use among the support and productions processes are found	Deciding which major energy-end-using processes that the participating companies have in terms of production and support processes is important. While a nonenergy-intensive mechanical engineering company can have an annual energy use for the support processes of 80%, the support process energy use of a chemical pulp mill is only a few percent

(*Continued*)

Table 17.3 (Continued)

Major important steps	**Comment**
Make an energy efficiency potential estimation	Using available international, national, and regional or local policy documents together with available scientific papers make an estimation of the energy efficiency potential of the program. If no figures are available, it is possible to make a pilot study asking companies of this. One such example by Backlund et al. (2012) showed that the stated energy efficiency potential for energy-intensive companies was 11% while for industrial SMEs, it was stated to be 33%.
Make an estimation of the expected deployment rate	A common mistake when making an impact estimation is that of assuming that all the energy efficiency potential is deployed, that is, implemented. Anderson and Newell (2004) found that approximately half of the suggested energy efficiency measures in the American IAC were implemented. A common figure can be to estimate the deployment rate to 50% or 0.5.
Quality control of your estimation	Make sure to make a quality control of your assessment both in terms of number of estimated energy efficiency measures implemented per company and also the expected energy efficiency per measures. A common number of measures from an audit can range considerably from only a few up to 10 or more per company Further, while average energy efficiency per measure from energy audits from industrial SMEs can be in the range of 50 MWh/measure and year or less, measure for production process intensive companies can be 1000 MWh/measure or even higher
Make an estimation of the additionality[a] of the deployment rate	A policy program's effect may also be due to other factors than the actual policy. An important estimation is that of additionality, that is, what was the additional contribution from the policy comparted with BaU and other active

(*Continued*)

Table 17.3 (Continued)

Major important steps	Comment
	policy programs and factors affecting the energy efficiency deployment rate. Using a baseline for expected effect of energy end-use development can be a good way of doing this, for example, including structural and autonomous effects in addition to policy effects. One example can be that a capacity building program for large companies within the EU on energy audits needs to take into account that all large companies' needs to conduct an energy audit every fourth year, so the actual net effect of the capacity building program is not the whole energy efficiency deployment from an energy audit but the additional measures implemented thanks to the capacity building program
Include nonenergy benefits or multiple energy benefits of the program	Make an estimation of other factors that positively affect the program. For an example of nonenergy benefits for public policy programs, see Nehler et al. (2018) and Johansson and Thollander (2019). These can, for example, be increased greening of companies, reduced maintenance cost for measures implemented among the companies
Scale up the expected net impact impacts in terms of energy efficiency and carbon dioxide improvements	When an average energy efficiency potential, deployment rate, additionality, and nonenergy-benefits have been estimated, it is now possible to provide an expected net impact of the policy program and related carbon dioxide emissions
Calculate the program's cost-effectiveness	Assume the relevant costs associated with the program. Primary public costs are energy audit costs, cost for marketing of the program, operationalization cost for staff in charge of the program, and cost for monitoring and evaluation

[a]Additionality means the extra or added contribution of the action in relation to already available programs and factors such as continuous replacement of technologies and processes. A subsidy for companies to conduct energy audits that have already undertaken such thus most likely have low additionality while a subsidy program for energy audits among companies with low experience in energy efficiency improvements and who have not undertaken an energy audit lately will most likely have a high additionality.

deployment rate is assumed to be 12% for this program distributed on an expected three measures per company from the audit and 32 MWh/measure. The additionality of the program is expected to be that no audits would have been undertaken without the program in place. If the audits would not be undertaken and provide the companies with specific information on the available energy efficiency measures, the SMEs will not undertake these measures. Eighty percent of the energy efficiency measures stem from electricity and 20% from natural gas.

These figures can now be scaled up which for energy efficiency means an estimated net impact of 96.0 GWh/year, 76.8 GWh/year for electricity, and 19.2 GWh/year for natural gas, for the participating 1000 companies.

17.6.5.1.2 Policy program costs

Major cost for the program is assumed to be 200,000 EUR per year for staff cost for the policy operator, 50,000 EUR for monitoring and evaluation, and audit costs of 2500 EUR per audit which for 1000 energy audits means 2.5 MEUR. Total costs are thus 2.75 MEUR.

17.6.5.1.3 Cost-effectiveness

The cost-effectiveness or more correctly in this case the subsidy effectiveness as the firm-specific costs are not included (e.g., investment costs for energy efficiency measures) is then 96,000 000 kWh to a cost of 2.75 MEUR or 0.020 EUR/kWh.

17.7 Challenges for industrial energy audit policy programs

17.7.1 Potential versus degree of implementation

Potential for energy efficiency is not synonymous with implementation of real measures. This has been shown in several scientific evaluations of various energy audit policy programs. An assessment of the US Industrial Assessment Centers (IAC), one of the world's largest industrial energy audit programs, involving more than 18,000 US SMEs, shows that about half of the proposed energy efficiency measures were implemented (Anderson and Newell, 2004). This can be compared with the evaluation of the first part of the Swedish regional energy audit program, Project Highlands, which showed that just over 40% of the proposed measures had been or were planned to be taken (Thollander et al., 2007).

17.7.2 Technology development and cost-effectiveness

Another challenge concerns the technology development degree of the national industries. Technology development follows a number of general development steps that end with a market where a few large players operate under strong price pressure, for example, the pulp and paper industry. The first technology development phase is characterized by a high degree of product innovation and a large number of small players. Finally, the technology reaches a so-called dominant design, and the degree of product innovation diminishes. During the next phase, the degree of process innovation increases, and as some players succeed less well with process innovation, their products become too expensive. In the third mature phase, the degree of product innovation also declines, and finally, a few large players are operating under strong price pressure. If a sector is in the third phase and has thus already undergone major efficiency improvements regarding its core business, the energy efficiency potential may of course be lower. In a Swedish study it was shown that companies' own estimates of their energy efficiency potential vary between just over 10% for energy-intensive companies to just over 30% for nonenergy-intensive companies (Backlund et al., 2012).

Research shows that the cost-effectiveness of the Swedish energy audit program, which is aimed toward Swedish SMEs, was between 200 and 400 kWh per EUR (Thollander and Dotzauer, 2010), where the majority of the measures were found in the support processes. For the Swedish VAP, the cost-effectiveness of about 10 kWh per EUR was found, where the majority of the measures are found in the production processes (Thollander and Rohdin, 2010).

17.7.3 Asymmetric energy policy shocks

Asymmetric energy policy shocks were first defined by Thollander et al. (2012), in their article on the challenges of Swedish energy policy formation to achieve the 2020 target of primary energy. An asymmetric energy policy shock is derived from the economic term asymmetric shock and describes that the potential for energy efficiency and the possibility of realizing this energy efficiency potential can vary within a region, state, or country. These differences or asymmetries mean that, for example, a target for primary energy cannot be considered symmetric. Thus, there is an asymmetry in the realization of the energy efficiency potential, given that a goal is set equal for a whole country or region. Asymmetric energy policy shocks may be due to the fact that a sector, a region, or a country, has already carried out extensive activities such as previous policies related to improved energy efficiency, which is why the energy efficiency potential is lower. Another reason is, for example, the region's climate. Heating or cooling demands due to the climate vary widely in some countries and regions. Asymmetric energy policy shocks can further be due to a country having a large proportion of energy-intensive industries, where the energy efficiency potential is generally always lower than for nonenergy-intensive industry. Since many national and international targets, apart from improved energy efficiency, also include CO_2 emissions reductions, a degree of caution has to be taken as achieving high shares of renewables in an energy system, means the energy efficiency potential may need to be justified as well, in order not to have conflicting targets.

17.8 Nonenergy benefits and energy policy programs

Results from studies on nonenergy benefits (NEB) and industrial policy programs showed that the most frequently mentioned NEBs were those noted in relation to energy efficiency networks, while companies participating in an energy audit program showed lower awareness of nonenergy benefits (Nehler et al., 2018; Johansson and Thollander, 2019). Moreover, the results also revealed that it is difficult to directly link NEBs to a specific policy program (Nehler et al., 2018). Further, the policy administrators seem to be more aware of available NEBs than the companies participating in the policy program (Nehler et al., 2018). The administrative law-based policy showed to have the lowest perceived NEBs available (Nehler et al., 2018).

The overall most commonly mentioned NEBs in one of the studies was related to production, for example, increased lifetime of equipment and more reliable production. However, participants from an energy audit program related the NEBs mainly to the technical installations, while participants of an energy efficiency networks program also perceived NEBs related energy management practices. Furthermore, one difference between perceived NEBs from respondents in companies participating in networks versus energy audit programs was the improved company environmental image, where energy efficiency network participants perceived this as an NEB while the energy audit program participants did not. Results of these studies indicate that the maturity of the respondents (in terms of how far they have matured in the in-house energy efficiency activities) naturally increases as the work with energy efficiency improvements is intensified. This in turn positively affect how companies perceive NEBs in relation to their company's energy efficiency work. A one-time law-based action or an energy audit carried out are not per see enough to create such high awareness, while an energy efficiency networks program seem to increase the chance for companies to mature in their energy management practices, and by so, more gains or NEBs are seen, related to the energy efficiency activities (Johansson and Thollander, 2019).

References

Anderson, S.T., Newell, R.G., 2004. Information programs for technology adoption: the case of energy-efficiency audits. Resource Energy Econ. 26 (1), 27–50.

Andrei, M., Thollander, P., Gindroz, B., Pierre, I., 2019. Impact assessment of industrial energy policy programs. Working paper.

Backlund, S., Broberg, S., Ottosson, M., Thollander, P., 2012. Energy efficiency potentials and energy management practices in Swedish firms. In: Proceedings of the ECEEE Industry Summer Study, September 11–14, 2012.

Backlund, S., Paramonova, S., Thollander, P., Rohdin, P., Karlsson, M., 2014. A regional method for increased resource-efficiency in industrial energy systems. In: Proceedings of ECEEE Industry Summer Study, Arnhem, 2–5 June.

Blomqvist, E., Thollander, P., 2015. An integrated dataset of energy efficiency measures. Energy Efficiency 8 (6), 1125–1147.

Carlén, A., Rosenqvist, M., Paramonova, S., Thollander, P., Municio, S., 2016. Energy effiency networks for small and medium sized enterprises-bosting the energy efficiency potential by joining forces. In: ECEEE Industry Summer Study, Berlin, 12–14 September.

EC (European Commission), 2012. Directive 2012/27/EU on energy efficiency, amending Directives 2009/125/EC and 2010/30/EU and repealing Directives 2004/8/EC and 2006/32/EC. Vol. 4: 1–56.

Farla, J.C.M., Blok, K., 1995. Energy conservation investment behaviour of firms: analysis of investments in energy efficiency in the Netherlands in the 1980s. In: Proceedings of the 1995 ACEEE Summer Study on Energy Efficiency in Industry, Washington, DC, USA.

Finman, H., Laitner, J.A., 2001. Industry, energy efficiency and productivity improvements. Proceeding of the 2001 Summer Study on Energy Efficiency in Industry: 561–570.

Hardell, 2006. Utvärdering av seminarieserie för tillverkande svenska SMF 2005-2006 (Evaluation of a series of seminars directed towards Swedish manufacturing SMEs 2005-2006). Swedish Energy Agency, Eskilstuna [in Swedish].

Johansson, I., Thollander, P., 2019. Non-energy benefits in energy audit and energy efficiency network policy programs for industrial SMEs. In ECEEE Summer Study, Belambra Presqu'île de Giens, 3–8 June.

Johansson, I., Mardan, N., Cornelis, E., Kimura, O., Thollander, P., 2019. Designing policies and programmes for improved energy efficiency in industrial SMEs. Energies 12 (7), 1338. Available from: https://doi.org/10.3390/en12071338.

Koewener, D., Mielicke, U., Jochem, E. 2011. Energy efficiency networks. for companies - concept, achievments and prospects, Karlsruhe, Germany: The Foundation of a Low-Carbon Society.

Nehler, T., Thollander, P., Fredriksson, L., Friberg, S., Nordberg, T., 2018. Non-energy benefits of Swedish energy efficiency policy instruments: a three-levelled perspective. In: ECEEE Industry Summer Study, Berlin, 10–13 June. pp. 139–149.

Paramonova, S., Ivner, J., Thollander, P., 2014. Outsourcing Industrial Energy Management - Industrial Energy Efficiency Networks Provided As an Energy Service. Nova Science Publishers, Inc. Page 71–98. Download at: https://www.researchgate.net/publication/268445085_Outsourcing_Industrial_Energy_Management_-_Industrial_Energy_Efficiency_Networks_Provided_As_an_Energy_Service.

Paramonova, S., Thollander, P., 2016. Energy-efficiency networks for SMEs: Learning from the Swedish experience. 65, 295–307.

Paramonova, S., Thollander, P., 2016. Ex-post impact and process evaluation of the Swedish energy audit policy programme for small and medium-sized enterprises. J. Clean. Prod. 135, 932–949.

Stern, P.C., Aronsson, E. (Eds.), 1984. Energy Use: The Human Dimension. The National Academies Press, Washington, DC.

Thollander, P., 2010. En översikt över hur ett energikartläggningsprogram kan utformas i en svensk context (an overview on how an energy audit program could be designed in a Swedish context). http://liu.diva-portal.org/smash/record.jsf?searchId = 1&pid = diva2:325174. [in Swedish].

Thollander, P., Dotzauer, E., 2010. An energy efficiency program for Swedish industrial small- and medium-sized enterprises. J. Clean. Prod. 18 (13), 1339–1346.

Thollander P., Rohdin P., 2010. Cost-effectiveness of energy programs involving energy audits – results from Sweden. In: Proceedings of the International Energy Program Evaluation Conference (IEPEC) 2010, Paris, France.

Thollander, P., Rohdin, P., Danestig, M., 2007. Energy policies for increased industrial energy efficiency: evaluation of a local energy programme for manufacturing SMEs. Energy Policy 35 (11), 5774–5783.

Thollander, P., Rohdin, P., Moshfegh, B., 2012. On the formation of energy policies towards 2020: challenges in the Swedish industrial and building sectors. Energy Policy 42, 461–467.

Thollander, P., Cornelis, E., Kimura, O., Morales, I., Zubizarreta Jiménez, R., Backlund, S., et al., 2014. The design and structure of effective energy end-use policies and programs towards industrial SMEs. In: Proceedings of ECEEE Industry Summer Study, Arnhem, 2–5 June.

Thollander, P., Paramonova, S., Cornelis, E., Kimura, O., Trianni, A., Karlsson, M., et al., 2015. International study on energy end-use data among industrial SMEs and energy end-use efficiency improvement opportunities. J. Clean. Prod. 104, 282–296.

Trygg, L., Karlsson, B., 2005. Industrial DSM in a deregulated European electricity market: a case study of 11 plants in Sweden. Energy Policy 33 (11), 1445–1459.

Väisänen, H. (co-ordinator), 2003. Guidebook for Energy Audit Programme Developers.

Index

Note: Page numbers followed by "*f*" and "*t*" refer to figures and tables, respectively.

D

L

Made in the USA
Monee, IL
13 September 2026

62977022R20210